# Angular Momentum and Mass Loss for Hot Stars

# NATO ASI Series

**Advanced Science Institutes Series**

*A Series presenting the results of activities sponsored by the NATO Science Committee, which aims at the dissemination of advanced scientific and technological knowledge, with a view to strengthening links between scientific communities.*

The Series is published by an international board of publishers in conjunction with the NATO Scientific Affairs Division

| | | |
|---|---|---|
| **A** | **Life Sciences** | Plenum Publishing Corporation |
| **B** | **Physics** | London and New York |
| | | |
| **C** | **Mathematical** | Kluwer Academic Publishers |
| | **and Physical Sciences** | Dordrecht, Boston and London |
| **D** | **Behavioural and Social Sciences** | |
| **E** | **Applied Sciences** | |
| | | |
| **F** | **Computer and Systems Sciences** | Springer-Verlag |
| **G** | **Ecological Sciences** | Berlin, Heidelberg, New York, London, |
| **H** | **Cell Biology** | Paris and Tokyo |

**Series C: Mathematical and Physical Sciences - Vol. 316**

# Angular Momentum and Mass Loss for Hot Stars

edited by

## L. A. Willson

Astronomy Program, Physics Department,
Iowa State University, Ames, U.S.A.

and

## R. Stalio

Dipartimento di Astronomia,
Università di Trieste, Trieste, Italy

**Kluwer Academic Publishers**

Dordrecht / Boston / London

Published in cooperation with NATO Scientific Affairs Division

Proceedings of the NATO Advanced Research Workshop on
Angular Momentum and Mass Loss for Hot Stars
Ames, Iowa, U.S.A.
October 23–27, 1989

**Library of Congress Cataloging in Publication Data**

```
NATO Advanced Research Workshop on Angular Momentum and Mass Loss for
  Hot Stars (1989 : Ames, Iowa)
    Angular momentum and mass loss for hot stars : proceedings of the
  NATO Advanced Research Workshop on Angular Momentum and Mass Loss
  for Hot Stars, Ames, Iowa, U.S.A., October 23-27, 1989 / edited by
  L.A. Willson and R. Stalio.
        p.   cm. -- (NATO ASI series. Series C, Mathematical and
  physical sciences ; vol. 316)
    "Published in cooperation with NATO Scientific Affairs Division."
    Includes indexes.
    ISBN 0-7923-0881-6
    1. Early stars--Congresses.  2. Stars--Evolution--Congresses.
  3. Mass loss (Astrophysics)--Congresses.  4. Angular momentum-
  -Congresses.   I. Willson, L. A. (Lee Anne)  II. Stalio, R.
  (Roberto)  III. Title.  IV. Series: NATO ASI series.  Series C,
  Mathematical and physical sciences ; no. 316.
  QB843.E2N37  1989
    523.8--dc20                                         90-41851
```

ISBN 0–7923–0881–6

Published by Kluwer Academic Publishers,
P.O. Box 17, 3300 AA Dordrecht, The Netherlands.

Kluwer Academic Publishers incorporates the publishing programmes of
D. Reidel, Martinus Nijhoff, Dr W. Junk and MTP Press.

Sold and distributed in the U.S.A. and Canada
by Kluwer Academic Publishers,
101 Philip Drive, Norwell, MA 02061, U.S.A.

In all other countries, sold and distributed
by Kluwer Academic Publishers Group,
P.O. Box 322, 3300 AH Dordrecht, The Netherlands.

*Printed on acid-free paper*

# *TABLE OF CONTENTS*

### VERY LUMINOUS AND VERY MASSIVE STARS

# PREFACE

Fundamental unsolved problems of stellar astrophysics include the effects of angular momentum on stellar structure and evolution, the nature and efficiency of the processes by which angular momentum is redistributed within and lost from stars, and the role that stellar rotation plays in enhancing or driving stellar mass loss. There appears to be a qualitative change in the nature and efficiency of these mechanisms near spectral type F0: hotter (more massive) stars typically retain more angular momentum at least until they reach the main sequence, while cooler stars typically spin down quickly. For the hotter stars, recent work suggests a strong link between the type of pulsation behavior, the mass loss rates, and the rotation velocity. If the same mechanisms are able to drive mass loss from the main sequence A stars, as has recently been proposed, then the current interpretations of a number of observations will be drastically affected: e.g. the ages of clusters may be incorrect by up to a factor of two, and the surface abundances of isotopes of He, Li and Be may no longer give constraints on cosmological nucleosynthesis. There are also effects on the evolution of the abundances of elements in the interstellar medium and on the general evolution of populations of stars. Thus the questions of the mechanisms of angular momentum and mass loss of stars more massive than the sun is important not only for stellar studies but for the foundations of much of modern astrophysics.

We and the Scientific Organizing Committee felt that a workshop on this topic would be timely because (1) computational codes incorporating stellar rotation into stellar models are beginning to be developed (for example at Yale and at Los Alamos); (2) there has not been a meeting recently concerning the angular momentum evolution of the more massive, more rapidly rotating stars -- recent and planned meetings incorporating a discussion of stellar rotation have concentrated on the heavily braked lower main sequence stars; (3) the importance of rotation as a factor in mass loss and in affecting pulsation, which in turn also affects mass loss, is just beginning to be appreciated; and (4) observational techniques (Doppler imaging) making use of rapid rotation to study (for example) pulsation properties of early type stars are also being developed.

The focus of this meeting was on the mass and angular momentum evolution of stars that on the main sequence have spectral types earlier than F, i.e. those stars that are mostly rapidly rotating. Because the mass loss and angular momentum history of a star determine the present mass and angular momentum, the topics included pre-main sequence, main sequence and post-main sequence stages of these stars. Goals of the conference, and some of the context, were discussed in introductory papers by Willson and Lamers. Kawaler and Catala reviewed the pre-main sequence stages, theory and observation. On the second day, Antonello, Stalio, Polidan, Friend, Cassinelli and Zickgraf discussed the nature and behavior of rotating stars on and near the upper main sequence; and on the third day, Gray, Perinotto, Sion and Langer took these stars from the main sequence through to their various ends. Finally, an inspiring theoretical summary was provided by the reviews by Maeder and Tassoul.

In preparing this book we have mostly followed the order of presentation of the papers, and have placed poster papers in groups following the reviews that most closely matched the topics.

Discussion was lively throughout the meeting, fueled in part by the ample coffee breaks and the stimulating posters contributed by many of the participants.  One issue that received recurring attention was the controversy over the nature of the WR stars:  are they evolved remnants of massive stars, or are they pre-main sequence objects?  As a result of the discussion of this point, Lamers et al. volunteered to write a summary of the "evolved remnant" point of view, and Underhill agreed to set forth her arguments in favor of these being pre-main sequence objects.  These papers appear in a special section at the end of the book.

The Holiday Inn Gateway Center provided excellent and unobtrusive service to the organizers and the participants in the conference.  A highlight of the meeting was the banquet with entertainment provided by the Musica Antiqua group of Ames, directed by Carl Bleyle.  The program was liberally spiced with astronomical references; and two very talented astronomers, "discovered" by the group, contributed to the success of the musical program.  The weather cooperated also, providing perfect fall conditions for the picnic (catered by a professor-emeritus of poultry science, Bill Marion) and tour to the Fick Observatory.

Numerous individuals contributed to the success of the meeting and to the assembly of this book.  We would like to first thank the scientific organizing committee and scientific advisory committee members, including G. H. Bowen, D. Baade, J. Castor, A. Hearn, M. Cohen, H. Lamers, Y. Kondo, A. Maeder, R. Polidan, C. Morossi, M. Smith, C. Catala, D. Gray and S. Kwok.  On the more practical side, Erlene Mooney, Physics Department secretary, took responsibility for that important detail, reimbursement for travel expenses.  Pnina Luban, graduate student, and Katsuyo Serizawa, undergraduate, minded the registration desk and worked extensively before and after the conference on the organization, as did Pam Marcum.  During the final preparation of the manuscripts, Kurt Rosentrater provided essential assistance with indexing, proofing and other important details.  ISU graduate students, postdocs and former students pitched in to assist with airport and social event transportation:  Brian Adams, Tom Beach, Joyce Guzik, Don Jennings, Roger Kirpes, Meg Lysaght, Don Luttermoser, Pam Marcum, and Heather Preston.  And of course the local organization was smoothly masterminded by the chair of the LOC, George Bowen, who now has two perfect meetings to his credit!

An important factor in the success of the meeting, and one particularly appreciated by the organizers as well as the participants, was the support provided by NATO and by NSF (through the US-Italy program) for participant travel, secretarial services and supplies. Iowa State University also contributed to make it possible for some participants from non-NATO countries to attend, and this support is also gratefully acknowledged.

Lee Anne Willson                    Roberto Stalio
Iowa State University               Università di Trieste

### _Participants_

| | |
|---|---|
| F. Adams | Center for Astrophysics, Massachusetts |
| E. Antonello | Osservatorio Astronomico di Brera, Milano |
| F. Araujo | Observatorio Nacional, Rio de Janeiro, Brazil |
| D. Baade | European Southern Observatory, Garching |
| L. A. Balona | South African Astronomical Observatory |
| T. Beach | Iowa State University |
| J. Bjorkman | University of Wisconsin |
| K. Bjorkman | University of Wisconsin |
| C. T. Bolton | David Dunlap Observatory, Toronto, Ontario |
| G. H. Bowen | Iowa State University |
| J. P. Cassinelli | University of Wisconsin |
| J. Castor | Lawrence Livermore National Laboratory California |
| C. Catala | Observatoire de Paris |
| E Covino | Osservatorio Astronomico di Capodimonte, Napoli |
| J. Cuypers | Koninklijke Sterrenwacht voan Belgie, Brussel |
| K. Davidson | University of Minnesota |
| M.-G. Franchini | Osservatorio Astronomico di Trieste |
| D. Friend | Weber State College, Utah |
| D. Gray | University of Western Ontario |
| J. Guzik | Los Alamos National Laboratory, New Mexico |
| J. Hakkila | Mankato State University, Minnesota |
| P. Harmanec | University of British Columbia |
| A. G. Hearn | Sterrekunding Institut, Utrecht |
| H. Henrichs | University Observatory, Munich |
| P. Judge | Joint Institute for Laboratory Astrophysics, Colorado |
| L. Kaper | University of Amsterdam |
| S. Kawaler | Iowa State University |
| H. J. G. L. M. Lamers | Space Research Laboratory, Utrecht |

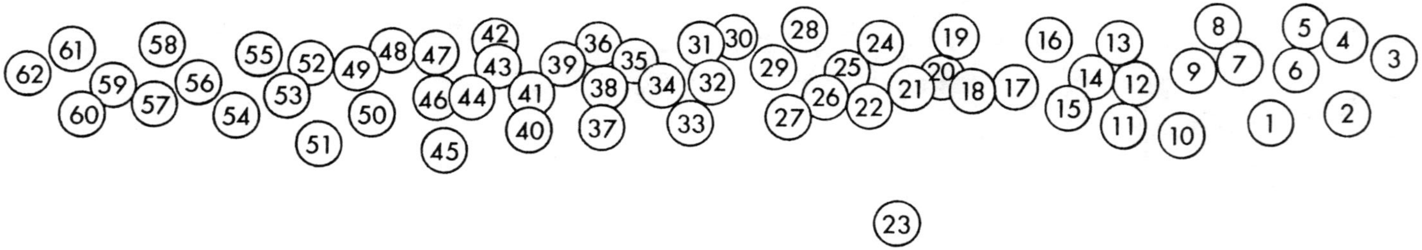

1. Willson
2. Kondo
3. Bowen
4. Stalio
5. Sion
6. Kawaler
7. Perinotto
8. Gray
9. Pinsonneault
10. Antonello
11. Waters
12. Lewis
13. Zickgraf
14. Mrs. Zickgraf
15. Pijpers
16. Friend
17. Harmanec
18. Hearn
19. Kaper
20. Prinja
21. Marlborough
22. Judge
23. Preston
24. Schmutz
25. Maeder
26. Balona
27. K. Bjorkman
28. Lamers
29. Luttermoser
30. Davidson
31. Baade
32. J. Bjorkman
33. Franchini
34. Smith
35. Adams
36. Cassinelli
37. Covino
38. Polidan
39. Morrison
40. Underhill
41. Vardya
42. Bolton
43. Hakkila
44. Morossi
45. M. Tassoul
46. Araujo
47. Castor
48. Waelkens
49. Beach
50. Guzik
51. Serizawa
52. Langer
53. Cuypers
54. Magalhaes
55. Rodriguez-Bell
56. J.-L. Tassoul
57. Vitello
58. Catala
59. Zorec
60. Peters
61. Henrichs
62. Talavera

WHY A MEETING ON ANGULAR MOMENTUM AND MASS LOSS FOR HOT STARS?

L. A. Willson
Astronomy Program
Physics Department
Iowa State University
Ames IA 50011 USA

ABSTRACT:   This is the second in a series of meetings organized jointly by
Iowa State University and the University of Trieste.   The theme for these
workshops is "dynamical phenomena in stars and circumstellar envelopes".   Here
I review the justification for the series and for the topic selected for this
meeting, and note some relevant recent developments in our understanding of
connections between angular momentum and mass loss for stars more massive than
the Sun.

1.  The series on Dynamical Phenomena

The Ames-Trieste series of workshops on dynamical phenomena in stars and
circumstellar envelopes is motivated by two factors:

> Dynamical phenomena produce time-variable observables -- light,
> velocity, color and spectral variations -- that provide important
> constraints on the nature of the systems involved.

> Dynamical modeling techniques are evolving rapidly, paralleling the
> development of ever-more-powerful computing methods.

The focus of this series of workshops is on physical phenomena.   It has
been traditional for astronomers to sort themselves by technique or wavelength, or by object (often interpreted narrowly, as one evolutionary stage of
one mass of star).   At many workshops one encounters the same group of people
discussing the same objects or techniques.   By selecting phenomena that occur
accross a broad class of objects, or that are important through several
observationally distinct evolutionary stages of a given class of objects, we
hope to stimulate new kinds of interactions and collaborations.

2. Objectives of the conference.

In materials distributed before the conference, we stated the objectives
as:  *To ascertain the present understanding of and stimulate further work
on the relationships between angular momentum and mass loss for stars that
are more massive than the sun, with an emphasis on the interpretation of
observations in terms of detailed physical mechanisms.*   Let us consider
what this implies:

*To ascertain the present understanding*:   We hope, by bringing together people
working on similar problems for different stages of a group of stars, to learn
what is known about rotation, angular momentum loss and mass loss.

*To stimulate further work*:  We are hoping that the speakers will emphasize
developments that are just becoming possible, problems rather than solutions.

1

*L. A. Willson and R. Stalio (eds.), Angular Momentum and Mass Loss for Hot Stars, 1–5.*
© *1990 Kluwer Academic Publishers. Printed in the Netherlands.*

*Relationships between angular momentum and mass loss*:  This phrase is deliberately ambiguous.  Most mechanisms for angular momentum loss require at least some mass loss, but also rapid rotation can cause or enhance mass loss, Rotation can also modify, for example, pulsation to which mass loss rates are sensitive -- the topic of the last Trieste-Ames meeting.  So we will look at angular momentum, at angular momentum loss, at mass loss, and at the connections among them.

*For stars that are more massive than the sun*:  there is a discontinuity in the rotational behavior of stars, probably tied to the decreasing importance of surface convection zones, for masses above 1 $M_\odot$.  Considerable effort has gone in recent years to extending our understanding solar-type magnetic/convective activity to other stars.  It is not obvious that the same mechanisms will be important in more massive stars without significant surface convection zones.  One should not, for example, infer from the existence of a corona that there is magnetic heating:  a corona can result from the thermalization of mechanical energy in a variety of forms.  An analogy:  malaria can cause a fever, but not all fevers are caused by malaria!

*With an emphasis on the interpretation of observations in terms of detailed physical mechanism*:  Astronomy is like a tree: observations are the leaves, converting photons to fuel; interpretation the trunk, connecting theory and observation; and theory is the roots, connecting astronomy to the rest of science and drawing nourishment from the contact.  Translation of the information contained in stellar spectra and time-variability into identification of the physical mechanisms makes it possible to construct models;  detailed models are needed to verify the identification of the physical mechanisms and derive quantitative information from the observations.  In the area of angular momentum and mass loss, modeling in sufficient detail to allow for this last step is at or beyond the edge of possibility.

3.  Angular momentum and mass loss: some questions

I'd like to focus discussion on three fundamental questions about angular momentum and mass loss:  (1)  What is the role of mass loss in removing angular momentum from stars of various masses?  (2)  What are the effects of rotation on stellar winds/mass loss?  and (3)  What can we learn from studying surface rotation and winds about the internal redistribution of angular momentum during stellar evolution?

*The effects of mass loss on stellar angular momentum.*  The angular momentum carried away by a wind can be expressed as $\jmath_{wind}\dot{M}$, where $\jmath = J/M$ is the angular mmomentum per unit mass, the specific angular momentum, in the wind far from the star.  Kawaler (1987) has recently re-examined the classical study by Kraft (1970); he finds that for stars earlier than about F0, $\langle v_{rot} \rangle \sim v_{crit}/3$, while G and K stars rotate much more slowly.  It is generally supposed that this "break" in the angular momentum vs. mass relation is caused by magnetic braking linked to the existence of sub-surface convection zones in the later type stars.  There is certainly substantial evidence that stars of spectral type G and K rotate more rapidly when they are young main sequence stars than they do later, so this magnetic activity must certainly be important.  However, some of the difference in $\jmath$ between high and low-mass stars can be seen well before they reach the main sequence -- most T Tauri stars appear to be rotating quite slowly, while a number of higher mass pre-main sequence stars are rotating more rapidly -- so magnetic braking on the main sequence can't be the whole story (see Kawaler's paper, this volume).

*The role of rotation in the mass loss process.* It is particularly evident in the case of the Be and B[e] stars that rotation enhances and alters the character of stellar winds: it can give rise to quite extreme differences between equatorial and polar outflows (see Zickgraf's paper, this volume). The details of the structure and nature of the winds of rotating stars are only beginning to be explored, and much theoretical and observational work remains to be done. It is, for example, theoretically possible to have a star with a relatively hot, low density, polar wind flow and a cool, relatively dense equatorial wind driven by a combination of rotation with pulsation and/or radiation pressure; it is also possible for this wind to alter its structure qualitatively (e.g. switching from cool to hot wind in the equatorial region) due to relatively small changes in the stellar pulsation parameters (Willson 1986; Willson and Bowen 1988).

*The link between internal and external (surface, wind) angular momentum.* There is an "anomaly" in the angular-momentum vs. spectral type (or mass) relation: among the A and early F stars there are not only many "normal" fast-rotating stars but also a population of slowly-rotating ones with peculiar spectra indicative of abnormal surface abundances. The currently popular assumption is that slow rotation means stable conditions in the outer envelope and atmosphere, and that this allows for the diffusion of elements into the star and the radiative levitation of other elements into the atmospheres, accounting for the anomalous abundances. The braking is presumably tidal (for the binary Am stars) or magnetic (for the Ap stars with strong global magnetic fields) -- see for example Wolff (1983). However, this leaves some big unanswered questions: Why do some A stars arrive on the main sequence with very strong magnetic fields, while others don't? Are the slowly rotating A stars rotating slowly throughout their interiors, or have only the surface layers been spun down? Do O, B, A or F stars lose angular momentum after they reach the main sequence, and if so, how?

One way to constrain the internal angular momentum of a class of stars is to look at later evolutionary stages when the material that on the main sequence may be relatively isolated in the interior of the star is on or more closely coupled to the stellar surface. Thus for example those planetary nebulae that show nitrogen overabundances and other indications of massive progenitors are also more likely to appear non-spherical (Kaler 1985) -- is this a result somehow of the higher specific angular momentum these stars have on the main sequence? White dwarfs are generally not rotating very rapidly (e.g. Sion, this volume): how and when do these stellar cores lose their primordial angular momentum?

4. Mass loss for main sequence A and F stars due to rotation and pulsation: the "main sequence mass loss hypothesis"

The possibility of a link between angular momentum and mass loss for A stars, particularly when enhanced by pulsation, is clearly a subject of particular interest to us here at ISU. We suggested a couple of years ago that such mass loss could be enough to alter substantially the evolution of these stars, and could remove up to a solar mass from stars that reach the main sequence among the early A types (Willson, Bowen and Struck-Marcell 1987). The first round exploration of the heretical hypothesis, that such mass loss is the rule for A stars, found no compelling reasons to reject it, and identified some problems that would be solved if A stars routinely lose large amounts of mass.

According to the main sequence mass loss hypothesis, stars reaching the main sequence within the instability strip and with moderately rapid rotation lose mass at rates of the order of $10^{-9}$ $M_\odot$/yr; this is sufficient to cause them to evolve down the main sequence until either/both the pulsation stops or magnetic braking reduces the rotation below some critical value.  The required mass outflow rates should be detectable with current instrumentation, depending on its excitation, ionization, and electron temperatures.  If the outflow is entirely ionized, as we originally proposed, and if it is flowing at a reasonable velocity (not more than a few times the stellar escape velocity) then VLA observations should be capable of placing interesting limits.  Brown et al (1990) have made such observations, and find an upper limit of about $10^{-10}$ $M_\odot$/yr; on the ionized mass loss rate.  If the outflow is not entirely ionized, then the outflow should be detectable from its influence on line profiles, particularly in the UV.  Several projects are underway to use IUE data to try to constrain the mass loss rates by this method, but definitive results are not yet available.  (Watch for reports from Brugel and Willson;  Rodriguez-Bell and York.)

One clear prediction is that there should be a deficit of A stars and a surplus of late F stars that should develop on a timescale of a few times $10^8$ to $10^9$ years.  A histogram of stars along the main sequence binned according to any temperature index shows a clear dip in the region of the instability strip, just as expected.  However, such a dip in the temperature distribution is also expected to result from the development of surface convection zones near the red edge of the instability strip, which alters the relation between spectral type or color and mass (Bøhm-Vitense and Canterna 1974).  Therefore one needs to look for a deficit in the appropriate mass range.  There are not very many main sequence A star binaries that have well-determined masses, so the relation between color or spectral type and mass is not very well calibrated observationally.  According to the best available calibrations for mass vs. spectral type, there is a deficit of stars in the mass range where the main sequence mass loss hypothesis predicts that there should be (Patten 1989).

Patten (1989) also looked for correlations between vsini and indications of past or present mass loss.  His vsini vs. spectral type curve is presented and discussed in his paper in this volume.  It shows interesting structure in the region of the A stars; however, the interpretation of this structure is far from unambiguous, as it is possible that rapidly rotating stars are assigned systematically earlier or later spectral types than more slowly rotating stars of the same mass and age.  Patten has also found that there is a significant correlation of IRAS color excesses with vsini for the A stars. This latter may indicate that the IR excess is tied to the star (rather than to a remnant protoplanetary disk); or it may indicate that even A stars, with no subsurface convection, can somehow spin down after reaching the main sequence.

We also noted in proposing the main sequence mass loss hypothesis that it could (1) provide an explanation for the blue stragglers found in many clusters and (2) allow for the possibility that we may be overestimating the ages of many clusters:  as the A star region is depleted of stars by their evolution down the main sequence, the main sequence turn off is mistakenly placed at the bottom of the "gap", and the slow rotators and/or late B stars left behind are seen as blue stragglers.  Guzik (1989 a,b) has computed isochrones for clusters of solar composition and of lower metallicity, using models that incorporate main sequence mass loss.  She finds that very normal-looking isochrones can be produced, and that errors up to a factor of two in age can be made by interpreting these (mass-losing model) isochrones with standard (mass-conservative) models.

If A and F stars lose substantial amounts of mass, and in particular if the sun lost a lot of mass after it arrived on the main sequence, then the standard interpretation of stellar surface Li and Be abundances as left over from the star formation won't work;  this Li and Be would be long gone, and the current surface layer would have been at sufficiently high temperature in the past that it would have destroyed its Li and Be.  Pantelaki is working on a new model for stellar surface Li and Be abundances that incorporates all known processes that may affect this abundance:  destruction at high temperature, diffusion into the star from the surface layers, mass loss with fractionation, and production by spallation.  Using an estimate for the X-ray losses expected to be associated with the spallation production of Li or Be in a solar flare environment from Ryter et al. 1970, assuming an energy spectrum that gives the right isotope ratios according to Walker, Mathews and Viola (1985), she finds substantial production of Li and Be in stellar flares is consistent with observed X-ray luminosities for stars between 1 and 1.5 solar masses.

The mechanism for this mass loss remains sketchy.  We now suspect that the wind is accelerated in part by the absorption of line photons (esp. Ly $\alpha$) produced in emission as the result of pulsation-induced shocks and escaping to the outer atmosphere thanks to the large Doppler shifts associated with the mass motions due to shocks in the atmosphere.  The modeling of pulsation and winds in rotating stars, of pulsation-enhanced winds in general, and of line-driven winds are all individually frontier areas, and the combination lies beyond what can be done now.  It is our hope that by bringing together this group of scientists working on problems in winds, rotation and pulsation we will accelerate progress in the direction of eventually making such models possible.

Acknowledgements:  I am grateful to NATO, NSF, CNR and ISU for their generous support for this meeting; to NASA for its support of our theoretical work (NAGW 1364) and use of IUE and IRAS data (NAG5-707 and NAG5-1187); and to all who have helped make this meeting happen.

BIBLIOGRAPHY

Brown, A., Veale, A., Judge, P., Bookbinder, J.A. and Hubeny, I  1990 Astrophys. J., *in press*.
Bøhm-Vitense, E. and Canterna, R. 1974, Astrophys. J. **194**, 629-635.
Guzik, J. A. 1989a,b:  in Proc. of 5th IAP Astrophysics Meeting on *Astrophysical Ages and Dating Methods*, eds J. Audouze, M. Cassé, and E. Vangioni-Flam, Paris (June 1989) *in press*.
Kaler, J. B. 1985, Annual Reviews of Astronomy and Astrophysics **23**, 1985.
Kawaler, S. D. 1987, Publ. Astron. Soc. Pacific **99**, 1322-1328.
Kraft, R. 1970, in *Spectroscopic Astrophysics*, ed. G. H. Herbig (Berkeley; University of California Press), p. 385.
Patten, B. N. 1989, MS Thesis, Iowa State University.
Ryter, C., Reeves, H. Gradsztajn, E., and Audouze, J. 1970, Astr. Ap. **8**, 389-397
Walker, T. P., Mathews, G. J., and Viola, V. E. 1985, Ap. J. **299**, 745-751
Willson, L. A. 1986, Pub. A. S. P.**98**, 37-40
Willson, L. A. and Bowen, G. H. 1988, in *Polarized Radiation of Circumstellar Origin*, G. V. Cyne *et al.* eds, Vatican Observatory, Vatican City State, p. 485.
Willson, L. A., Bowen, G. H. and Struck-Marcell, C. 1987, Comments on Astrophysics **12**, 17-34.
Wolff, S. 1983, *The A-Stars:  Problems and Perspectives*, NASA SP-463.

# THE EFFECTS OF ROTATION ON STELLAR STRUCTURE AND EVOLUTION


Jean-Louis Tassoul,
Département de Physique, Université de Montréal



ABSTRACT. Following a brief review of some basic concepts, I discuss the effects of rotation in main-sequence stars having a radiative envelope and a convective core. Single stars and detached components in close binaries are considered in turn. (The effects of rotation are so strikingly different in single and double stars that they are worth discussing in the same review.) For single stars I summarize in non-technical terms what is known about the interaction between rotation, meridional circulation, and the ever-present smaller-scale motions. Applications are made to the chemically peculiar stars and the magnetic stars. I also discuss the properties of a newly discovered meridional flow that pervades the interior of a nonsynchronous binary component. These transient, mechanically-driven currents  -- which are much faster than the steady, thermally-driven Eddington-Vogt currents --  are of direct relevance to the problem of synchronization and orbital circularization in the early-type binaries.


## 1. INTRODUCTION

The study of stellar rotation began about 1610, when Galileo recognized sunspots as being associated with the visible surface of the Sun and measured the rotation rate of this star by observing their motions across the solar disc. Yet, it is to Scheiner that credit belongs for showing, circa 1630, that the solar photosphere does not rotate as a solid at a uniform rate, but rather as a fluid with a rotation period depending upon heliocentric latitude. In 1667, Bouillaud argued that the variability in light of some stars was the direct consequence of axial rotation, the spinning bodies showing alternately their bright (unspotted) and dark (spotted) hemispheres to the observer. This idea was popularized in Fontenelle's <u>Entretiens sur la pluralité des mondes</u> -- a highly successful introduction to astronomy that went through many revised editions during 1686-1742 (Brunet 1931, Fontenelle 1973). Although this explanation for the variable stars did not withstand the passage of time, it is nevertheless worth mentioning because it shows the interest and fascination that stellar rotation has aroused since its inception.

7


*L. A. Willson and R. Stalio (eds.), Angular Momentum and Mass Loss for Hot Stars, 7–32.*
© 1990 *Kluwer Academic Publishers. Printed in the Netherlands.*

In fact, it is not until 1909 that Schlesinger found convincing observational evidence that other stars also rotate. In the thirties, Struve and his associates measured the rotation rate of many single and double stars by the broadening of their spectral lines. In particular, Struve and Elvey (1931, p. 673) found that there exists a sharp decline in rotational velocities in the middle F's along the main sequence, and that in late F-type stars and later types rapid rotation occurs only in close spectroscopic binaries. However, it is Kreiken (1935) who first noticed that there is a marked tendency of the components of close, early-type binaries to rotate more slowly than single stars of the same spectral type. Definite evidence of synchronism or quasi-synchronism in the majority of short-period, early-type binaries was originally found by Swings (1936).

Although large efforts have since been devoted to the study of rotating stars, we are still far from understanding all of their properties. Previous work performed prior to 1977 has been reported in my book (Tassoul 1978, hereafter T.R.S.). Given the fact that the literature on rotating stars is growing at a rate of about one hundred papers per year (or more), it is not surprising that many chapters of that book have become largely out-of-date!

On the observational side, much progress has been made in the measurements of ultra-low rotational velocities by making use of high-resolution spectra and Fourier-transform techniques. In the case of spotted stars, ultra-low rotational velocities have also been obtained by determining the modulation frequency of the star's light due to the motion of starspots across its surface. (Bouillaud's original idea was not that wrong, after all!) The determination of the internal rotation of the Sun by helioseismology is also a quite remarkable achievement. On the theoretical side, thanks to the advent of very large computers, much progress has been made in understanding the early moments of a star's lifetime. These three-dimensional calculations throw some light on the formation of double (and multiple) stars and planetary systems, although it is often difficult to reconcile the results obtained on the basis of different codes. Another, but much less conspicuous progress is the recent finding that a small amount of turbulence is required to obtain a satisfactory description of the motions within the radiative zone of a star. As we shall see in this review paper, this has noticeable consequences for both the single stars and the components of close binaries.

Admittedly, this paper is not an exhaustive review of the literature. Rather, I shall try to review and summarize in nontechnical terms the theoretical work I'm pursuing in collaboration with my wife. Following a brief discussion of some basic concepts, in Section 4 we shall discuss the properties of the so-called Eddington-Vogt currents in the radiative envelope of a slowly rotating, early-type star (Tassoul and Tassoul 1982-1989b; hereafter Papers I-X). These thermally-driven motions are of direct relevance to the chemically peculiar stars, since they are most likely to provide the missing link that was needed to explain the correlation between slow rotation and abnormal spectrum in the A-type stars. In section 5 we shall discuss the properties of a newly discovered meridional flow that pervades the interior of a nonsynchronous binary component (Tassoul 1987-1990, Tassoul and Tassoul 1990; hereafter Papers A-D, respectively).  As we shall see, these transient mechanically-driven currents are of direct relevance to the problem of synchronization and orbital circularization in the close (and not-so-close) early-type binaries.


## 2. GENERAL CONSIDERATIONS

In principle, by making use of the basic equations of stellar hydrodynamics, one should be able to calculate at every instant the angular momentum distribution within a star, no matter whether it is on the main sequence or whether it is rapidly expanding or contracting.

To be specific, if we consider an axisymmetric star, it is convenient to write the velocity field in the form

$$\vec{v} = \Omega\omega\vec{1}_\phi + \vec{u}, \tag{1}$$

where $\Omega(\omega,z,t)$ is the angular velocity of rotation and $\vec{u}(\omega,z,t)$ is the velocity in meridian planes passing through the rotation axis. (We shall make use of the cylindrical coordinates $\omega$, $\phi$, $z$ and the spherical coordinates $r$, $\Theta$, $\phi$.)  Thence, it is a simple matter to show that the whole rotational history of a star is actually described by the azimuthal component of the Navier-Stokes equations.  In an inertial frame of reference, one has

$$\frac{\partial}{\partial t}(\Omega\omega^2) + \vec{u}.\text{grad}(\Omega\omega^2) = F_\phi + L_\phi, \tag{2}$$

where $F_\phi$ is proportional to the $\phi$-component of the frictional force and $L_\phi$ is proportional to the $\phi$-component of the Lorentz force (in case there is a prescribed magnetic field).  If one has $F_\phi = L_\phi = 0$, equation (2) merely expresses the fact that the specific angular momentum, $\Omega\omega^2$, is preserved as one follows the motion of each fluid parcel.  In general, if the motions are turbulent rather than laminar, $\vec{v}$ denotes a mean velocity and $F_\phi$ depends on suitable coefficients of eddy viscosity.  Any loss of mass and, hence, of angular momentum must be prescribed as a boundary condition on equation (2), mass loss affecting also the equation of continuity.

Equation (2) illustrates the basic difficulty one has to face when discussing phases of expansion and contraction: because $\vec{u}$, $F_\phi$, $L_\phi$ and the mass-loss rate are not known in advance, it is impossible to calculate from first principles alone the rotational history of a star. This is of direct relevance to the very late phases of stellar evolution.  Indeed, one knows that the fast implosion of a star's core may lead -- by conservation of angular momentum -- to the formation of a rapidly spinning neutron star.  On the contrary, the fact that the majority of white dwarfs are slow rotators indicates that these stars have gradually lost most of their angular momentum during the pre-white-dwarf phase of evolution.  Obviously, unless detailed information is provided by observations, no one can make any firm theoretical

statement about the rotational history of an old collapsing star.  The
same arbitrariness is encountered during all the pre-main-sequence
phases of stellar evolution.

In principle at least, the discussion of equation (2) is made
somewhat simpler in the case of a main-sequence star.  But then, because stellar rotation is basically a problem of fluid dynamics, it is
important to distinguish between those stars having an outer convective
zone from those having a subphotospheric envelope in which radiative
equilibrium prevails.

Nothing is known with any certainty about the rotation rate in a
convective core.  (In practice, it is often assumed that the angular
velocity is uniform throughout the core.)  As is well known, the actual
motions in a convective envelope (such as the Sun's) result from intri-
cate interactions between turbulent convection, rotation, meridional
circulation, dynamo-generated magnetic fields, and mass loss.  All
theories that can be found in the literature thus depend, in one way or
another, on free parameters; fortunately, in the case of the Sun, they
can be adjusted to the observational data.

Now, it is often believed that the rotation in a radiative zone
(either a core or an envelope) is a much simpler problem, because the
claim has frequently been made that laminar motions always prevail
outside a convective zone.  As we shall see, this is far from being
true, the actual motions in a radiative zone being also a complex
combination of rotation, meridional circulation, and small-scale
eddy-like and/or wave-like motions.  To understand these matters, let
us first discuss the various rotationally-driven instabilities that
beset barotropes and baroclines.


## 3. BAROTROPES AND BAROCLINES

By definition, a <u>barotrope</u> is a circulation-free configuration for
which the angular velocity is constant on cylinders centered on the
rotation axis, i.e., $\Omega = \Omega(\omega,t)$.  As explained in T.R.S. (Sec. 4.3),
these highly idealized bodies are characterized by the following properties:

(i) the effective gravity  -- i.e., the gravitational attraction
modified by the centrifugal force --  can be derived from the potential

$$\Phi = V - \int^{\omega} \Omega^2(s)\, s\, ds, \tag{3}$$

where $V(\omega,z,t)$ is the gravitational potential,

(ii) the level surfaces ($\Phi$ = constant), the isobaric surfaces (p=
constant), and the isopycnic surfaces ($\rho$ = constant) always coincide,

(iii) when the chemical composition is homogeneous, these three
families also coincide with the isothermal surfaces (T = constant).

In actual practice, however, a barotrope is a rather crude approximation to a stellar radiative zone, whose structure is much closer
to that of a <u>barocline</u>.  By definition, a barocline is a configuration

for which one has $\Omega = \Omega(\omega,z,t)$. In the case of a realistic stellar model, one can thus make the following statements:

(i) the effective gravity cannot be derived from a potential, so that there are no level surfaces,

(ii) the isobaric surfaces and the isopycnic surfaces are in general inclined to each other by a finite angle,

(iii) even when the chemical composition is homogeneous, there exist temperature variations over each surface of constant pressure.

## 3.1. The Rotationally-Driven Instabilities

Catalogs of instabilities can be found in the literature (e.g., Smith 1987). My experience with these matters shows that it is convenient to classify the instabilities according to their action on the global structure of a star: there is a violent one that can virtually destroy the system, there are two mild ones that generate a permanent spectrum of baroclinic eddies and/or waves, and there are numerous feeble ones that can be neglected altogether. All these instabilities depend either on the specific angular momentum $j = \Omega\omega^2$ or on the gradient Richardson number $Ri = N^2/S^2$, where N is the buoyancy frequency and S is the shear in the linear velocity $\Omega\omega$. (Since we consider radiative zones only, one has $N^2 > 0$ everywhere.)

An instability with respect to axisymmetric disturbances occurs whenever the j-distribution decreases outward on the surfaces of constant specific entropy. (In a barotrope, this condition reduces to $dj/d\omega < 0$.) In geophysics, this is called the condition for symmetric instability, and it merely generalizes the Rayleigh criterion for an incompressible fluid (e.g., T.R.S., Sec. 7.3). It is a barotropic instability in the sense that it draws its energy mainly from the kinetic energy of the rotation; it may therefore occur in barotropes and baroclines alike. As was shown by Lorimer and Monaghan (1980), it is a violent one because, given an adverse j-distribution, the system will at once generate three-dimensional motions in the nonlinear regime, the resulting flow becoming chaotic with a very slow trend to equilibrium. It must be avoided at any cost, therefore.

Contrastingly, the shear-flow instability and the baroclinic instability are mild ones in the sense that they merely generate small-scale turbulent motions that are superposed on the mean flow. These two instabilities are quite distinct concepts, however, although they both develop from nonaxisymmetric disturbances. The former is a barotropic instability that occurs wherever $Ri \lesssim 1/4$, that is, wherever the shear is large enough to overcome the stabilizing influence of the density stratification. The latter draws its energy from the potential energy of the basic stratification; it occurs in baroclines only, wherever one has $Ri \gtrsim 1$ (Paper I, pp. 343-347). As we shall see in Section 4.2, strict radiative equilibrium compels the isothermal and isobaric surfaces to be inclined to each other by a finite angle; it follows at once that baroclinic instability must necessarily occur almost everywhere in a stellar radiative envelope. (Because this instability is caused by the temperature variations over the isobaric surfaces, it is sometimes called "slender convection.") Unfortunately, there is still much confusion about this concept in the astronomical literature (Paper VI, n. 3, and Paper VII, n. 6). Yet, the recent work of Fujimoto (1987) clearly shows that baroclinic instability cannot be ignored altogether in the bulk of a stellar radiative zone. And since it occurs wherever one has Ri $\gtrsim$ 1, there is thus no need to invoke large shears in the rotation rate to generate a <u>permanent</u> spectrum of eddies and/or waves in a radiative zone.

For completeness, I shall also mention the <u>thermal instabilities</u> which develop from axisymmetric disturbances in a <u>barocline (Goldreich</u> and Schubert 1967, Shibahashi 1980). These are the so-called GSF and ABCD instabilities. (They are well known since they have their roots in the astronomical literature.) These instabilities are feeble ones in the sense that their time scales are often larger than the age of the systems. Hence, they are necessarily overshadowed by the ever-present shear-flow and baroclinic instabilities, which are dynamical ones (Paper I, pp. 341-343).

In my opinion, there is no need to discuss these matters any further, because we cannot relate the turbulent motions that are generated by the dynamical instabilities to suitable coefficients of <u>eddy</u> <u>viscosity</u>. Attempts to do so have been made, of course, thus adding to the confusion that already exists in the phenomenological literature on stellar rotation. My viewpoint is quite simple. The geophysicists -- who have all a genuine background in fluid mechanics and can make direct measurements -- openly admit that they cannot calculate their eddy coefficients from first principles alone. If so, then, how could one expect to make such a calculation in stellar astronomy -- a field in which the problems are often ill-defined and much more complex than those encountered in geophysics? Of course, one can always write a coefficient of eddy viscosity, $\mu_t$ (say), in the form $\rho L_c V_c$, where $L_c$ is a typical length and $V_c$ is a typical speed of the turbulent motions. Unfortunately, it must be borne in mind that it is impossible to calculate $L_c$ and $V_c$ separately from first principles alone. In fact, it is even beyond our ability at this time to calculate the <u>product $L_c V_c$</u>. These matters have been also discussed by Charbonneau <u>and Michaud</u> (1990) in a somewhat different spirit.


## 4. THE THERMALLY-DRIVEN CURRENTS

Almost from the start, in the twenties, the problem of rotation and meridional circulation in a radiative zone was ill-formulated. Hence, it is not inappropriate to review the classical papers before presenting a consistent solution. (The present section and the <u>Appendix</u> supersede T.R.S., Sec. 8.2-8.4.)

### 4.1. <u>The Classical Papers</u>

<u>Milne</u> (1923) - The general equations of a rotating star in radiative equilibrium are formulated. Solutions are obtained on the basis of truncated expansions in the small parameter $\varepsilon = \Omega^2 R^3/GM$, which is the ratio of the centrifugal force to gravity at the equator. It is

shown that a star will appear hotter at the poles than at the equator.

von Zeipel (1924) - It is properly demonstrated that the conditions of mechanical and radiative equilibrium are, in general, incompatible in a uniformly rotating barotrope.

Vogt (1925) and Eddington (1925) - In order to resolve the so-called von Zeipel paradox, they point out independently that the small departures from spherical symmetry in a rotating star lead to unequal heating along the polar and equatorial radii. This, in turn, causes a large-scale flow of matter in meridian planes passing through the rotation axis.

Eddington (1929) - The time scale of the thermally-driven currents is claimed to be of the order of the Kelvin-Helmholtz time, $t_{KH}$ = $GM^2/RL$, where L is the total luminosity of the star. This incorrect result has misled the people for more than twenty years.

Krogdahl (1944) - It is shown that, to $O(\varepsilon)$, the angular velocity $\Omega$ and the circulation velocity $\vec{u}$ take the form

$$\Omega = \Omega_0 \ (\omega_0 + \varepsilon\omega_1 + \ldots) \quad \text{and} \quad \vec{u} = \varepsilon\vec{u}_1 + \ldots, \tag{4}$$

where $\Omega_0$ is a constant. (For the sake of simplicity, he merely let $\omega_0$ = 1.) Unfortunately, in attempting to solve analytically equation (2), with $\partial/\partial t = L_\psi = 0$, he prescribed an overly restrictive condition on the circulation pattern. Yet, this is the first consistent formulation of the problem, because it shows that some kind of viscous action must be retained so as to obtain self-consistent solutions.

Gratton (1945) - It is claimed that the circulation pattern in a uniformly rotating, inviscid radiative envelope consists of two distinct cells separated by the particular level surface over which $\Omega^2$ = $2\pi G\rho$. This artificial result, which does not apply to realistic stellar models, was independently found by Öpik (1951). The Gratton-Öpik paradox is discussed further in the Appendix.

Schwarzschild (1947) - The von Zeipel paradox is resolved by choosing the rotation law so that the vector $\vec{u}$ identically vanishes in the models. As was shown by Aikawa (1970), these steady, inviscid models are somewhat arbitrary because their rotation law depends on a free constant. Moreover, one must bear in mind that there is no obvious reason to expect rotating stars to select zero-circulation configurations. In fact, because these inviscid baroclinic models are always unstable with respect to nonaxisymmetric motions, the slightest disturbance will generate three-dimensional motions and, as a result, a large-scale meridional circulation will commence.

Sweet (1950) - He was the first to show that the circulation caused by a slow solid-body rotation is quadrupolar in structure -- e.g., $u_r = \varepsilon u(r)P_2(\cos \Theta)$, where $P_2$ is the Legendre polynomial. The

14

meridional flow consists of a single cell, with interior upwelling at the poles which is compensated by interior downwelling at the equator. He also showed that the time scale of these currents, $t_{ES}$, is of the order of $t_{KH}/\varepsilon$, thus correcting Eddington's (1929) wrong estimate. Unfortunately, one also finds that

$$u_r \propto 1 \quad\text{and}\quad u_\Theta \propto 1/(R-r), \tag{5}$$

as $r \to R$ near the free surface, and

$$u_r \propto 1/(r-R_c) \quad\text{and}\quad u_\Theta \propto 1/(r-R_c)^2, \tag{6}$$

as $r \to R_c$ near the core-envelope interface. Hence, this inviscid solution does not satisfy the essential boundary condition

$$\vec{n}.\vec{u} = 0, \tag{7}$$

at the two boundaries. ($\vec{n}$ is the outer normal.) This condition merely expresses the fact that, without mass loss, the currents must flow along the boundaries. Moreover, with $\partial/\partial t = F_\psi = L_\psi = 0$ and $\Omega = $ constant, it is evident that one cannot satisfy equation (2), the transport of specific angular momentum by the meridional currents remaining always unbalanced (Randers 1941).

Mestel (1953) - It is argued (without adequate proof) that the presence of chemical inhomogeneities will considerably restrain the circulatory currents, unless the star is on the verge of equatorial break-up (see also Paper VII, n. 4). This suggestion is quite plausible since, as we know from empirical evidence, most stars do not mix extensively in their radiative interior as they evolve away from the main sequence. It is also suggested that (laminar) viscous boundary layers could form near the core and the surface, thus preventing the unwanted singularities in equations (5) and (6). Unfortunately, neither at the core nor at the surface was a satisfactory boundary-layer analysis made.

Baker and Kippenhahn (1959) - When the prescribed rotation law is nonuniform, it is found that, instead of equation (5), one has

$$u_r \propto 1/(R-r)^n \quad\text{and}\quad u_\Theta \propto 1/(R-r)^{n+1}, \tag{8}$$

as $r \to R$ near the free surface. (n is the effective polytropic index.) As we shall see in Section 4.3, this result is a mathematical property that has no physical content. The important point to remember is that condition (7) is not satisfied, and that the transport of angular momentum always remains unbalanced because these authors have also let $\partial/\partial t = F_\psi = L_\psi = 0$ in equation (2).

## 4.2. <u>Thirty Years Later</u>

The newly proposed solution rests essentially on a dynamical linkage between eddy-like and/or wave-like motions (which may be called "anisotropic turbulence" because they are predominantly two-dimensional) and the mean flow (i.e., the differential rotation and concomitant meridional currents). To be more specific, because strict radiative equilibrium prevents a rotating star from being a barotrope, the main idea is that the chemically homogeneous parts in a radiative envelope are filled with small-scale transient motions that are caused by the ever-present barotropic-baroclinic instabilities. This anisotropic turbulence, in turn, generates thin thermo-viscous boundary layers so that the circulation velocities do not become infinite and may satisfy condition (7) at the boundaries of the radiative envelope. Simultaneously, the turbulent friction $F_\phi$ acting on the differential rotation can be made to balance the transport of angular momentum in equation (2). Thus, by taking into account the so-called <u>eddy/wave-mean flow interaction</u> in a stellar radiative zone, one can combine into a single, <u>coherent framework</u> the far-reaching but incomplete contributions that were originally made by Krogdahl (1944), Sweet (1950), and Mestel (1953).

As we have seen, it is the transport of radiation in a non-spherical body that causes the slow but inexorable Eddington-Vogt currents in a stellar radiative zone. Accordingly, thermally-driven currents also exist in a tidally distorted star, as well as in a magnetic star, since the tidal interaction with a companion and the Lorentz force both generate small departures from spherical symmetry in a star (Papers II-III-IIIa and Papers IX-IXa-X). Similar currents also exist in a cooling white dwarf (Paper V). Here we shall describe the state of motion in the chemically homogeneous envelope of a single, nonmagnetic, early-type star that does not deviate greatly from spherical symmetry (Papers I-IV-VI-VIa). Hence, it is quite appropriate to expand about hydrostatic equilibrium in powers of the small parameter $\varepsilon$, neglecting all terms of $O(\varepsilon^2)$ or smaller. (In a realistic main-sequence model in almost uniform rotation, $\varepsilon$ does not exceed the value $\varepsilon_c \approx 0.4$, at which point equatorial break-up is likely to occur; in the early- type stars, one has in general $\varepsilon \approx 0.01 - 0.10$.) Letting $\omega_0 = 1$ in equation (4), one finds that the overall rotation rate, of $O(\varepsilon^{1/2})$, generates a large-scale meridional flow, which is of $O(\varepsilon)$ because the centrifugal force is of that order; these motions, in turn, react back on the driving mechanism, thus bringing a correction of $O(\varepsilon^{3/2})$ to the overall rotation rate. By making use of equation (2), with $L_\phi = 0$, one obtains

$$\Omega = \Omega_0 \left( 1 + \varepsilon \, (A + B \sin^2 \Theta) \right), \tag{9}$$

where the functions A and B depend on r and t. The problem of evaluating the meridional flow is thus neatly separated from that of evaluating the departures from solid-body rotation in equation (9).

16

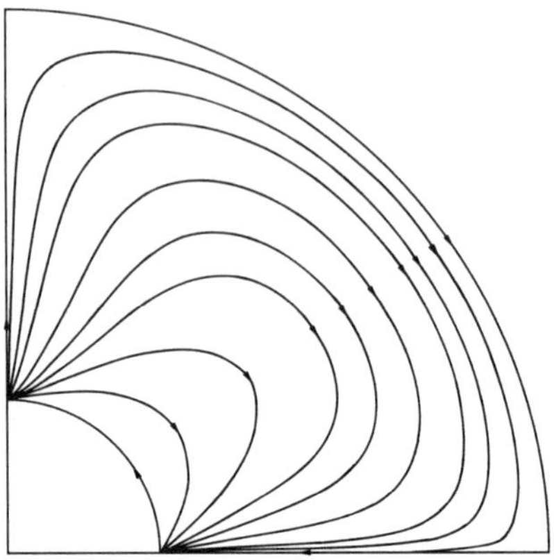

Fig. 1 - Streamlines of meridional circulation in a Cowling
point-source model, with electron-scattering opacity, M =
3 $M_\odot$, and N = 6.


Fig. 1 illustrates the streamlines of the meridional flow in a
slowly rotating, Cowling point-source model. The circulation pattern
consists of a single cell extending from the convective core (which it
does not penetrate) to the free surface, with rising motions at the
poles and sinking motions at the equator. In fact, the turbulent friction acting on the circulation is utterly negligible in the bulk of the
radiative envelope. Because of the presence of thin thermo-viscous
boundary layers, there are no singularities in the meridional flow, and
condition (7) is naturally satisfied at both boundaries. Figures 2 and
3 illustrate the function y, which is the radial part u(r) of $u_r$ (in
normalized units), as a function of x, which is the distance from the
boundary divided by the thickness of the corresponding boundary layer.
(As explained in Paper VIa, these thin layers, which combine the properties of the equations of motion with those of the energy balance,
are quite distinct from an ordinary Ekman layer!) Note especially how
the actual solutions depart from the inadequate, inviscid solutions
defined in equations (5) and (6), and which are depicted by dashed
curves in Figures 2 and 3. The circulation velocities thus remain
uniformly small everywhere in the radiative zone. Of course, in the
boundary layers these velocities depend directly on the vertical coefficient of eddy viscosity $\mu_t$. Fortunately, because they depend,
respectively, on $\mu_t^{1/7}$ and $\mu_t^{1/10}$ in the core and surface boundary
layers, the dependence on this poorly known parameter is appreciably
reduced. The typical speed of these currents is of the order of
$\varepsilon R/t_{KH}$, thus confirming the idea that the Eddington-Sweet time, $t_{ES} =
t_{KH}/\varepsilon$, is indeed the characteristic time of the meridional flow in the
chemically homogeneous parts in a radiative zone.
The derivation of the rotation law is a much more uncertain ven-
ture, however, because it is impossible at this time to perform a mea-
ningful evaluation of the coefficient $\mu_t$. In principle, the functions

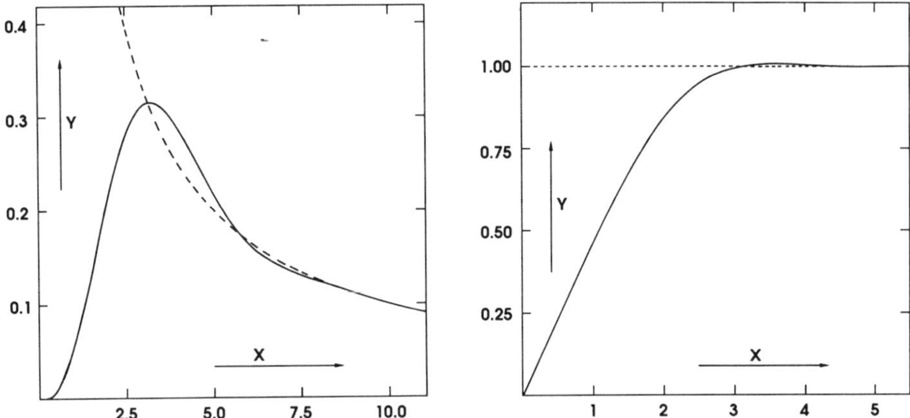

Fig. 2 (left) - The radial part of the radial component $u_r$ in the core boundary layer. Sweet's inviscid solution, $y = 1/x$, is indicated by a dashed curve.

Fig. 3 (right) - The radial part of the radial component $u_r$ in the surface boundary layer. Sweet's inviscid solution, $y = 1$, is indicated by a dashed line.

A and B in equation (9) can be obtained from equation (2) -- with $L_\phi = 0$ and, e.g., $\partial/\partial t = 0$. However, because the viscous force $F_\phi$ is directly proportional to $\mu_t$, there is no hope to calculate these functions with any accuracy. (A similar difficulty occurs in the theories of solar rotation; but, then, it is possible to adjust the theoretical rotation law to the observed surface rotation rate!) By making use of equation (2), one can at least find under what condition the asymptotic expansion (4) for $\Omega$ converges. This condition, $\varepsilon|\omega_1| < 1$, requires that one has $t_v < t_{ES}$, where $t_v$ is the viscous time scale. If we let $\mu_t = 10^N \mu_r$, where $\mu_r$ is the radiative viscosity, this result implies that our steady solutions remain valid as long as $\varepsilon 10^{6-N} < 1$ in the models. Since one has typically $\varepsilon \approx 10^{-2}$, one must thus let $N \gtrsim 4\text{-}5$ (say) -- a lower limit which is quite similar to those encountered in geophysics. To be specific, if one lets $10^N = \mu_t/\mu$, where $\mu_t$ is the vertical eddy viscosity and $\mu$ is the molecular viscosity (of air and water, successively), one has $N \approx 6\text{-}7$ in the Earth's atmosphere and $N \approx 3\text{-}5$ in the oceans! These "large" values are quite reasonable, because huge natural systems can always sustain a wide spectrum of small-scale motions with large Reynolds numbers. Laboratory experiments with rotating fluids, for which one generally finds that $N \approx 1$ or 2, are therefore not at all comparable.

## 4.3. Comments and Criticisms

What happens if one has $t_\nu \gtrsim t_{ES}$ in the models? In this case, as explained in Paper VIa (p. 313), one can no longer make use of an expansion in powers of $\varepsilon$ for the angular velocity, i.e., the full nonlinea-
rity of equation (2) must be retained in the calculations. Although
this will alter the solution for $\Omega$, there is no reason to believe that
the corresponding solution for $\vec{u}$ will be much different from that presented in Section 4.2. Anyhow, this is quite a challenging problem for
the theoretician. A preliminary attempt to formulate the full nonlinear problem has been made by Sakurai (1986, Sec. 6); no practical
results are yet available.

What happens if we let $\omega_0 \neq 1$ in equation (4), so that equation
(5) must be replaced by equation (8)? This problem has already been
considered in Paper I (pp. 347-350). It is a simple matter to show
that there always exists a surface boundary layer that prevents the $1/\rho$-
singularity in the interior, inviscid solution (see Fig. 10, p. 349, in
Paper I). Parenthetically note that one has $u_r \propto 1$ near the surface
when $\omega_0 = 1$ because, in this case only, the centrifugal potential can
be written, to $O(\varepsilon)$, as a combination of the Legendre polynomials $P_0$
and $P_2$. Whether one has $u_r \propto 1$ or $u_r \propto 1/\rho$ is quite irrelevant since
it is a mere artifact of the inviscid solution, which must always be
replaced by a boundary-layer solution near the free surface.

The claim has also been made that we should have found a double-
cell pattern in the meridional flow. As explained in the Appendix,
such a flow pattern does not exist when realistic assumptions are made.
Consequently, because the $1/\rho$-factor does not appear in the second-
order solutions, the often-quoted formula $u_r \propto \varepsilon^2/\rho$ is also quite irrelevant. In fact, to all orders in $\varepsilon$ there exist thermo-viscous boundary layers that allow the fluid to flow along the rotationally distorted
surface.

Since boundary condition (7) does not apply when there is mass
loss from the star's surface, one may also argue that the inviscid
solutions satisfying either equation (5) or equation (8) will remain
valid in this case. This is also incorrect because mass loss modifies
the formulation of the problem at two distinct places: (i) conservation
of the total mass requires the presence of a sink-term -- $s_m(r,\Theta,t)$,
say -- in the equation of continuity, and (ii) condition (7) must be
replaced by the requirement that $\vec{n}.\vec{u}$ is always equal to the mass-loss
speed -- $u_m(r,\Theta,t)$, say -- at the outer surface. Both functions $s_m$
and $u_m$ must be prescribed in an ad hoc manner. Obviously, there is no
reason to believe that the standard inviscid solutions, which do not
depend on mass loss, will satisfy this new constraint for any given
mass loss.

It has also been suggested that there exist other means to remove
the mathematical singularities in the meridional flow: a magnetic
field, a gradient of mean molecular weight, a nonlocal equation for
radiative transport, the inertial terms $\vec{u}.\mathrm{grad}\,\vec{u}$, etc. It is a simple
matter to show that none of these suggestions works! For example, let
us consider the so-called inertial boundary layer that was discussed by
Smith and Roxburgh (1977), and which is sometimes proposed as an

alternative solution to the surface thermo-viscous boundary layer
presented in Section 4.2.  A mere inspection of their paper shows
that their solution is not consistent because the circulation
velocity normal to the free surface does not vanish there, as it
is required by condition (7).  This is not surprising since it has
been known for some time that only a viscous boundary layer within
the inertial boundary layer could fulfill this requirement (e.g.,
Charney 1955).

## 4.4. Grounds for Turbulence in a Stellar Radiative Zone

It is immediately apparent from the newly proposed solution that
some kind of viscous action is needed for the following three
reasons: (i) the ever-present barotropic-baroclinic instabilities
in a rotating star generate a whole spectrum of anisotropic turbulent motions, (ii) the circulation velocities must always satisfy
the kinematical boundary condition (7), with $|\vec{u}|$ finite, and (iii)
in equation (2), a viscous force must be present to balance the
inexorable transport of specific angular momentum by the meridional flow.  (As explained in the Appendix, purely magnetic action
cannot do the job!)  Because the microscopic (molecular and radiative) viscosity is negligibly small in a star, it follows that
allowance must be made for eddy-like and/or wave-like motions in a
stellar radiative zone.  These rational arguments are undoubtedly
sufficient to convince any theoretician who is familiar with the
theory of real fluids.  Yet, because most astronomers seem to
belong to another school of thinking, it may not be inappropriate
to review some practical justifications for turbulence in a stellar radiative zone.  (This kind of turbulence, which is probably
not directly observable, should not be confused with turbulence in
a convective zone or with convection itself!)

The oblique-rotator model for the magnetic stars is a good
case to point.  Indeed, although there is no general agreement
about these matters, most advocates of this model believe that the
observed distribution of the obliquities is at most a marginally
nonrandom one.  Yet, if one assumes strict laminar, inviscid motions in a fully ionized radiative envelope, it is known that the
thermally-driven currents will rapidly convert a small-obliquity
field into one with an apparently large obliquity.  As was shown
in Paper X, the picture is quite different if one assumes that the
magnetic star is a less-than ideal body, having a wide spectrum of
small-scale motions that coexist with the largest-scale flow in
its radiative envelope.  In this case, because the field lines can
more easily diffuse through the body, it is found that the
Eddington-Vogt currents are by far too inefficient to produce a
perpendicular rotator over the main-sequence lifetime of a typical
magnetic star (see Fig. 1, p. 810, in Paper X).  In other words,
by making allowance for a moderate amount of dissipation in the
models, one can easily prevent the formation of too many perpendicular rotators within their main-sequence lifetime.  This requires
the presence of turbulence.

20

The nicest feature of the newly proposed solution is that
the meridional flow does not depend on turbulent friction in the
bulk of a radiative envelope, and that it is almost independent of
the eddies and/or waves in the surface boundary layer. Accordingly, the circulation velocities presented in Papers I and II are
more than adequate to discuss the influence of large-scale meridional streaming on the gravitational sorting of the elements in
the (single and double) early-type stars (Charbonneau et al.
1988a,b, 1989). Since they have made their point clearly, I shall
not repeat the many arguments that are in favor of both the diffusion model and the newly proposed solution for the meridional
currents in a radiative envelope. Again, consistency requires
some turbulence.

The case for the turbulent diffusion of the elements in the
A-type stars is not so clear, however, because we do not know the
magnitude of the vertical coefficient of eddy diffusivity with any
degree of certainty (Sec. 3.1, Charbonneau and Michaud 1990).
Yet, various measurements made in the laboratory and in the
Earth's atmosphere indicate that the eddy diffusivity of matter
decreases more drastically than that of momentum as the gradient
Richardson number increases in a stably stratified system (Paper
VIIIa, p. 400, and references therein). In other words, the fact
that there is some turbulent friction acting in the surface layers
of an early-type star does not imply that the concomitant turbulent diffusion of matter should be equally efficient in these
layers. Anyhow, whichever model is finally corroborated by observations will require some turbulence in the surface layers.

The recent observational data also indicates that much of
the Sun's radiative interior is rotating at a rate close to that
of the surface equatorial belt, while the hydrogen-burning core is
apparently rotating more rapidly than the chemically homogeneous
parts of the core. As was pointed out in Papers VIII and VIIIa,
the Sun's interior differs in two important respects from the
radiative envelope of an early-type star: (i) the Eddington-Vogt
currents are utterly negligible in the solar core because the
circulation time is much larger than the Sun's age, and (ii) angular momentum is continuously transferred away from the solar convective envelope to outer space. Accordingly, there must exist a
very efficient mechanism of angular momentum transport that keeps
the inner and outer parts of the radiative interior rotating nearly uniformly in spite of the inexorable solar-wind torque. Again,
it has been shown that a moderate amount of turbulence in the
chemically homogeneous parts of the radiative core is sufficient
to reproduce the present quasi-solid inner and outer rotation
rates of the Sun. Moreover, the constraints on the vertical eddy
viscosity derived from the turbulent diffusion of momentum are in
good agreement with those derived from the turbulent mixing of
material and the solar surface abundances (Pinsonneault et al.
1989, Paper VIIIa). This is "nonstandard" stellar evolution theory.

## 5. THE MECHANICALLY-DRIVEN CURRENTS

In detached close binaries, tidal interaction will continuously change the axial rotation of the components as well as the eccentricity of their orbits. Unless the binary components rotate in perfect synchronism with a circular orbital motion, each star senses a variable external gravitational field, thus becoming liable to oscillatory motions which can be described as an "equilibrium tide" and a "dynamical tide". The former one is the instantaneous shape obtained by assuming that strict mechanical equilibrium prevails; the latter one refers to the dynamical response of the star to the tidal forcing of its natural modes of oscillation.

As far back as 1879, Darwin pointed out that viscous dissipation produces a slight misalignment of the tidal bulges with respect to the lines joining the two centers of mass; this small tidal lag, in turn, produces a torque which tends to synchronize the axial and orbital motions. Unfortunately, because turbulent viscosity is quite small in a stellar radiative envelope, this braking mechanism is much too weak to account for the high degree of synchronism or pseudo-synchronism (i.e., synchronism at periastron) that is observed in the early-type binaries. Following Cowling's (1941) pioneering work, Zahn (1984) has shown that the retardation process caused by the tidal forcing of the gravity modes is also a quite inefficient one. To be specific, this mechanism for synchronization remains effective up to the distance ratio $d/R \approx 6$, whereas pseudo-synchronization extends up to $d/R \approx 20$ in the early-type binaries (Giuricin et al. 1984b).

Attempts to patch up Zahn's (1984) calculations have already been made. Goldreich and Nicholson (1989) have pointed out that the retardation process caused by the tidal forcing of the gravity modes proceeds from the outside toward the inside of a star. According to Rocca (1989), the proper inclusion of the Coriolis force can also enhance the process at the beginning of the tidal evolution. Although these suggestions might somewhat improve the efficiency of the dynamical-tide theory, they do not change the inescapable fact that this mechanism in a <u>short-range</u> one, since its characteristic time for synchronization is proportional to $(d/R)^{8.5}$.

Following Papers A and D, I shall present another braking mechanism, which is much more efficient than the classical ones, but has hitherto escaped notice. It involves a large-scale meridional flow, superposed on the motion around the rotation axis of the tidally distorted component. These transient, mechanically-driven currents -- which are much faster than the steady, thermally-driven Eddington-Vogt currents -- cease to exist as soon as synchronization has been achieved in the star. The newly proposed mechanism is quite effective because the corresponding synchronization time depends primarily on $(d/R)^{4.125}$. This fact makes it a rather <u>long-range</u> mechanism, as compared to the braking caused by the tidal forcing of the gravity modes. It also explains the observations quite naturally.

22

## 5.1. The Hydrodynamical Spin Down

We consider a system of two rotating stars in circular orbits about their common center of mass, with separation d and orbital period $P_0 = 2\pi/\Omega_0$. The rotation axes are perpendicular to the orbital plane. We take as the origin of our system of coordinates the center of mass of the primary -- an early-type star of mass M and radius R. The z-axis is parallel to the rotation axis, and the x-axis points to the secondary -- a mass-point M'. Since we are not interested in the two classical retardation mechanisms, we shall thus neglect both the small tidal lag and the gravity modes.

Now, if synchronization has not yet been achieved, it is evident that the primary is not at rest with respect to the frame corotating with the orbital angular velocity $\Omega_0$. To be specific, if $\Omega_i$ is a typical value of the initial rotational angular velocity of the primary, then the velocity $\vec{v}$ in the chosen frame is of the order of $(\Omega_i - \Omega_0)\, \omega \vec{1}_\psi$. (In this corotating frame, thus, a state of perfect synchronism corresponds to $\vec{v} = 0$.) Yet, because of the presence of the secondary, each fluid particle moving on the surface of the primary is forced to move along an ellipse, with slight accelerations and decelerations along its trajectory. (Recall that the tidal bulges are steady in the corotating frame, the external gravitational attraction thus acting as if it was a solid container!) This is illustrated on Figure 4, where the four arrows indicate the tidal attraction corrected for the tidal attraction at the center of mass of the primary. These small variations in the azimuthal direction imply that one has $\partial v_\psi/\partial\psi \neq 0$ so that, by virtue of mass conservation, one also has $v_r \neq 0$ and $v_\Theta \neq 0$. In other words, a large-scale meridional flow must necessarily come into existence in a nonsynchronous binary component. Obviously, these time-dependent currents vanish altogether when the primary has reached a state of hydrostatic equilibrium in the corotating frame, that is, when synchronism has been attained.

As explained in Paper D, boundary-layer theory can be used to describe the general features of these time-dependent motions. To make a long story short, one writes

$$\vec{v} = \sum_k \vec{v}_k \ (r,\Theta,\psi) \ \exp\ (-\alpha_k t), \qquad (10)$$

where the $\vec{v}_k$'s can be expanded in terms of radial functions and spherical harmonics (k = 1,2,...). This expansion must satisfy the basic equations of the problem, as well as appropriate boundary conditions. In particular, one must prescribe that the vector $\vec{v}$ remains everywhere parallel to the outer boundary. Thus one has, at every instant,

$$\vec{n}.\vec{v} = 0, \qquad (11)$$

on the tidally and rotationally distorted surface of the primary. The stress vector must also vanish on this surface. These bounda-

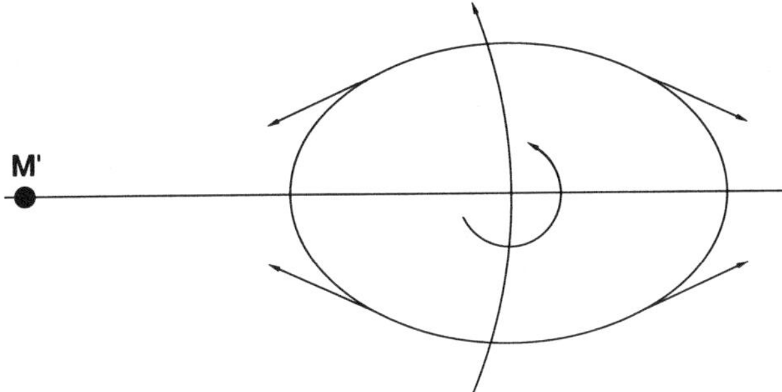

Fig. 4 - The relative tidal attraction due to the mass-point M' at four places in the equatorial belt of the primary. The rotation axis is perpendicular to the plane of this schematic drawing. The vertical arrow indicates the sense of the orbital motion. (The small tidal lag is not represented.)

ry conditions are essential because they define the permissible values for the $\alpha_k$'s in equation (10). The lowest eigenvalue $\alpha_1$ is the most important one, since it defines the e-folding time of the decaying motions, $t_{sd} = 1/\alpha_1$, which is also the spin-down (or spin-up) time of the newly proposed braking mechanism.

Detailed mathematical calculations show that there always exists a thin Ekman-type <u>suction</u> layer that induces a large-scale flow of matter within the practically inviscid interior of a non-synchronous binary component. In the case of a spin down ($\Omega_i > \Omega_0$), these motions correspond to a quadrupolar circulation pattern that is weakly dependent on longitude, the fluid entering the boundary layer in the equatorial belt and returning with decreased angular momentum to the poles. (The reverse phenomenon occurs when $\Omega_i < \Omega_0$.) If the departure from synchronism is small (i.e., if $\xi = |\Omega_i - \Omega_0| / \Omega_0 < 1$), this time-dependent advection of matter and, hence, of angular momentum -- which is regulated by the free boundary layer -- brings effective synchronization within a time of the order of

$$t_{sd} = \frac{1}{4\pi} P_0 \left(\frac{\delta}{R}\right)^{-1} \frac{M}{M'} \left(\frac{d}{R}\right)^3 , \qquad (12)$$

where $\delta$ is the boundary-layer thickness. Typically, the radio $\delta/R$ lies in the range of $10^{-4}$ - $10^{-3}$ in a realistic stellar model.

Following Paper A, one can also derive this time scale by means of a simple physical argument. Indeed, one readily sees that $|\partial v_\phi/\partial \phi|$ is of the order $\eta |\Omega_i - \Omega_0| \omega$ in the surface boundary layer, so that the equation of continuity implies a mass exchange between this suction layer and the inviscid interior with a velo-

24

city of magnitude $(\delta/R)\eta|\Omega_i-\Omega_0|\omega$, where $\eta = (M'/M)(R/d)^3$.  Since R
is the typical dimension of the radiative envelope, it follows at
once that one has $|\vec{u}| \approx (\delta/R)\eta|\Omega_i-\Omega_0|\omega$ in this region.  Now, be-
cause the specific angular momentum is approximately conserved in
the bulk of the inviscid interior, a fluid particle moving outward
the distance $\xi\omega/2$ will acquire the lower angular velocity $\Omega_0$; that
is, it will be at rest in the corotating frame.  Letting D be this
distance, one readily sees that the spin-down time must be of the
order of $D/|\vec{u}|$ , which is the value defined in equation (12).

## 5.2.  Comments and Criticisms

Physically, this spin-down (or spin-up) process is quite similar
to the braking mechanism that is responsible for the decay of a
cyclonic depression in the Earth's atmosphere (Charney and
Eliassen 1948).  It bears also much resemblance to the decay of
the whirling motion created when a cup of tea is stirred
(Greenspan and Howard 1963).  In this case, near the bottom of the
cup viscosity slows down the motion, so that a radial inflow of
matter takes place in a thin boundary layer.  By continuity, this
radial inflow in the bottom boundary layer requires upward motion
along the axis and a slow compensatory outward radial flow
throughout the remaining inviscid parts of the cup.  By virtue of
equation (2), this slow outward motion approximately conserves
specific angular momentum; hence, by replacing high angular-velocity fluid by low angular-velocity fluid, it serves to spin down
the whirling motion far more rapidly than could mere viscous friction.  Visualization of this transient meridional flow is provided
by the tea leaves, which are always observed to cluster near the
center at the bottom of the fluid (e.g., Holton 1972).
    The claim has been made that the spin-down mechanism will
not work in stars because their boundaries are free and not solid.
This remark is incorrect because, in both cases, a viscous boundary layer is needed to satisfy the appropriate boundary conditions,
either the no-slip condition near a solid boundary, or condition
(11) in a binary component.  In fact, all these problems have one
essential feature in common:  a variation in the azimuthal velocity near the boundary.  And it makes no difference whether this
azimuthal variation is due to friction on a solid wall, or whether
it is the external gravitational attraction that forces the fluid
particles to move nonuniformly along fixed ellipses on the free
surface.
    The claim has been made that the spin-down mechanism will
not work in stars for, otherwise, a tidally-induced boundary layer
would be observed on Earth, at the free surface of the oceans.
This is also an incorrect remark, because the newly proposed mechanism is operative only in bodies that are fluid all the way
from the center to the surface.  Any comparison with the Earth-
Moon system is therefore irrelevant since the oceans overlying the
Earth's crust do not penetrate into the mantle.  Another confusion
may arise because there is also an Ekman boundary layer at the top

of the oceans. This layer, which is caused by the winds blowing over the seas, is a "passive" one because it is merely needed to satisfy the surface boundary conditions. On the contrary, the boundary layer discussed in Section 5.1 is an Ekman-type suction layer. It is an "active" one in the sense that, because $\overline{\delta v_\phi/\delta\phi} \neq 0$, it regulates the exchange of matter between the surface and the bulk of the radiative envelope. There is thus no need to have a wind blowing over the star's surface to enforce the so-called Ekman-pumping mechanism!

Since this Ekman pumping does not contribute much to the Sun's retardation process, one may argue that it is not very efficient in the double stars either. Such a comparison is irrelevant because the two problems are quite different. In the double-star problem, a nonsynchronous component is never in hydrostatic equilibrium in our corotating frame so that, near the surface, the Coriolis force and the viscous force must already play a role to $O(\eta)$ in the mechanical balance. In other words, it is this lack of mechanical equilibrium that produces an Ekman-type suction layer in a tidally distorted star. By contrast, because the Sun can readjust itself continuously to mechanical equilibrium in spite of the solar-wind torque, the Ekman-type suction mechanism is an effect of $O(\varepsilon^2)$ in the small rotational parameter $\varepsilon$. Hence, it can be neglected in the Sun whereas it is of paramount importance in the close binaries.

Parenthetically note that Hyun (1983) has recently considered the spin up from rest of a thermally-stratified fluid in a finite, circular cylinder. The most important consequence of this vertical thermal stratification is that the inviscid meridional flow is confined to a region closer to the Ekman layers, the impulsive spin up being controlled in the bulk of the cylinder by advection of matter and viscous diffusion. On the basis of this hydrodynamical result, one may therefore argue that the efficiency of the newly proposed mechanism for the close binaries is also much reduced. Again, such a comparison is irrelevant for the following reasons. First, in a distant star, one always measures the rotation rate of its surface layers, which coincide with the Ekman-type suction layer. In fact, all relevant studies refer to synchronism between the surface rotation of a star and its orbital motion. Second, in a stellar radiative envelope, there always exist thermally-driven Eddington-Vogt currents that take care of its thermal stratification. The effects of thermal stratification in a star and in the above experiment are not comparable, therefore.

The claim has been made that a gradient of density will inhibit the mechanically-driven currents in a tidally distorted star. As explained in Paper D, this is also incorrect because well-defined meridional currents do exist in all cases, no matter whether there is a large or a small density gradient. (One should not confuse density gradient for gradient of mean molecular weight!) In a realistic stellar model, with electron-scattering opacity, the boundary-layer thickness is proportional to $\mu_t^{1/4}$, rather than to $\mu_t^{1/2}$ as in an incompressible fluid. Moreover, the meridional currents are probably confined within the radiative envelope, the core-envelope interface acting as an effective barrier. This is of little concern to us, since only the surface rotation rates can be measured. On theoretical grounds, however, a concomitant braking of the convective core by turbulent diffusion of momentum is quite plausible.

Finally, let us also note that many authors have attempted to calculate the three-dimensional velocity field in a tidally distorted star (e.g., Scharlemann 1981, Sec. III and references therein, Campbell and Papaloizou 1983). According to these authors, the main effect of turbulent viscous friction is that, as a result of energy losses, the magnitude of the tidal velocity decreases with time -- synchronism being then achieved in a time of the order of $t_v(d/R)^6$. Unfortunately, because these authors have ignored the essential boundary condition (11) altogether, they all missed the unsteady velocity field (10) that brings effective synchronization in a time much shorter than the time provided by viscous dissipation alone. In fact, there are three time scales in the problem: (i) the dynamical time over which the Ekman-type suction layer develops, (ii) the spin-down time (12) over which synchronization is effectively achieved, and (iii) the dissipation time, $t_v(d/R)^6$, over which the small residual motions are eventually dissipated by turbulent friction.

In my opinion, the newly proposed mechanism has hitherto escaped notice because too much attention was paid to the (slow) viscous dissipation of vorticity in a tidally distorted star, whereas no attention was being paid at all to the (much faster) advection of angular momentum by the tidally-induced meridional flow. This is another way of saying that the formulation of a hydrodynamical problem remains incomplete until one has properly prescribed the boundary conditions.

## 5.3. Comparison Between Theory and Observation

Because we have assumed that the difference between the (unknown) initial angular velocity $\Omega_i$ and the (known) orbital angular velocity $\Omega_0$ is small, equation (12) gives no more than a lower limit on the actual synchronization time, $t_{syn}$ (say), in a real binary. Tentatively, in Papers A-D we have made the reasonable assumptions that $t_{syn}$ is one order of magnitude larger than $t_{sd}$, and we took the eddy viscosity as a free parameter. Fortunately, because $\delta$ is proportional to $\mu_t^{1/4}$, the dependence of the time $t_{syn}$ on the eddy viscosity is considerably reduced; its dependence on the nonlinearity when $|\Omega_i - \Omega_0| >> \Omega_0$ cannot be ascertained at this time, however.

Detailed comparison between theory and observation has been made in Papers A-D. Here I shall only summarize the salient features of these semi-quantitative discussions.

As was shown in Papers A and B, the hydrodynamical spin-down mechanism is a very efficient one because, by virtue of equation (12) and Kepler's third law, $t_{syn}$ depends primarily on $(d/R)^{4.125}$

or $(P_0)^{2.75}$. This fact makes it a long-range mechanism that can easily explain the high degree of synchronism (or pseudo-synchronism) that is observed up to $d/R \approx 20$ in the early-type binaries (Giuricin et al. 1984b). It also explains why the longer period, A-type binaries have much reduced rotational velocities, effective synchronization having not been achieved in these stars (Abt and Levy 1985).

The circularization process that is concomitant with this very efficient braking mechanism has been discussed in Papers B and C. Because the corresponding time -- $t_{cir}$ (say) -- is proportional to $(d/R)^{6.125}$ or $(P_0)^{4.083}$, the newly proposed mechanism is also effective in inducing orbital circularization in the close (and not-so-close) early-type binaries. In this case, one can even compare the data for the massive close binaries, with circular or almost circular orbits up to $P_0 \approx 30$ d (e.g., Massey 1982, p. 258), and the less massive ones, with circular orbits up to $P_0 \approx 2$ d (Giuricin et al. 1984a). If $t_{cir}$ was dependent on $P_0$ only, the theory would not be in agreement with both groups of binaries. Fortunately, because $t_{cir}$ also depends on the ratio $M^2/R^5$ which is rapidly varying along the upper main sequence, the theory is able to account for the large difference between these two upper period limits. As can readily be seen from Zahn's (1984, p. 385) Table 1, the dynamical-tide theory is unable to account for the existence of circular orbits among the wider early-type binaries. To be specific, this theory is effective only up to $P_0 \approx 1-2$ d in the mass range 2-10 $M_{\odot}$, and it is quite ineffective beyond $P_0 \approx 3$ d in the more massive binaries. It is a short-range mechanism, therefore, as compared to the newly proposed one.


## 6. CONCLUDING REMARKS

Although this review paper was devoted to radiative zones almost exclusively, I hope I made it clear that the modelling of a rotating star is primarily an exercise in physical fluid dynamics. That is to say, in order to build a consistent model, one must solve all the basic equations together with all the boundary conditions, thus retaining all the pertaining forces. Magnetic fields are optional, however, because hydrodynamics provides seven equations for the seven unknowns ($p$, $\rho$, $T$, $V$, and $\vec{v}$), the number of boundary conditions being just sufficient to prescribe completely the solutions. By contrast, the new results presented in Sections 4 and 5 clearly indicate that the large-scale dynamics in a rotating star always demands some frictional forces to be present.

The thermally-driven Eddington-Vogt currents and concomitant differential rotation are a good case in point since, obviously, one cannot satisfy equations (2) and (7) without making allowance for some turbulent dissipation in the models. In this case, however, it is the transport of self-generated radiation in a non-spherical star that causes the large-scale currents. Accordingly, except in "passive" boundary layers where turbulent friction merely allows the circulation velocities to satisfy boundary condition (7), the meridional flow can be obtained from the conservation of energy. This explains why the formulae derived by Sweet (1950) and Baker and Kippenhahn (1959) have withstood the passage of time in spite of the fact that these strictly inviscid solutions must be replaced by viscous ones near the outer boundaries. As explained in Section 4.3, when there is mass loss, turbulent friction is also required to obtain acceptable solutions in the outermost layers of an early-type star.

On the contrary, the large-scale motions in a nonsynchronous binary component are primarily caused by the requirement that the fluid must flow along the free surface of the triaxial body. Thus, in this case, it is the surface boundary condition (11) that controls the motions in the tidally and rotationally distorted component. This leads to the formation of an "active" boundary layer that exchanges mass and angular momentum with the inviscid interior of the star. This transient flow, in turn, will tend to synchronize the axial and orbital motions in a quite efficient manner. Evidently, one cannot calculate these braking currents unless one has properly streamlined the non-spherical surface of the model. As explained in Section 5.1, this requirement leads at once to an eigenvalue problem which does not seem to have been considered before. In my opinion, too much reliance upon standard potential theory and stellar-pulsation calculations are probably among the reasons for this neglect.


APPENDIX: THE GRATTON-OPIK PARADOX

It is often believed that the circulation pattern in a stellar radiative envelope consists of two distinct cells separated by a particular level surface. In order to show the incorrectness of this statement, I shall closely scrutinize Mestel's (1966) proof because it is quite general and does not depend on approximate expansions.

We consider a uniformly rotating, chemically homogeneous body in which the viscous forces can be neglected altogether. As was pointed out in Section 3, these approximations imply that there exist level surfaces, which coincide with the isobaric, isopycnic, and isothermal surfaces. Conservation of energy then reduces to

$$\rho A(\Phi)\vec{u}.\mathrm{grad}\Phi = f(\Phi)(4\pi G\rho - 2\Omega^2) + f'(\Phi)g^2, \qquad (A1)$$

where $\vec{u}$ is the circulation velocity, $f'(\Phi) = df/d\Phi$, and $g = d\Phi/dn$ is the magnitude of the effective gravity. $A(\Phi)$ and $f(\Phi)$ take constant values on each level surface; their definition is irrelevant for the present discussion. Dividing equation (A1) by $g$ and integrating over a level surface, one obtains

$$f(\Phi)(4\pi G\rho - 2\Omega^2) \ <g^{-1}> \ + f'(\Phi) \ <g> \ = 0, \qquad\qquad (A2)$$

since in a steady state there can be no flux of matter across a closed surface. (A bracket designates a mean value over a level surface.) Combining next equations (A1) and (A2), one can write

$$\rho A(\Phi)\vec{u}.\mathrm{grad}\Phi = f'(\Phi) \ (g^2 \ - <g> / <g^{-1}> \ ). \qquad\qquad (A3)$$

If the function $f'(\Phi)$ vanishes for a value $\Phi^*$ (say), this equation implies that the circulatory flow does not cross the corresponding level surface. From equation (A2), $f'(\Phi)$ vanishes on that level surface with density $\rho^*(\Phi^*)$ given by $\Omega^2 = 2\pi G\rho^*$. This concludes the proof that there apparently exists a double-cell pattern in an inviscid, uniformly rotating radiative envelope.

Let us first consider a configuration for which one can neglect the inertial terms $\vec{u}.\mathrm{grad}\ \vec{u}$ in the equations of motion. In this case, it is immediately apparent that the velocity $\vec{u}$ is present only in the equations expressing conservation of mass and energy. From the $r$-and $\Theta$-components of the equations of motion, the equation of state, and Poisson's equation, one can thus calculate unequivocally the pressure $p$, the density $\rho$, the temperature $T$, and the potential $V$ -- all these functions being independent of the velocity $\vec{u}$. Hence, the potential $\Phi = V - \Omega^2\omega^2/2$ is also completely determined, and it does not depend on the circulation velocity either. Since equation (A2) is independent of the four equations that were used to calculate the potential $\Phi$, there is thus no reason to believe that the constraint (A2) will be satisfied by the functions $<g>$ and $<g^{-1}>$ that one has derived from the known potential $\Phi$. Thus, in this case, the double-cell pattern is a mere paradox.

If one cannot neglect the inertial terms in the equations of motion, then the functions $p$, $\rho$, $T$ and $V$ actually depend on the vector $\vec{u}$, so that the four families ($p$ = constant, ...) do not necessarily coincide. Accordingly, even if one assumes that they do coincide approximately, the dependence on $\vec{u}$ implies at once that equation (A3) is no longer an explicit expression for the velocity $\vec{u}$: it has become a complex integro-differential equation, both sides depending on the velocity $\vec{u}$. This, in turn, implies that the particular level surface $\Phi^*$ = constant no longer plays a particular role. Again, one cannot claim that there exists a double-cell pattern.

If these arguments are not sufficient to convince the reader that we are dealing here with a mere paradox, then we may ask the following question: is it actually possible to have a strictly inviscid body that has all the properties of a barotrope? Recall that Mestel's (1966) proof rests on the fact that the inviscid body remains strictly steady and barotropic in spite of the inexorable Eddington-Vogt currents. If not, then the system is a barocline, i.e., the density and the temperature are <u>not</u> in general constants on the isobaric surfaces. This, in turn, implies that the quantities A and f are <u>not</u> constants on the isobaric surfaces,

30

so that one can no longer write the simple relation (A2) from
which one has deduced the double-cell pattern.

Now, the claim has often been made (without any proof) that
there exists an inconspicuous magnetic field that can enforce <u>by
itself</u> almost uniform rotation in a stellar radiative zone.
Strictly without dissipation, Mestel (1961) has obtained the exact
solution

$$\Omega = \frac{\alpha + \eta\beta/\rho\omega^2}{1-4\pi\eta^2/\rho} \qquad (A4)$$

and a similar expression for the toroidal magnetic field. Here $\alpha$,
$\beta$ and $\eta$ are constants on the poloidal field lines, which coincide
everywhere with the streamlines of the meridional circulation.
Obviously, Mestel's (1961) so-called isorotation law is singular
on the axis, since $\Omega \propto 1/\omega^2$ as $\omega \to 0$. As explained in Paper IXa,
<u>with little or no turbulent friction</u>, the rotation rate in a ra-
diative envelope must necessarily tend toward a solution which is
similar to equation (A4) and has, in general, a large gradient in
the angular velocity near the rotation axis. That is to say,
unless one includes some kind of viscous action  -- which then
counters the local nonuniformities in the rotation rate that are
due to the magnetic field --  it is impossible to obtain a quasi-
solid rotation rate, <u>all the way from the rotation axis to the
free surface.</u>

To sum up, we have shown that the double-cell pattern is a
mere consequence of an excessively large number of conflicting
assumptions that cannot be met in a realistic stellar model.
Although we have scrutinized Mestel's (1966) proof only, it is a
simple matter to show that the other proofs based on equation (A2)
and truncated expansions are also incorrect (Gratton 1945, Opik
1951, Pavlov and Yakovlev 1978). Furthermore, Maheswaran's (1968)
derivation, which is not based on equation (A2), contains alge-
braic errors. When these errors are corrected, one finds that the
second-order expansions no longer contain the factor $(1 - \Omega^2/2\pi G\rho)$
which leads to the double-cell pattern and the inadequate formula
$u_r \propto \varepsilon^2/\rho$ near the outer surface.


REFERENCES


Abt, H.A., Levy, S.G. 1985, Ap.J. Suppl., 59, 229.
Aikawa, T. 1970, Sci. Reports Tôhoku Univ., I, 53, 21.
Baker, N., Kippenhahn, R. 1959, Z. f. Ap., 48, 140.
Brunet, P. 1931, L'introduction des théories de Newton en France
        au XVIIIe siècle, pp. 223-228 (Genève: Slatkine Reprints,
        1970).
Campbell, C.G., Papaloizou, J. 1983, M.N.R.A.S., 204, 433.
Charbonneau, P., Michaud, G. 1988a, Ap.J., 327, 809.
    ------. 1988b, Ap.J., 334, 746.
    ------. 1990, Ap.J., 352, in press.

Charbonneau, P., Michaud, G., Proffitt, C.R. 1989, Ap.J., 347, in
     press.
Charney, J.G. 1955, Proc. Natl. Acad. Sci. USA, 41, 731.
Charney, J.G., Eliassen, A. 1948, Tellus, Vol. 1, No. 2, p. 38.
Cowling, T.G. 1941, M.N.R.A.S., 101, 367.
Eddington, A.S. 1925, The Observatory, 48, 73.
     -----. 1929, M.N.R.A.S., 90, 54.
Fontenelle, B. 1973, Entretiens sur la pluralité des mondes, p.
     116 (Verviers: Marabout Université).
Fujimoto, M.Y. 1987, Astr. Ap., 176, 53.
Giuricin, G., Mardirossian, F., Mezzetti, M. 1984a, Astr. Ap.,
     134, 365.
     -----. 1984b, Astr. Ap., 135, 393.
Goldreich, P., Nicholson, P.D. 1989, Ap.J., 342, 1079.
Goldreich, P., Schubert, G. 1967, Ap.J., 150, 571.
Gratton, L. 1945, Mem. Soc. Astr. Italiana, 17, 5.
Greenspan, H.P., Howard, L.N. 1963, J. Fluid Mech., 17, 385.
Holton, J.R. 1972, An Introduction to Dynamic Meteorology, pp.
     88-92 (New York: Academic Press).
Hyun, J.M. 1983, Geophys. Ap. Fluid Dyn., 23, 127.
Kreiken, E.A. 1935, Z. f. Ap., 10, 199.
Krogdahl, W. 1944, Ap.J., 99, 191.
Lorimer, G.S., Monaghan, J.J. 1980, Proc. Astr. Soc. Australia, 4,
     45.
Maheswaran, M. 1968, M.N.R.A.S., 140, 93.
Massey, P. 1982, IAU Symposium 99, Wolf-Rayet Stars: Observations,
     Physics, Evolution, eds. C.W.H. de Loore, A.J. Willis (Dor-
     drecht: Reidel), p. 251.
Mestel, L. 1953, M.N.R.A.S., 113, 716.
     -----. 1961, M.N.R.A.S., 122, 473.
     -----. 1966, Z. f. Ap., 63, 196.
Milne, E.A. 1923, M.N.R.A.S., 83, 118.
Opik, E.J. 1951, M.N.R.A.S., 111, 278.
Pavlov, G.G., Yakovlev, D.G. 1978, Astr. Zh., 55, 1043.
Pinsonneault, M.H., Kawaler, S.D., Sofia, S., Demarque, P. 1989,
     Ap.J., 338, 424.
Randers, G. 1941, Ap.J., 94, 109.
Rocca, A. 1989, Astr. Ap., 213, 114.
Sakurai, T. 1986, Geophys. Ap. Fluid Dyn., 36, 257.
Scharlemann, E.T. 1981, Ap.J., 246, 292.
Schwarzschild, M. 1947, Ap.J., 106, 427.
Shibahashi, H. 1980, Publ. Astr. Soc. Japan, 32, 341.
Smith, B.L., Roxburgh, I.W. 1977, Astr. Ap., 61, 747.
Smith, R.C. 1987, IAU Colloquium 92, Physics of Be Stars, eds. A.
     Slettebak, T.P. Snow (Cambridge: Cambridge Univ. Press), p.
     123.
Struve, O., Elvey, C.T. 1931, M.N.R.A.S., 91, 663.
Sweet, P.A. 1950, M.N.R.A.S., 110, 548.
Swings, P. 1936, Z. f. Ap., 12, 40.

Tassoul, J.L. 1978, Theory of Rotating Stars (Princeton:
    Princeton Univ. Press) (T.R.S.).
  -----. 1987, Ap.J., 322, 856 (Paper A).
  -----. 1988, Ap.J. Letters, 324, L71 (Paper B).
  -----. 1990, preprint (Paper C).
Tassoul, J.L., Tassoul, M. 1982a, Ap.J. Suppl., 49, 317 (Paper I).
  -----. 1982b, Ap.J., 261, 265 (Paper II).
  -----. 1982c, Ap.J., 261, 273 (Paper III).
  -----. 1983a, Ap.J., 264, 298 (Paper IV).
  -----. 1986a, Ap.J., 310, 786 (Paper IX).
  -----. 1986b, Ap.J., 310, 805 (Paper X).
  -----. 1986c, Geophys. Ap. Fluid Dyn., 36, 303 (Paper VIa).
  -----. 1989a, Astr. Ap., 213, 397 (Paper VIIIa).
  -----. 1990, preprint (Paper D).
Tassoul, M., Tassoul, J.L. 1983b, Ap.J., 267, 334 (Paper V).
  -----. 1983c, Ap.J., 271, 315 (Paper VI).
  -----. 1984a, Ap.J., 279, 384 (Paper VII).
  -----. 1984b, Ap.J., 286, 350 (Paper VIII).
  -----. 1988, M.N.R.A.S., 232, 481 (Paper IIIa).
  -----. 1989b, Ap.J., 345, 472 (Paper IXa).
Vogt, H. 1925, Astr. Nachr., 223, 229.
von Zeipel, H. 1924, M.N.R.A.S., 84, 665.
Zahn, J.P. 1984, IAU Symposium 105, Observational Tests of the
    Stellar Evolution Theory, eds. A. Maeder, A. Renzini (Dor-
    drecht:  Reidel), p. 379.

# NEW EVOLUTIONARY ASPECTS OF MASS LOSS AND ANGULAR MOMENTUM


*André Maeder*
*Geneva Observatory*
*CH-1290 Sauverny, Switzerland*



ABSTRACT. After a brief summary of the main effects of mass loss on massive star evolution, the interest is focused on several topical problems, likely promised to interesting future developments. In particular, the effects of metallicity on the evolution of massive stars and on their very different distributions in galaxies are firstly considered. We then review the case of the vibrational pulsations theoretically predicted in Wolf–Rayet stars and their possible relation with the high mass loss rates of these objects. The possible origin of the instabilities of the Luminous Blue Variables (LBV) and of their huge mass ejections is also discussed. Finally, a few possible evidences for rotationally or tidally induced mixing in massive O–stars are presented, suggesting that some very massive stars may evolve close to homogeneity.


## 1   INTRODUCTION

Over the last decade there has been a large number of papers and reviews devoted to the evolution with mass loss and to some effects of stellar rotation . Thus, rather than making one more review on these subjects, I shall only give here a condensed summary of the main evolutionary effects of mass loss and then make a progress report on a few selected topics, which are the areas of great future potential developments. These selected topics are:

- The metallicity effects on the mass loss and evolution of massive stars and on their distributions in galaxies.

- The yet undiscovered vibrational instabilities of WR stars in relation with their high mass loss rates.

- Physical considerations on the instabilities and mass ejection of LBV stars.

- The rotational mixing and possible homogeneous evolution of a fraction of the massive O–stars.

These are certainly among the most uncertain problems related to mass loss and rotation. This is precisely why they need to be examined and discussed in this very useful astrophysics conference.

33


*L. A. Willson and R. Stalio (eds.), Angular Momentum and Mass Loss for Hot Stars, 33–51.*

## 2 BASIC CONTEXT OF MASS LOSS AND STELLAR EVOLUTION

The main effects of mass loss on the inner stellar evolution and their observational consequences have been discussed by Chiosi and Maeder (1986). Further points have also been considered by Maeder and Meynet (1987, 1989) and by Langer (1989ab).

Table 1 summarizes very briefly the main effects of mass loss on massive star evolution. For the currently observed mass loss rates, the consequences of stellar winds on main sequence (MS) evolution are small. The lifetime is $t_{MS} \sim q_c\, M/L$. The mass of the convective core $q_c\, M$ decreases with increasing mass loss, while the luminosity L does the same. Thus, the net result for $t_{MS}$ is small: at most an increase by 5 – 10%, associated to moderately small differences in the tracks.

On the contrary, mass loss effects on the He–burning phase are enormous. The lifetime $t_{He}$ in the He–burning phase is usually shared, for massive stars, between the blue supergiant, the red supergiant and the WR phases: $t_{He} \simeq t_{BSG} + t_{RSG} + t_{WR}$. The sharing and balance between these three stages depend very sensitively on the mass loss rates $\dot{M}$. Typically, for an initial 30 $M_\odot$ star, in case of no mass loss, $t_{He} \simeq t_{BSG}$: the whole He–phase is spent in the blue. This is mainly due to the large intermediate convective zone (cf. Stothers and Chin 1978; Maeder 1981) which keeps a large zone more homogeneous both in composition and temperature, thus the blue location. For increasing $\dot{M}$–rates, the time fraction spent in the blue declines, and thus the star moves to the red supergiant stage earlier and the time fraction spent in this phase is thus increased. However, the larger the mass loss rates in the red supergiant phase, the earlier the star will be pealed off and evolve to the stage of a bare core, generally identified with the WR stage: henceforth, the surprising behaviour of $t_{RSG}/t_{He}$ with $\dot{M}$ illustrated in Table 1 and the very understandable growth of $t_{WR}$ with mass loss rates.

The above effects lead to different evolutionary sequences according to the ranges of initial stellar masses (for given $\dot{M}$–rates).

| | | |
|---|---|---|
| for $M \geq M_1$ | : | O–Of–BSG–LBV–WR–SN |
| $M_1 > M \geq M_2$ | : | O–BSG–YSG–RSG–WR–SN |
| $M_2 > M$ | : | O–RSG (with or without cepheid loop) – SN |

BSG, YSG and RSG mean blue, yellow and red supergiants respectively. LBV stands for luminous blue variables, WR for Wolf–Rayet stars and SN for supernova. The limiting masses between these sequences very much depend on the exact value of the mass loss rates. For the currently observed $\dot{M}$–rates, $M_1 = 50 M_\odot \pm 10 M_\odot$, $M_2 = 35 M_\odot \pm 5 M_\odot$.

Due to convective dredge–up and mass loss, the surface abundances may change during stellar evolution and these effects have been studied in several works (cf. Maeder 1987). The most sensitive abundance ratio is C/N, which may change from about 4 (cosmic ratio) to 0.02, i.e. by a factor of 200. The O/N ratio also undergoes a maximum change by a similar factor, although these changes occur later and more progressively in the evolution than for C/N. Along the evolutionary sequences described above the various surface abundance ratios H/He, C/N, O/N, $^{12}C/^{13}C$, $^{14}N/^{15}N$, $^{16}O/^{17}O$, $^{17}O/^{18}O$ are progressively changing. The future comparisons between models and observations should ideally not only bear on the HR diagram of clusters, but simultaneously they should include a close comparison between observed and predicted abundances at a given location in the HR diagram. Some first comparisons have even shown that some nuclear cross sections, e.g. for $^{17}O\,(p,\alpha)\,^{14}N$ are undoubtedly wrong (cf. Maeder 1987).

**TABLE 1.  Summary on mass loss effects in massive star evolution.**

MAIN SEQUENCE

* $M_{core}$ ↓ , $q_c = \dfrac{M_{core}}{M}$ ↑ , semi-convection ↓

* L ↓ , L/M ↑

* MS lifetime $t_H$ ↑

* moderate $\dot{M}$ : MS widening
  very high $\dot{M}$ : MS narrowing (quasi-homogeneous evolution)

HE - BURNING PHASE

* Large effects in HRD/Very small in log Tc vs. log ρc.
  (central conditions)

* 3 evolutionary sequences according to $\dot{M}$ and $M_{initial}$

* $t_{He} \cong t_{BSG} + t_{RSG} + t_{WR}$ , sharing varies with $\dot{M}$

* *BLUE SUPERGIANTS (BSG):*

  no $\dot{M}$ : $t_{He} \cong t_{BSG}$
  with $\dot{M}$ : $\left\{ \begin{array}{l} \text{He - phase moves to red} \\ \text{Blue loops reduced} \end{array} \right\}$ $t_{BSG}$ ↓

* *RED SUPERGIANTS (RSG):*

  moderate $\dot{M}$ => $\dfrac{t_{RSG}}{t_{OBA}}$ ↑
  (for low $\dot{M}$ : lack of RSG)

  high $\dot{M}$ => $\dfrac{t_{RSG}}{t_{OBA}}$ ↓

  $\dot{M}$ ↑ => $\dfrac{t_{RSG}}{t_{OBA}}$ ∧

* *WOLF-RAYET STARS (WR):*

  $\dot{M}$ increases $t_{WR} / t_{OBA}$
     lowers threshold mass for
     forming WR stars (most
     from $M_{initial} \geqslant 40\ M_\odot$)

  $\dot{M}$ ↑ => $\dfrac{N_{WR}}{N_{OBA}}$ ↑↑

  $\dot{M}$ ↑ => $N_{WR}/N_{RSG}$ ↑↑

  Mass - luminosity relation      $\log \dfrac{L}{L_\odot} = 3.8 + 1.5 \log \dfrac{M}{M_\odot}$
  for WR stars

The identification of WR stars with bare cores has been confirmed in IAU Symposium 99 (cf. de Loore and Willis, 1982). The abundances of WNL (late), WNE (early), WC and WO stars are consistent with a progression in the exposition of nuclear products, as shown by many authors (cf. Smith and Willis 1983; Nugis 1982; Conti et al. 1983; Maeder 1983; Smith and Hummer 1988; Torres and Conti 1987).

A major distinction between the constraints imposed by the abundances in WN and WC stars must be made. The abundances in WN stars (particularly in WNE stars) are CNO equilibrium values, which are very unsensitive to the model properties, and thus tell us very little about them. The WNE abundances, however, constitute a marvelous test of the nuclear cross–sections for the CNO cycles. This is the only place in Nature where the abundances characteristics of these cycles reveal themselves, independently of any dilution factor. At the opposite, the abundances in WC stars are products of partial He–burning. The values of these abundances ($^{4}$He, $^{12}$C, $^{16}$O, $^{22}$Ne) depend very much on the stage of nuclear processing at which they are revealed. Thus, in addition to the nuclear cross–sections, they also very much depend on the structural properties of the models and on effects such as overshooting or mixing.

Finally, let us point out that, while the surface properties (location in the HR diagram, abundances) are greatly influenced by mass loss, the central properties (e.g. $\log T_c$, $\log \rho_c$ and thus nucleosynthesis) for long only show a negligible dependence on mass loss. This is true as long as the mass $M_\alpha$ of the He–core is not reduced by mass loss (cf. Maeder and Lequeux 1982). However, when the mass $M_\alpha$ is reduced, very substantial changes also occur. Such a situation may well be more frequent than thought before, if the mass loss rates in the WR stage depend on the actual mass of the WR stars, as suggested by Langer (1989b) and several other authors (cf. § 3 below). In this case, the entry in the WNE phase is marked by heavy mass loss rates, which rapidly lead the star to the WC stage where it experiences a reduction of $M_\alpha$ and consequent changes in central properties. Such an effect is expected to affect significantly the nucleosynthetic yields of massive stars and its study is now being undertaken.

## 3 METALLICITY EFFECTS ON MASSIVE STAR EVOLUTION AND DISTRIBUTIONS IN GALAXIES

In Table 2 the number ratios of Wolf–Rayet stars (WR) to O stars, Wolf–Rayet stars of subtypes C and N, and of M supergiants to WR stars are shown for various galactic and extragalactic sites of different metallicities Z given in the second column. SN 6–7.5 kpc means stars in the solar neighbourhood with galactocentric distances of 6–7.5 kpc. These data originate from several sources (Arnault et al., 1989; Azzopardi et al. 1988; Breysacher 1986; van der Hucht et al. 1988; Meylan and Maeder 1983; Smith 1988).

The large variations of these number ratios are quite impressive as they amount to factors 10, 20 and 45 respectively. There have been great debates in the past on the origin of these variations. The point of view supported by Maeder, Lequeux and Azzopardi (1980) was that metallicity influences the mass loss rates, which in turn affect the lifetimes in the various considered stages, i.e. an interpretation in terms of stellar evolution. Some authors have invoked effects related to the initial mass function (IMF) or the star formation rate (SFR) (e.g. Bertelli and Chiosi, 1981). However, Meylan and Maeder (1983) have shown that the galactic gradient of WR stars is much steeper than that of their massive O–

progenitors. Besides, we may notice that Table 2 concerns number ratios, and that the observed changes imply differences in the various galactic gradients.

Metallicity, of course, influences stellar opacities, but in the interior of massive stars the effect is quite small because the main opacity source is electron scattering. In the very outer stellar layers, however, different metallicities will result in different opacities, which may in turn result in different mass loss rates. Thus the main effect of differences in Z in massive star evolution is likely to be due to differences in the mass loss rates, as already proposed by Maeder et al. (1980).

Grids of models of massive stars were constructed for initial masses 15, 20, 25, 40, 60, 85 and 120 $M_\odot$ and for initial metallicities Z = 0.040, 0.020, 0.005 and 0.002. Appropriate (O/Fe) and ($\alpha$–nuclei/Fe) abundance ratios were taken in the opacity tables for different metallicities. A total of more than 40.000 individual stellar models has been computed (cf. Maeder, 1990). The sets with Z = 0.020 and Z = 0.002 are illustrated in Figs. 1 and 2. For the solar composition the $\dot{M}$–rates by de Jager et al. (1988) have been used. What is the situation at other metallicities? The models by Brunish and Truran (1982ab) had effectively larger mass loss rates at low metallicities and such a behaviour is rather doubtful. Schaller (1986) has used a linear scaling of the mass loss rates with metallicity Z, as predicted by the stellar winds models by Abbott (1982). However, his models predicted no red supergiants for the metallicity of the Small Magellanic Cloud, while many are existing. This was interpreted (Schaller 1986) as indicating that the mass loss rates in the SMC should not be as small as given by the scaling $\dot{M} \sim Z^\zeta$, with $\zeta = 1$.

TABLE 2. Ratios of star number in various galaxies.

| | Z | $\dfrac{WR}{O}$ | $\dfrac{WC}{WN}$ | $\dfrac{M}{WR}$ |
|---|---|---|---|---|
| M31 | .035 | – | 2.82 | – |
| SN 6–7.5 kpc | .029 | .21 | 2.85 | .53 |
| SN 7.5–9.5 kpc | .020 | .10 | 1.22 | 1.4 |
| SN 9.5–11 kpc | .013 | .033 | 1 | 13.5 |
| M33 | .007–.02 | .06 | .71 | – |
| LMC | .0057 | .04 | .26 | 8 |
| NGC 6822 | .0045 | .03: | 0/2 | – |
| SMC | .0021 | .015 | .14 | 24 |
| IC 1613 | .002 | .02 | – | – |

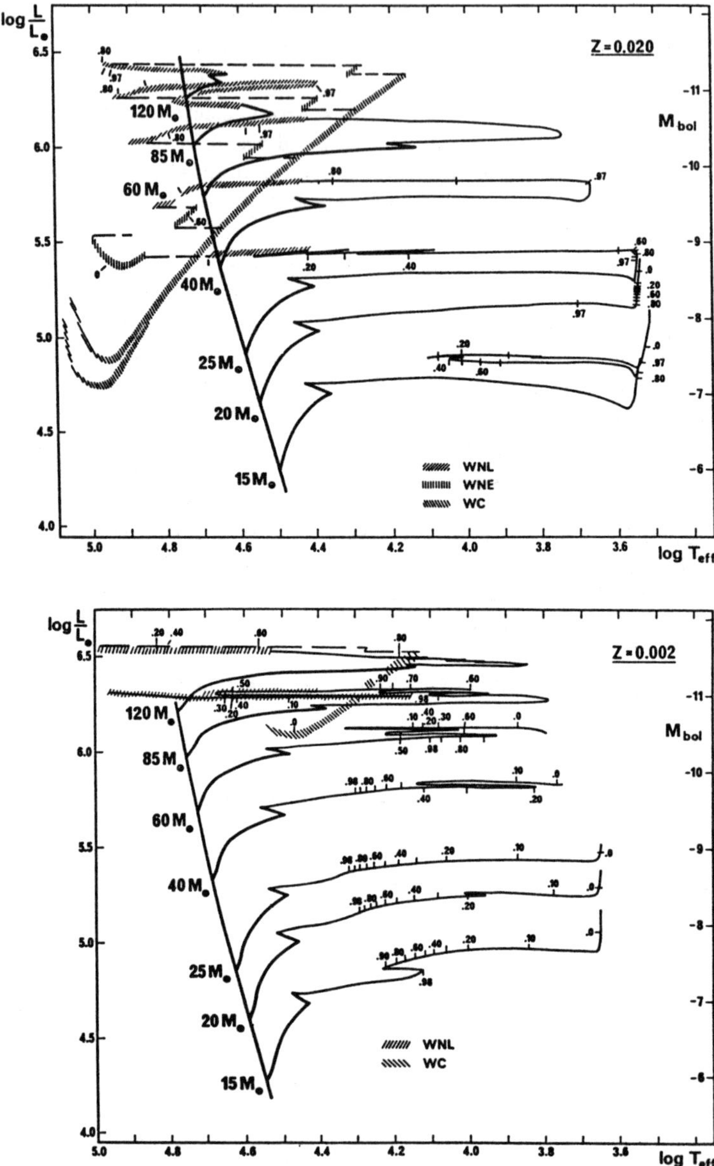





## Figures 1 and 2

Evolutionary tracks in the HR diagrams for Z = 0.02 and 0.002. The data along the tracks indicate the values of the central helium contents $Y_c$ during the He–burning phase. The WR phases are indicated by hatched lines; the corrections for the optical thickness of the wind have been applied.

For the other metallicities we have taken the data by de Jager et al. (1988), scaled according to a relation $(\dot{M}/\dot{M}_{\odot}) = (Z/Z_{\odot})^{\zeta}$. The published stellar wind models of O–stars by Kudritzki et al. (1987) give a value $\zeta \simeq 0.5$ (see also the discussion by Conti, 1988). For supergiants and especially red supergiants, there is yet no reliable information (and probably not before long) on the way the $\dot{M}$–rates may change with Z (cf. Kwok 1987). Thus Kudritzki's 1987 value of $\zeta = 0.5$ was chosen in the above grids of models. There is no doubt that further progresses in models and observations will lead to tighter constraints on the $\zeta$–values and lead us to distinguish the $\zeta$–values in the blue and red supergiants and O–stars.

The mass loss rates of WR stars are important parameters too. Up to now, the models generally used the most straightforward procedure, that is to take the average observed mass loss rates of the WR stars, especially more than these rates do not exhibit any clear dependence with the WR subtypes (cf. Conti 1988). However, as shown by Schmutz et al. (1989), this leads to the troublesome result that the predicted WR luminosities are much higher than the observed values. Since there is a mass–luminosity relation for WR models (cf. Maeder 1983; Langer 1989a), this implies that the actual masses of most of the Wolf–Rayet stars are much smaller than predicted. This in turn indicates that the mass loss rates should be larger at least in some parts of the WR stages.

Besides the above evidence, there has been a number of other indications in favour of mass loss rates for WR stars depending on the actual masses of WR stars. Chiosi (1982), Chiosi and Maeder (1986) pointed out that WNL stars (WN with H still present to adopt a short definition) may evolve directly into WC stars (products of partial He burning visible) and proposed that the WNL stars become highly vibrationally unstable when entering the WNE stage (WN with no H present in general). The instability, either by mixing or by removing the He–rich external layers, would soon suppress the WNE stage and reveal the products of the He–burning reactions (WC phase). Indeed, the models of vibrational instability (cf. Maeder 1985) have effectively shown that the more massive a WR star (WNE or WC stage), the more unstable it should be. From a simplified analytical development, a relation between $\dot{M}$ and M was predicted for the WR stars.

Abbott et al. (1986; cf. also Conti 1988) by observing the $\dot{M}$–rates in five binary systems found a relation of the form $\dot{M}_{\mathrm{WR}} \sim M^{2.3}$. However, it was not clear whether all WR stars should obey such a relation. Smith and Maeder (1989) found, for the ensemble of WR stars (including WNL), a relation $\dot{M}_{\mathrm{WR}}$ vs. M relation with a slope of the order of 1. Langer (1989b) considers the effects of a $\dot{M}_{\mathrm{WR}} \sim M^{\alpha}$ relation on an ensemble of observable quantities, such as the masses and luminosities of WR stars, the WR lifetimes, the WC/WNE ratios. He concludes that an $\alpha$–value of 2.5 would best represent these observations. Interestingly enough, he shows that by removing the WNL from the sample by Smith and Maeder (1989) one also obtains a value of $\alpha = 2.6$. Although it is not so clear what would be causing the WR phenomenon in WNL stars having still substantial H–surface layers, there are good theoretical reasons from the models of vibrational instability (cf. Maeder 1985) to consider the WNL separately. Thus, for the WNE and WC stars the expressions given by Langer (1989b), i.e. $\dot{M}_{\mathrm{WR}} = (0.60 - 1.00) \cdot 10^{-7}(M_{WR}/M_{\odot})^{2.5}$ in solar mass per year have been taken in the above–mentioned grids, while for WNL stars a constant value $\dot{M} = 4 \cdot 10^{-5} M_{\odot} y^{-1}$ was used (cf. Conti 1988).

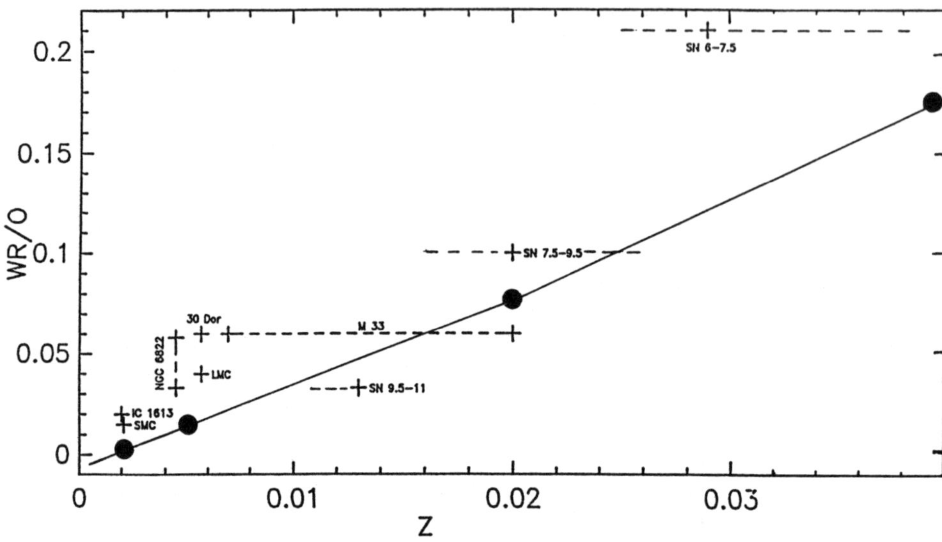



**Figure 3**

The WR/O number ratios in function of metallicity Z (cf. Arnault et al. 1989) compared to model predictions.

The lifetimes predicted by these models at various Z were integrated over a mass spectrum $dN/dM \sim M^{-2.7}$ (cf. Scalo 1986). In this way various number ratios could be obtained, for example the number ratio WR/O and WN/WC. Figure 3 compares the observed WR/O number ratios in various galactic and extragalactic sites (cf. Table 2) with the model predictions at various metallicities. We notice an interesting agreement both in the general trend and the detailed values. It is most remarkable that the big changes in star distributions just appear to be an evolutionary effect resulting from the connection $Z \rightarrow \dot{M} \rightarrow$ lifetimes in the WR stage. The good agreement in Fig. 3 also appears as a support of the employed relation between $\dot{M}$ and Z. Similar tests have been performed and they also lead to an excellent agreement. To what extent the very differentiated galactic distributions of some WR subtypes can also be explained (for example, WC subtypes) is now being examined.

We have discussed here some nice consequences of mass loss on the distributions of stars in galaxies, but it is also clear that other interesting effects also exist regarding, for example, nucleosynthesis and the mass of supernovae progenitors. Indeed, the ratio of elements ejected in the winds (typically $^4$He, $^{22}$Ne, $^{25}$Mg, $^{26}$Mg, s–elements) to those ejected in supernova explosions (typically $^{16}$O, $^{20}$Ne, $^{24}$Mg, $^{28}$Si) is much larger at high metallicity Z than at low Z. This means, for example, that the relative helium to metal enrichment $\Delta Y/\Delta Z$ is much lower at low Z than at high Z. Such effects are now under investigation.

# 4 WOLF–RAYET INSTABILITIES AND MASS LOSS

As repeatedly emphasized in this conference, some Wolf–Rayet stars have high values of the momentum ratio $\dot{M}v_\infty\ c/L$ up to more than 10, for which the theory of radiatively driven wind has not yet provided any convincing explanation. Therefore, it is worth here to examine the case of the instabilities of WR stars in relation with their high mass loss rates.

The observational status of the variability of WR stars has been reviewed by Moffat and Shara (1986) and by Abbott and Conti (1987). Variability of a few hundredths of magnitude are found in many cases, but the nature of the pulsations (binary motions, non radial, radial, etc. ...) remains uncertain. The identification of the pulsation mode is made very difficult by the fact that in WR stars the instabilities cannot be observed directly: WR stars are surrounded by optically thick winds, which do not allow us to observe the photosphere directly.

Let us examine here the theoretical status about the pulsations of WR stars and start with the non radial case, which has been claimed by Vreux (1985) to be responsible for some observed line variations. The driving of non radial oscillations by central nuclear energizing is in a very unfavourable situation, because the amplitudes $\delta T/T$, $\delta\rho/\rho$ tend to be zero at the stellar center and no efficient pumping of energy can occur there (cf. Simon 1957). This explains why Kirbiyik et al. (1984), in an investigation of non radial oscillations for WR models, found these stars to be stable for the low harmonics $\ell$; they noticed however the appearance of instabilities for high degrees of harmonics ($\ell = 15$). Noels and Scuflaire (1986) interestingly showed that, while there is no driving of non radial oscillations in WR stars to be expected from He–cores, the H–burning shells may produce some efficient driving of non radial pulsations. The periods found are of the order of a few hours. There is, however, a limitation: the H–shell is, if any, only present for a very short time in WR stars. In the case studied by Noels and Scuflaire it lasts only 6000 years. Indeed, most WR models (cf. Maeder, 1981) even do not exhibit an H–burning shell. This is obvious if we remember that WNE as well as WC stars no longer have hydrogen at their surface; thus only a fraction of the WNL stars could have an efficient driving from the H–shell. Studying various WR models, Cox and Cahn (1988) found no unstable g–type non–radial modes. The non–radial modes have large amplitudes outside the convective core where they might be expected to be driven by the $\epsilon$–mechanism in the rare cases where there would be an H–burning shell. But even there Cox and Cahn (1988) find the WR models to be stable, with respect to non–radial oscillations, because of the large radiative damping in the outer layers. Thus, on theoretical grounds, non–radial instabilities in WR stars seem to be very unlikely.

At the opposite, WR models evolved with mass loss at the observed rates were found to be unstable in the fundamental mode of radial oscillations (Maeder 1985). Stability analyses in 32 models at various evolutionary stages in massive stars with initial masses 85 and 120 $M_\odot$ were performed. The rate $L_{PN}$ of gain of pulsation energy from nuclear sources was calculated, as well as $L_{PH}$, the rate of loss of pulsation energy by heat leakage. If the value of $L_{PN}$ over the whole stars is larger than $L_{PH}$, the star is unstable. Interestingly, during the evolution towards the WR stage, the stability of the model strongly declines. The physical reason is that as the external H–containing layers become thinner, the stellar radius decreases as a result of lower opacity. The ratio $\rho_c/\bar{\rho}$ of central and average densities decreases, and thus the ratio $A_s/A_c$ of surface to central amplitudes does the same. This

42

means that the heat leakage in the outer layers becomes relatively less important with respect to the pumping of pulsational energy at the center. Thus, it was found that the relative importance of nuclear energizing with respect to radiative damping is higher by 4 orders of magnitude in a WR star than in a MS star. Fig. 4 shows the radiative damping and nuclear energizing in an initial 120 $M_\odot$ during MS evolution; at each point along the fractional radius r/R, the integrated works $L_{PH}$ and $L_{PN}$ from the center to the considered point are shown. We notice that this model is highly stable. Fig. 5 shows the same kind of data for a typical WR star. WR stars are generally unstable due to the nuclear energizing of pulsations, and they are probably the only stars where the so–called $\epsilon$–mechanism of Eddington is at work.

These instabilities were confirmed by Cox and Cahn (1988). They found that a hydrogen–free 50 $M_\odot$ model, evolved from an original MS 120 $M_\odot$ model, has a pulsationally unstable radial fundamental mode. However, all radial overtones were found stable. The same conclusion is now obtained by Schaller and Maeder (1990), who are using, like Cox and Cahn, the non–adiabatic approximation, while the first calculations by Maeder were made in the so–called quasi–adiabatic approximation.

The exploration of WR models by Maeder (1985) show that the unstable regime is entered only when the surface H–content nearly vanishes ($X_s = 0.07$ in some example). The instability persists during the whole stage of a bare core, which is usually identified with a WR star. Only at the very end of the WR phase, when the WR mass has decreased to $M \leq 8\ M_\odot$, the instability declines or disappears.

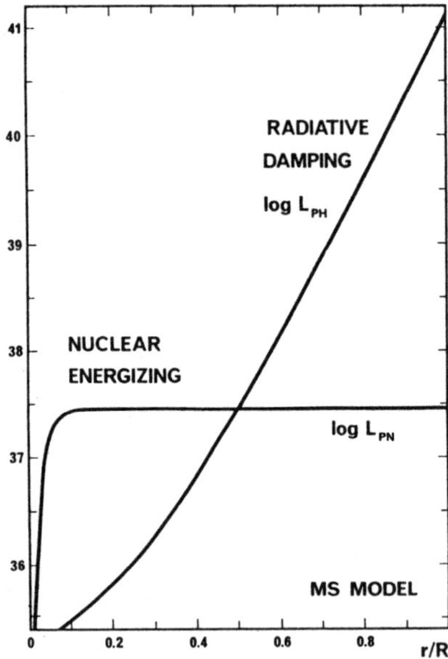


**Figure 4**

The integrated values of nuclear energizing and radiative damping in a MS model of an initial 120 $M_\odot$ star at the time when the central H–content is $X_c = 0.24$. We notice that integrated over the whole star (see the values at r/R = 1.0) the radiative damping is 3 to 4 oder of magnitude larger than the nuclear energizing. Thus, the star is highly stable.

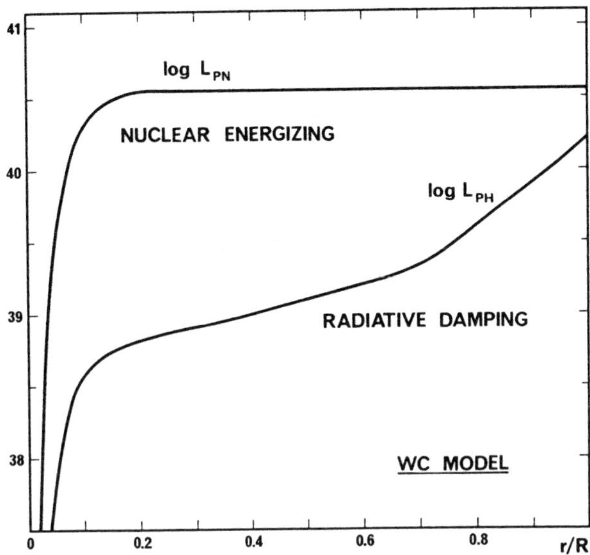


**Figure 5**

Same as Fig. 4, but for a WR star. Initial mass 120 $M_\odot$, actual mass 34 $M_\odot$, central and surface helium contents $Y_c = .37$, $Y_s = .45$. There we see that at r/R = 1.0, i.e. over the whole star, $L_{PN}$ is larger than $L_{PH}$, thus the star is unstable.

A very interesting point concerns the relation between mass loss and instability. If, in an evolutionary sequence, rather low mass loss rates are taken with respect to the observed average values, the evolution proceeds more inhomogeneously. This means that larger concentration ratios $\rho_c/\bar{\rho}$ are present and this prevents the instability to show up. This is certainly the reason why Noels and Gabriel (1981) found WR models to be stable during most of the WR stage; the low $\dot{M}$–rates they used never bring the star to the state of quasi-homogeneous evolution that high $\dot{M}$–rates produce and which is thought (cf. Maeder, 1983; cf. also §3) to correspond to WR stars.

Let us now consider a sequence of quasi–homogeneous WR models, obtained with the observed mass loss rates. An interesting feature appears there (cf. Schaller and Maeder, 1990). If suddenly the $\dot{M}$–rate is enlarged by a high factor, the instability declines due to the fact that central T and $\rho$ are slightly reduced. At the opposite, reduction of the $\dot{M}$–rate leads to an enhancement of the instability. This numerical result well supports the view (cf. Maeder 1985) that WR stars evolve keeping at the edge of vibrational instability. Indeed, it was also shown analytically that if WR stars evolve quasi–homogeneously and keep at the edge of vibrational instability with $M_\mu^{-2} \simeq$ const, the pulsations would be able to sustain the observed mass flux.

The unstable WR models occupy a well defined mass–luminosity relation $\log L/L_\odot = 3.8 + 1.5 \log M/M_\odot$ (for $M \geq 10\ M_\odot$), which is characterized by a large overluminosity with respect to the main sequence. At present we cannot assume that vibrational instability is the only process responsible for the WR phenomenon, but it is a very good candidate.

Radiation is also likely to contribute to the high mass loss of WR stars, in particular since WR stars may have higher $T_{eff}$ than previously considered (cf. Abbott and Conti 1987). For now we do not know whether radiation or pulsation is the main effect responsible for the WR phenomenon. However, I want to emphasize that the very high L/M ratio of WR stars is likely to be at the origin of both pulsation and radiative effects. Thus, the very large L/M ratio is probably the deep physical reason for the WR phenomenon.

## 5 PHYSICAL CONSIDERATIONS ON THE INSTABILITIES AND EJECTIONS IN LBV STARS

The observed properties and astrophysics of Luminous Blue Variables (LBV), i.e. extreme supergiants at the top of the HR diagram, have been recently discussed in IAU Colloquium 113 (cf. Davidson et al. 1989). The origin of the instabilities and of the huge mass ejections (up to 1 $M_\odot$ is a few centuries) characteristic of these stars is far from being properly modellized and completely understood. This is why they have to be discussed. A few evidences start emerging, along which we may possibly go on, and it may be useful to emphasize here the main points.

To my knowledge, all modelmakers performing models of very massive stars (M > 60 $M_\odot$) have got numerical difficulties in their models as the star evolves away to the red from the MS. The outer and inner solutions become difficult to match. Generally, a large density inversion also occurs somewhere in the envelope (cf. Bisnovatyi–Kogan and Nadyozhin 1972; Stothers and Chin 1973). In our models (cf. Maeder 1987) this density inversion typically appears for stars with log $T_{eff} \simeq 3.9$ and it grows very fastly at lower $T_{eff}$. The $\rho$–inversion occurs at a distance corresponding to $10^{-4}$ to $10^{-6}$ in mass faction from the surface; it lies in the layers where is located the opacity peak due to partial H–ionisation and where L happens to be larger than the local Eddington luminosity (account being properly given to the various opacity sources of the models). Physically, the origin of the $\rho$–inversion is due to the creation of a large opacity peak as $\rho$ and T go down during the envelope expansion. This opacity peak produces supra–Eddington luminosities, and thus a corresponding density inversion, as the convective flux is close to zero and unable to wash the inversion out.

The occurrence of $\rho$–inversion in the models does not necessarily mean that $\rho$–inversions effectively occur in real stars. Rather, the occurrence of a strong $\rho$–inversion, which is a consequence of the local supra–Eddington luminosity, must be considered as the sign that *hydrostatic models are no longer valid in the upper right corner of the HR diagram and that massive outflows occur.* The absence of hydrostatic solutions in the upper right corner of the HR diagram is well indicated by the fact that the matching of the Henyey interior solutions with the outer solutions can progessively no longer be achieved as the star moves to the right. We may also point out that $\rho$–inversions occur in the subsonic regime of convective velocities, i.e. for $V_{conv}/V_{sound} \leq 0.1$; thus, the instability revealed by the $\rho$–inversion is not the same as that due to turbulent pressure and studied by de Jager (1984). Indeed, on a redwards track the $\rho$–inversions are met before sonic velocities are reached. The breakdown of hydrostatic solutions is due to supra–Eddington luminosities in the region of the opacity peak, and is thus the continuation for stars of lower $T_{eff}$ of the instability proposed by Appenzeller (1986) and Lamers (1986; cf. also Lamers and Fitzpatrick 1988) from atmospheric studies of hotter stars.

The departure of hydrostatic equilibrium revealed by the $\rho$–inversions which can reach a factor of 10 near $\log T_{\text{eff}} \simeq 3.9$, is quite large and therefore heavy mass loss is to be expected. Rates of mass loss as high as 0.5 $M_\odot \, y^{-1}$ have been obtained from non–static solutions (cf. Bisnovatyi–Kogan and Nadyozhin 1972). We can roughly estimate the above figure by simple analytical arguments. Let us call $\tau_{\dot{M}} = M/\dot{M}$ the characteristic time of mass loss and $\tau_\rho$ the characteristic time for the growth of the $\rho$–inversion in the models. If $\tau_\rho < \tau_{\dot{M}}$, the non–contraried $\rho$–inversion would grow without any limit. At the opposite, $\tau_{\dot{M}}$ cannot be much shorter than $\tau_\rho$, since the mass loss is driven by the instability announced by the $\rho$–inversion. Thus $\tau_{\dot{M}}$ and $\tau_\rho$ must be of the same order of magnitude, which means that $\dot{M} \simeq M/\tau_\rho$. For an initial 60 $M_\odot$ model in the LBV stage, we have typically M $\simeq$ 40 $M_\odot$ and $\tau_\rho \simeq 10^2 y$; thus we get $\dot{M}$–rates of the order of magnitude of several 0.1 $M_\odot \, y^{-1}$. The above value is only a rough estimate of the order of magnitude, but it tells us that *the strong departure from hydrostatic equilibrium in the upper right corner of the HR diagram must result in phases of intense mass loss* since the growth rate of the instability is extremely short.

Why is matter ejected in violent outbursts rather than in the form of a steady outflow? What determines the amount of mass in the ejected shell? These questions are critical ones, which could by the way also be asked for other kinds of variable stars. Indeed, if only the matter in and above the $\rho$–inversion would be expelled in one episode, the amount of matter in the ejected shell would be at most $\sim 10^{-4}$ of the stellar mass; this is far less than observed.

Some understanding may come from considerations about the timescales involved in the problem. The timescale for thermal adjustment is very short in the outer layers of an LBV. Thus, during the time of the ejection, which lasts for more than the dynamical time scale (free fall time), the opacity peak is able to go down in the star by a substantial amount. For example, in a time as short as the dynamical time scale, the opacity peak can move inwards by as much as 0.06 $M_\odot$ in a star of 41.2 $M_\odot$ (from an initial 60 $M_\odot$) at $\log L/L_\odot = 6.05$ and $\log T_{\text{eff}} = 3.89$. Since any ejection event will last in any case for more than the dynamical time, this means that the ejection process will be sustained for a while by the fact that the blasting front (the opacity peak) has time to deeply move inwards during the explosion event itself. But, due to the heavy mass loss the star cannot keep the same radius and it soon moves bluewards in the HR diagram, on timescales which depend on the ejection parameters as illustrated in Fig. 6 below.

During the bluewards motion the opacity decreases in the critical layers which are then no longer characterized by large supra–Eddington luminosities and the huge and catastrophic ejection ceases. The hydrostatic solutions are again acceptable approximations until interior secular evolution again brings the star to its instability limit, after a certain recovery time.

A number of simulations have been performed to obtain the horizontal displacements in the HR diagram and the corresponding light curves in B magnitude for various amounts $\Delta M$ of mass ejected and for various mass loss rates. Only the consequences of the thermal response of the LBV to a mass ejection have been considered. A more complete description would require an hydrodynamic code, as well as the inclusion of the effects of the wind and dust on the light curve (cf. Davidson 1987). At this stage already, several interesting conclusions can be reached from the calculations made:

- For $\dot{M}$–rates equal or lower than $10^{-3}$ $M_\odot$, secular evolution dominates over the effects of readjustments in external layers and there is little variability in B magnitude. For ejections made at rates larger than a few $10^{-3}$ $M_\odot\,y^{-1}$ (the exact value depends on the stellar mass), the thermal adjustment overpasses the effects of secular evolution, and the net result is a shrinkage of the stellar radius and thus a bluewards evolution in the HR diagram. The overall luminosity undergoes only limited changes. The main effect for the light curves in B or V magnitudes results from the change of the appropriate bolometric corrections as $T_{eff}$ varies during the horizontal oscillations in the HR diagram.

- The peak–to–peak amplitude of the B or V light curve is mainly a function of the total amount $\Delta$ M lost in the shell ejection: the larger $\Delta M$, the larger the amplitude.

- The $\dot{M}$–rates during the outburst essentially determine the initial slope of the variations, i.e. the rapidity of the initial changes: the higher $\dot{M}$, the faster the initial variation.

- The recovery time after an ejection mainly depends on the total amount of mass ejected.

- The light curve in B–magnitude during an outburst also depends on the temperature before the ejection, since it determines the bolometric correction before the outburst.

From these first simulations we see that a proper analysis of the LBV light curve might give an insight on the parameters of the shell ejections. Future models should include the dynamical effects and those from the optically thick wind, in order to provide a more complete description of the LBV.


# 6   ON ROTATION AND MIXING IN MASSIVE STARS

Rotation can influence stellar evolution and structural stellar properties in four different ways.

### a. Effect of rotation on stellar equilibrium

The centrifugal term in the equation of mechanical equilibrium leads to reductions of central pressure and temperature, and thus of total luminosity. The maximum effect in case of uniform rotation is about 7–8% (cf. Faulkner et al. 1968). As the effect of rotation on the Schönberg–Chandrasekhar limit $q_c$ amounts to a maximum of 3% (cf. Maeder 1971), this means that the modification of stellar equilibrium due to rotation has only a small effect on the MS lifetime $t_{MS} \sim q_c\,M/L$, i.e. a maximum increase by a few percents.

### b. Change of atmospheric shape

The equipotential surfaces are modified by rotation. The equatorial radius is increased and thus over the stellar surface the local gravity and temperature are varying from place to place. This leads to changes of the emergent flux. Atmospheric models including this effect were constructed (cf. Maeder, Peytremann 1970, 1971; cf. also Zorec, this meeting). For most orientations, the star looks cooler and thus stellar ages from colour–magnitude diagrams may be somewhat overestimated. A procedure to correct for this effect has been devised (cf. Maeder 1971). Also and contrarily to a current belief, the equivalent widths of spectral lines are modified due to the spheroidal geometry of a rotating star.

### d. Rotational mixing

By far the potentially most significant effect of rotation on stellar evolution is that caused by rotational instabilities inducing transport of angular momentum and mixing of chemical elements in stellar interiors. Several excellent recent reviews have been made on the rotational instabilities and mixing (for example Zahn 1983; Spruit 1989; Tassoul 1989 and this meeting). We shall not consider these mechanisms in detail here, especially more that there seems to be no general consensus about the main mixing mechanism.

For the construction of evolutionary mixing the diffusion coefficient D associated to the mixing processes should be known. Schatzman and Maeder (1981) have constructed solar models with a coefficient D of the form $D = Re^* \nu$, where $\nu$ is the viscosity, radiative plus molecular, and $Re^*$ the so–called critical Reynolds number, which is a pure number. Solar oscillations, $^3$He and $^7$Li abundances in the Sun impose $Re^* \leq 25$ in the Sun. In massive stars it is expected that the D coefficient is much larger since the (radiative) viscosity is about $10^6$ times larger than in the Sun. Of course, $Re^*$ can be used as a trial parameter, but it is very desirable to have it from theoretical principles. Such an approach has been made by Zahn (1983) and has been applied to massive star evolution by Maeder (1987).

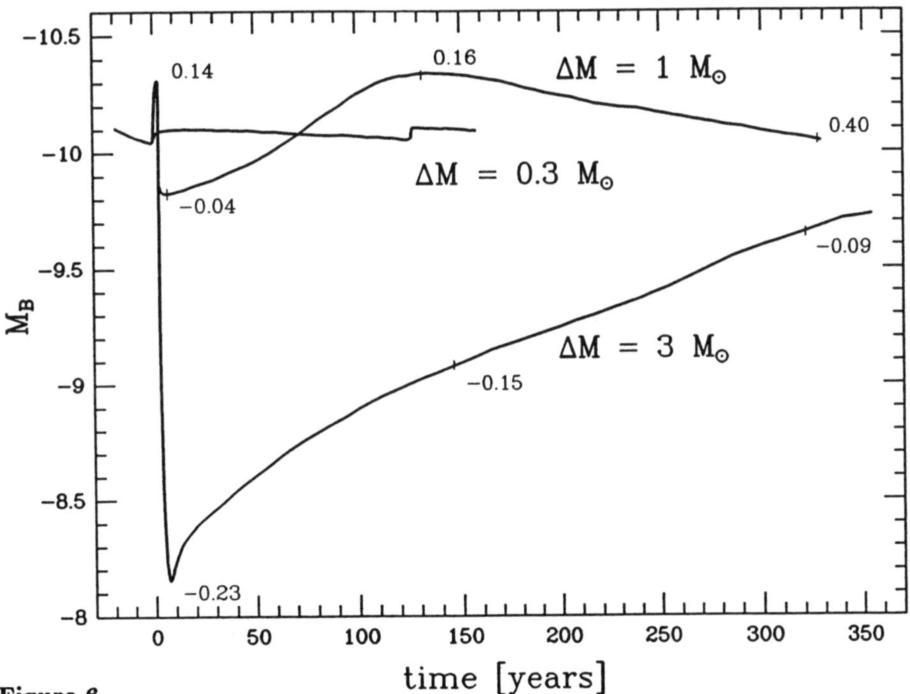


**Figure 6**

Example of the light curves in B–magnitude resulting from shell ejections of various shell masses $\Delta M$ at mass loss rate 0.5 $M_\odot$ y$^{-1}$. Before the ejection the model had 46.9 $M_\odot$, log $L/L_\odot = 6.04$ and log $T_{eff} = 3.82$; the initial mass was 60 $M_\odot$. The values of (B–V) are indicated along the curves.

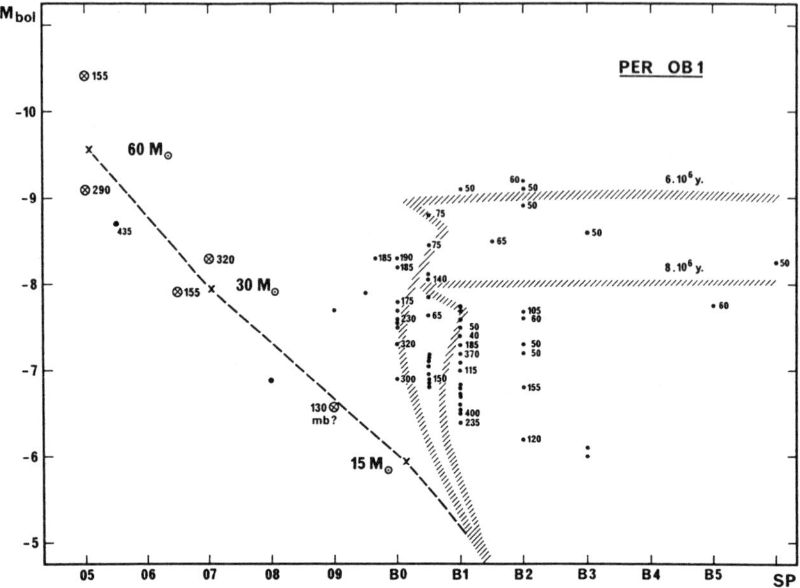



## Figure 7

The $M_{bol}$ vs. spectral type diagram for the association Per OB1 according to the data by Humphreys (1978). The axial rotational velocities are indicated. The known N-enriched stars are indicated by an open circle with a cross. The zero age main sequence is represented by a broken line. Two isochrones are also shown by hatched lines.

As far as evolution is concerned, the main effects are the following ones. In case of low and moderate rotation, the stabilizing effect of the $\mu$-gradient severely limits the extension of turbulent diffusion and all models nearly follow the same tracks close to the classical ones. The evolution is essentially inhomogeneous; the MS lifetime and the surface CNO abundances undergo only limited changes. For fast rotation, the stabilizing effect of the $\mu$-gradient is unable to prevent mixing. This evolution resembles the homogeneous one giving bluewards tracks in the HR diagram. Processed CNO elements rapidly appear at the stellar surface. The luminosity firstly increases strongly. In view of its composition, the star becomes a WR star and then undergoes a decline in luminosity due to the rapid decrease of its mass. Mainly because more nuclear fuel is available, the MS lifetime is larger by about 55%. The switching from the classical regime to that of homogeneous evolution occurs quite rapidly in terms of the rotational parameter.

The comparisons with observations suggest that ON-stars, or at least those which are blue stragglers, may correspond to the above picture of homogeneous evolution. Indeed, it was noticed by Kudritzki (1985) that some ON stars are located close to the zero-age sequence, which is impossible for standard evolution. The comparison with the young cluster Per OB 1 supports the idea of homogeneous evolution for a fraction ($\leq$ 15%) of massive O stars. In this cluster there is a well defined branch of O-blue stragglers, which are either rapidly rotating stars or binaries. Most of them show N–enrichments characteristic of internal mixing. Indeed, tidal interaction in binaries as well as rotation can generate instabilities and mixing, leading to homogeneous evolution and the case of ON–blue stragglers is likely to correspond to this picture.

On the whole, the evolution of massive stars appears to lie at the crossroad of many fascinating new research lines, such as the distribution of massive stars in galaxies, the stellar nucleosynthesis, the huge mass ejection of LBV and mass loss of WR stars, the status of supernovae progenitors, etc.... The many remaining unsolved problems contribute to make stellar evolution a topical research subject both observationally and theoretically, as well as an enthusiastic and healthy exercise.

# 7  REFERENCES


Appenzeller, I.: 1986, in *Luminous stars and associations in galaxies*, IAU Symp. 116, Ed. C. de Loore and A.J. Willis, p. 139

Appenzeller, I.: 1987, in *Instabilities in luminous early type stars*, Ed. H. Lamers and C. de Loore, Reidel Publ. Co., p. 55

Abbott, D.: 1982, Astrophys. J. **259**, 282

Abbott, D.C., Bieging, J.H., Churchwell, E., Torres, A.V.: 1986, Astrophys. J. **303**, 239

Abbott, D.C., Conti, P.S.: 1987, Ann. Rev. Astron. Astrophys. **25**, 113

Appenzeller, I.: 1986, in *Luminous stars and associations in galaxies*, IAU Symp. 116, Ed. C. de Loore and A.J. Willis, Reidel Publ. Co., p. 139

Arnault, P. Kunth, D., Schild, H.: 1989, Astron. Astrophys. **224**, 73

Azzopardi, M., Lequeux, J., Maeder, A.: 1988, Astron. Astrophys. **189**, 34

Bertelli, G., Chiosi, C.: 1981, in *The most massive stars*, Ed. S. D'Odorico et al., ESO Garching, p. 211

Bisnovatyi–Kogan, G.S., Nadyozhin, D.K.: 1972, Astrophys. Space Sci. **15**, 353

Breysacher, J.: 1986, Astron. Astrophys. **160**, 185

Brunish, W.M., Truran, J.W.: 1982a, Astrophys. J. **256**, 247

Brunish, W.M., Truran, J.W.: 1982b, Astrophys. J. Suppl. **49**, 447

Chiosi, C.: 1982, in *Wolf–Rayet stars: observations, physics, evolution*, IAU Symp. 99, Ed. C. de Loore, A.J. Willis, Reidel Publ. Co., p. 323

Chiosi, C., Maeder, A.: 1986, Ann. Rev. Astron. Astrophys. **24**, 329

Conti, P.S.: 1988, in *O stars and Wolf–Rayet stars*, Ed. P.S. Conti and A.B. Underhill, NASA SP–497

Conti, P.S., Leep, E.M., Perry, D.N.: 1983, Astrophys. J. **268**, 228

Cox, A.N., Cahn, J.H.: 1988, Astrophys. J. **326**, 804

Davidson, K.: 1987, Astrophys. J. **317**, 760

Davidson, K., Moffat, A.F.J., Lamers, H.J.G.L.M.: 1989, *Physics of luminous blue variables*, IAU Coll. 113, Kluwer Acad. Publ.

Faulkner, J., Roxburgh, I.W., Strittmatter, P.A.: 1968, Astrophys. J. **151**, 203

van der Hucht, K.A., Hidayat, B., Admiranto, A.G., Supelli, K.R., Doom, C.: 1988, Astron. Astrophys. **199**, 217

Humphreys, R.: 1978, Astrophys. J. Suppl. **38**, 309

de Jager, C.: 1984, Astron. Astrophys. **138**, 246

de Jager, C., Nieuwenhuijzen, H., van der Hucht, K.A.: 1988, Astron. Astrophys. Suppl.

Ser. **72**, 259

Kirbiyik, H., Bertelli, G., Chiosi, C.: 1984, in 25th Liège Colloquium, Ed. A. Noels, M. Gabriel, p. 126

Kudritzki, R.P.: 1985, in *Production and distribution of CNO elements*, ESO workshop, Eds. I.J. Danziger, F. Matteucci, K. Kjär, p. 277

Kudritzki, R.P., Pauldrach, A., Puls, J.: 1987, Astron. Astrophys. **173**, 293

Kwok, S.: 1987, Physics Reports **156**, 111

Lamers, H.: 1986, in *Luminous stars and associations in galaxies*, IAU Symp. 116, Eds. C. de Loore and A.J. Willis, Reidel Publ. Co., p. 157

Lamers, H., Fitzpatrick, E.L.: 1988, Astrophys. J. **324**, 279

Langer, N.: 1989a, Astron. Astrophys. **210**, 93

Langer, N.: 1989b, Astron. Astrophys. **220**, 135

de Loore, C., Willis, A.J.: 1982, *Wolf–Rayet stars: observations, physics, evolution*, IAU Symp. 99, Ed. C. de Loore, A.J. Willis, Reidel Publ. Co.

Maeder, A.: 1971, Astron. Astrophys. **10**, 354; **14**, 354

Maeder, A.: 1981, Astron. Astrophys. **99**, 97

Maeder, A.: 1983, Astron. Astrophys. **120**, 113

Maeder, A.: 1985, Astron. Astrophys. **147**, 300

Maeder, A.: 1987, Astron. Astrophys. **173**, 287; **178**, 159

Maeder, A.: 1990, Astron. Astrophys. Suppl. Ser. in press

Maeder, A., Lequeux, J.: 1982, Astron. Astrophys. **114**, 409

Maeder, A., Lequeux, J., Azzopardi, M.: 1980, Astron. Astrophys. **80**, L17

Maeder, A., Meynet, G.: 1987, Astron. Astrophys. **182**, 243

Maeder, A., Meynet, G.: 1989, Astron. Astrophys. **210**, 155

Maeder, A., Peytremann, E.: 1970, Astron. Astrophys. **7**, 120

Maeder, A., Peytremann, E.: 1971, Astron. Astrophys. **21**. 279

Meylan, G., Maeder, A.: 1983, Astron. Astrophys. **124**, 84

Moffat, A.F.J., Shara, M.M.: 1986, Astron. J. **92**, 952

Noels, A., Gabriel, M.: 1981, Astron. Astrophys. **101**, 215

Noels, A., Scuflaire, R.: 1986, Astron. Astrophys. **161**, 125

Nugis, T.: 1982, in *Wolf–Rayet stars: observations, physics, evolution*, IAU Symp. 99, Ed. C. de Loore, A.J. Willis, Reidel Publ. Co., p. 127

Scalo, J.: 1986, Fundamentals Cosmic Phys. **11**, 1

Schaller, G.: 1986, Diploma Work, Geneva Observatory

Schaller, G., Maeder, A.: 1990 in preparation

Schatzman, E., Maeder, A.: 1981, Astron. Astrophys. **96**, 1

Schmutz, W., Hamann, W.-R., Wessolowski, K.: 1989, Astron. Astrophys. **210**, 236

Simon, R.: 1957, Bull. Acad. Roy. Belgique, Cl. Sci. **43**, 610

Smith, L.F.: 1988, Astrophys. J. **327**, 128

Smith, L.F., Hummer, D.: 1988, MNRAS **230**, 511

Smith, L.F., Maeder, A.: 1989, Astron. Astrophys. **211**, 71

Smith, L.J., Willis, A.J.: 1982, MNRAS **201**, 451

Spruit, H.C.: 1989, in *Inside the sun*, IAU Coll. 121, Eds. G. Berthomieu, M. Cribier, Kluwer Acad. Press, in press

Stothers, R., Chin, C.W.: 1973, Astrophys. J. **179**, 555

Stothers, R., Chin, C.W.: 1979, Astrophys. J. **233**, 267

Tassoul, J.L.: 1989, Course given in *3e Cycle Belge en astronomie et astrophysique*, in press

Torres, A.V., Conti, P.S., Massey, P.: 1986, Astrophys. J. **300**, 379

Vreux, J.M.: 1985, Publ. Astron. Soc. Pac. **97**, 274

Zahn, J.P.: 1983, in *Astrophysical processes in upper MS stars*, 13th Saas–Fee course, Eds. B. Hauck and A. Maeder, Geneva Obs. p. 253

Zahn, J.P.: 1987, in *The internal solar angular velocity*, Eds. B.D. Durney and S. Sofia, Reidel Publ. Co., p. 201

MASS LOSS DURING THE EVOLUTION OF MASSIVE STARS


Henny J.G.L.M. Lamers
Dept of Astronomy, University of Wisconsin, Madison, USA
SRON Laboratory for Space Research, Utrecht, The Netherlands


ABSTRACT. Mass loss during various phases in the evolution of massive stars ($M_* > 30\ M_o$) has been discussed recently in various reviews.

## 1. MASS LOSS FROM EARLY TYPE STARS

Methods of mass loss determinations of hot stars based on UV line profiles, H$\alpha$ emission, IR-excess and radio-flux have been reviewed by Lamers (1988). The resulting mass loss rates and their dependence on the stellar parameters have been discussed by Cassinelli and Lamers (1987), Lamers (1988) and Garmany (1988). The variability of stellar winds has been reviewed by Henrichs (1988).

The most succesfull theory for explaining the mass loss rates from hot stars is the radiation driven wind theory, which was reviewed by Kudritzki et al (1988). However, the variability of the winds is not explained by this theory. The observed degrees of ionization in the winds of hot stars do not agree with the predictions of the radiation driven wind theory (review by Lamers and Groenewegen, 1990)

## 2. MASS LOSS FROM LUMINOUS BLUE VARIABLES

Mass loss from Luminous Blue Variables (LBV's) was recently reviewed by Lamers (1989) for the quiescent phases and the outbursts. The overall mass loss from LBV's is dominated by the large outbursts.

## 3. MASS LOSS FROM WOLF RAYET STARS

Mass loss from Wolf Rayet stars was recently reviewed by Willis and Garmany (1987) on the basis of the UV observations. Conti (1988) reviewed the different mass loss determinations and the dependence of mass loss on the stellar parameters.

## 4. MASS LOSS FROM LATE TYPE SUPERGIANTS

Mass loss from late type supergiants was discussed recently by Goldberg (1986) and by Dupree and Reimers (1987).



*L. A. Willson and R. Stalio (eds.), Angular Momentum and Mass Loss for Hot Stars, 53–54.*
© 1990 *Kluwer Academic Publishers. Printed in the Netherlands.*

## 5. MASS LOSS AND ROTATION

The mass loss from the rapidly rotating Be stars is a clear indication
that rotation can enhance the mass loss rates and produce equatorial
excretion disks. The mechanism is not very well known. The mass loss
rates from UV observations were reviewed by Snow and Stalio (1987) and
the mass loss determinations from the IR-excess were discussed by Lamers
(1986).

## 6. CONCLUSIONS

Stars with initial masses higher than 40 $M_O$ lose about 10 to 15 percent
of their mass during the H core-burning phase. After this phase they
evolve into Luminous Blue Variables and lose an additional 5 to 20 $M_O$,
mainly during the outbursts. When their mass has dropped below a
critical limit, i.e. when the mass of the star is about 1.4 times the
mass of the processed core, the star contracts to a Wolf Rayet star.

Stars with initial massses lower than 40 $M_O$ lose less than 10 percent
of their mass during the main-sequence phase. They will evolve into red
supergiants and suffer high mass loss in that phase. When the mass of
the star is less than 1.3 times the mass of the core the star will
contract and become a Wolf Rayet star.

The high mass loss rates continue during the WR phase, until the WR
star explodes as a supernova.

## REFERENCES


in *O Stars and Wolf Rayet Stars*, 1988, eds P.S. Conti and A.B.
Underhill, NASA SP-497:

    Conti. P.S. 1988, p.168
    Garmany, C.D. 1988, p 160
    Henrichs, H. 1988, p 199
    Kudritzki, R.P., Pauldrach, A., Puls, J. 1988, p.173

in *Exploring the Universe with the IUE Satellite*, 1987, eds Y. Kondo et
al, Reidel, Dordrecht:

    Cassinelli, J.P., Lamers, H.J.G.L.M. 1987, p.139
    Dupree, A.K., Reimers, D. 1987, p.321
    Snow, T.P., Stalio, R. 1987, p.183
    Willis, A.J., Garmany, C.D. 1987, p.157

*other reviews:*

Goldberg,L. 1986, in *The M-type Stars*, eds H. Johnson and F. Querci,
    NASA SP-492, p.245
Lamers, H.J.G.L.M. 1986, in *Physics of Be stars*, eds T. Snow  and A.
    Slettebak, Cambridge University Press p.219
Lamers, H.J.G.L.M. 1988, in *Mass Outflows from Stars and Galactic
    Nuclei*, eds L. Bianchi and R. Gilmozzi, Kluwer Acad Publ. p.39
Lamers, H.J.G.L.M. 1989, in *Physics of Luminous Blue Variables*,  eds K.
    Davidson, A. Moffat, H.J.G.L.M. Lamers, Kluwer Acad Publ  p.135
Lamers, H.J.G.L.M., Groenewegen, M. 1990 in *Intrisic Properties  of O-
    stars*, ed C. Garmany, ASP Conf Series (in press)

ANGULAR MOMENTUM LOSS IN PRE-MAIN SEQUENCE OBJECTS AND THE
INITIAL ANGULAR MOMENTUM OF STARS

Steven D. Kawaler
Department of Physics
Iowa State University
Ames, IA  50010 USA

ABSTRACT.  Several lines of observational evidence point to angular momentum loss from
pre-main sequence stars between the end of the dynamical collapse phase and arrival on the
main sequence.  The degree of angular momentum loss is a strong function of mass, with low
mass stars ($M<1.3M_\odot$) experiencing the most angular momentum loss.  This angular
momentum loss prevents the use of low mass main sequence stars in studies of primordial
stellar angular momentum.  However, intermediate mass stars ($1.3M_\odot<M<{\sim}6M_\odot$) have
apparently lost little angular momentum and can be used to probe the initial distribution of
angular momentum.  While the angular momentum distribution is broad for a given spectral
type, the mean angular momentum is a well determined function of the stellar mass.  This
relationship, derived for intermediate mass main sequence stars, is consistent with the measured
rotation velocities of pre-main sequence stars of lower mass as well.

## I INTRODUCTION

In studying stellar rotation, we must always be concerned with the issue of whether the
observed rotation rate of a star reflects its primordial angular momentum or it is the result of
evolutionary processes.  Similarly, when we study the distribution of stellar rotation for stars of
a given type, we need to know whether the observed *distribution* is primordial or evolutionary.
Only after these fundamental questions are resolved can we use the observed stellar rotation
properties as probes of stellar evolution.

For the purposes of this review, we will consider the "initial" angular momentum as the
angular momentum that remains following the dynamical collapse phase, and the appearance of
the star on the Hayashi track.  Estimates of the mass dependence of the mean initial angular
momentum $J_0(M)$, and the distribution of angular momenta for stars of a given mass $\sigma_J(M)$,
come from the observed rotation rates of main sequence stars.  A smooth trend of decreasing
mean angular momentum with mass exists for intermediate mass main sequence stars (the
"Kraft" curve).  Main sequence stars less massive than $1.3M_\odot$ rotate much more slowly than
this trend, suggesting that they lose angular momentum with time.  Observations of the rotation
of very young stars provide additional estimates of the initial angular momentum.

The observed rotation velocities of stars in young clusters of different ages allow us to
map out the history of angular momentum loss by magnetic braking in low and intermediate
mass stars.  With the help of these clusters, the process of magnetic braking is now reasonably



*L. A. Willson and R. Stalio (eds.), Angular Momentum and Mass Loss for Hot Stars, 55–63.*
© 1990 *Kluwer Academic Publishers. Printed in the Netherlands.*

56

well understood, at least for low mass stars.  Application of models of magnetic braking to intermediate mass stars indicates  that they lose little angular momentum by this process.  Hence the main sequence rotation velocities of intermediate mass stars should reflect their initial angular momenta.

## II INITIAL ANGULAR MOMENTA

### A) The Angular Momentum "Problem"

Can we obtain the function $J_0(M)$ from first principles, and then refine the theory with appropriate observations?  Consider an interstellar cloud that will eventually collapse to form a star.  One would expect the cloud to co-rotate with the Galaxy; at the position of the Sun, this corresponds to an angular rotation frequency of $10^{-15}$ s$^{-1}$.  If the cloud is spherical and of uniform density, then it is easy to show that

$$J_{cloud} \approx 5 \times 10^{55} \text{ g cm}^2 \text{ s}^{-1} \left(\frac{M}{M_\Theta}\right)^{5/3} \left(\frac{\rho_0}{10^{-24}\text{gcm}^{-3}}\right)^{-2/3} \qquad (1)$$

is the angular momentum of a cloud with density $\rho_0$ and mass M.  Main sequence stars have *much* less angular momentum.  Using representative values for M, R, and $\Omega$ ($3M_\Theta$, $3R_\Theta$, $\Omega \approx 7 \times 10^{-5}$s$^{-1}$) gives a typical value for the angular momentum of intermediate mass main sequence stars of $3 \times 10^{50}$ g cm$^2$ s$^{-1}$, which is six orders of magnitude smaller than the parent cloud!

The angular momentum of a star in its diffuse protostellar state is therefore several orders of magnitude too large to allow collapse to stellar dimensions; rotational support would halt protostellar collapse in the equatorial plane once the cloud collapses to about 0.1 parsec!  The cloud's solution to this problem may involve fragmentation, magnetic fields, disk formation, accretion, and planetary system formation and evolution. By the time it appears as a true stellar object near the Hayashi track, it can only retain a tiny fraction of the primordial angular momentum.  Yet it is this residual spin, which we refer to as the initial stellar angular momentum, manifests itself in the observed rotation, magnetic field dynamics, and other key properties of stars. Clearly, it is as yet impossible to determine $J_0(M)$, or $\sigma_J(M)$, from theoretical arguments alone.  We are therefore forced to seek an empirical determination of the angular momentum retained by a protostar that has reached the Hayashi track and emerged as a star after its dynamical collapse phase.

### B. Main Sequence Rotation Velocities

It is well known that stars less massive than about $1.3M_\Theta$ lose angular momentum with time. Evidence for this is provided by the observed dependence of stellar rotation velocity on mass, as first determined by Kraft (1970).  With the assumption that stars rotate as solid bodies, this implies that the mean angular momentum scales as $M^2$ for stars more massive than $1.3M_\Theta$. Below this mass, the angular momentum drops dramatically with further decreasing mass. Intermediate mass stars ($M>1.3M_\Theta$) have extremely shallow surface convection zones and weak magnetic fields, and there is no observational evidence that their rotation velocity is a function of age.  These factors together imply that these stars suffer little angular momentum loss near the main sequence.

For stars of a given mass, the range in rotation velocities covers at least an order of magnitude.  However, as shown by Kraft (1970, see also Kawaler 1987), the *mean* rotation velocities follow a power law in mass.  In fact, the mean main sequence rotation velocities are approximately 1/3 breakup over the range from 1.5 to 6 solar masses (Tarafdar and Vardya 1971, Kawaler 1987). Assuming solid body rotation, the empirical relationship between the mean angular momentum and mass is

$$J_0(M) = (9.0 \pm 0.6) \times 10^{49} \left(\frac{M}{M_\odot}\right)^{2.09 \pm 0.05} \mathrm{g\ cm^2\ s^{-1}}, \qquad (2)$$

as determined by Kawaler (1987) using the data of Fukuda (1982) in an approach similar to Kraft (1970). The tightness of this relation is easily understood if one accepts the observed fact that the mean rotation velocity is 1/3 of the breakup velocity.  Then the theory of stellar structure sets the rules; the breakup velocities, moments of inertia, and radii of main sequence models all follow simple power laws in mass that can be derived from the basic equations of stellar structure (see Kawaler 1987).

While the mean rotation velocity is a well behaved function of stellar mass, $\sigma_J(M)$ is a broad distribution ranging from near zero to near breakup.  $J_0(M)$ as determined above is only a low order moment of the distribution; the shape can change significantly from one spectral type to the next (Wolff *et al.* 1982).  On the average, the distribution looks somewhat Maxwellian, with a peak at or slightly below $J_0(M)$.

The outer layers of stars contribute most to the total moment of inertia.  Even if there is a large degree of internal differential rotation, then, stars cannot store significant amounts of angular momentum in their interiors.  Estimates of the total angular momentum that are based on the assumption of solid body rotation are therefore not as bad as they might sound.  The paper by Marc Pinsonneault (these proceedings) examines the consequences of this property on the later evolution of intermediate mass stars.   Therefore, the observed rotation velocities of stars are indeed a direct window on $J_0(M)$ and $\sigma_J(M)$.

## C. Rotation Velocities of Young Stars

The Kraft curve may represent the dependence of the mean initial angular momentum on stellar mass, but only if those stars lose little angular momentum when evolving from the Hayashi track to the main sequence.  Measuring rotation velocities for intermediate mass pre-main sequence objects such as T Tauri stars provides a more direct measurement of their initial angular momentum.

We reluctantly turn to the T Tauri stars when studying the early history of stellar angular momentum; reluctantly because it is extremely difficult to assign accurate masses to individual objects. This is a particularly difficult job because stars that populate the Hayashi track have no obvious observational mass discriminator; they lie along a line of roughly constant effective temperature, with the luminosity determined both by the mass of the star and its age.  Another complicating factor is that the rotation velocities themselves can only be determined if greater than 5 to 10 km/s. In addition, T Tauri stars display vigorous winds, and these winds can carry angular momentum away from the star.  Thus the rotation of T Tauri stars only provide lower limits to the initial stellar angular momentum.  In any case, these young objects provide a direct way to estimate how much angular momentum stars retain following protostellar collapse.

The first systematic observations of pre-main sequence rotation velocities was the work of Vogel and Kuhi (1981).  Smith, Beckers, and Barden (1983) observed rotation velocities in the Orion Ic complex, and rotation velocities for T Tauri stars were measured by Hartmann *et al.*(1986), Bouvier *et al.* (1986), and Hartmann and Stauffer (1989).  Recently, McNamara (1989) re-examined the data on NG2264 and Orion with updated memberships, and was able to reconcile the distribution of rotation velocities observed by Vogel and Kuhi (1981) and Smith, Beckers, and Barden (1983).

In general, PMS rotation velocities are binned into "high mass" and "low mass" ($M>1.3M_\odot$ and $M<1.3M_\odot$) bins.  Figure 1 shows the data for T Tauri stars taken from Bouvier *et al.* (1986) and Hartmann *et al.* (1986).  The high mass distribution is peaked at low velocities, but has a tail that extends beyond 100 km/s.  The mean rotation velocity for this sample is roughly 40 km/s.  Assuming a mean mass of about $1.5M_\odot$ this corresponds to an angular momentum of $3\times10^{50}$ gcm$^2$s$^{-1}$, which is about 50% higher than the value obtained from the Kraft curve.  The low mass distribution shown in Figure 1 shows a sharp peak at low rotation velocities; the mean rotation velocity is only 14km/s.  With a mean mass of $0.7M_\odot$, this velocity implies a mean angular momentum of $3\times10^{49}$ gcm$^2$s$^{-1}$, or about 70% of the Kraft value.

Hence, within the uncertainties associated with assigning masses to pre-main sequence objects, T Tauri stars fall within a factor of about 1.5 of the Kraft relation for mean angular momentum as a function of stellar mass.  The distribution of $J_0(M)$ appears to be different for stars of high and low mass, but there is a significant mass range in each "bin".  Within each mass grouping, a mass range of a factor of 2 could mean a range in $<J>$ of a factor of 4 if the Kraft relationship indeed reflects the initial angular momenta.

The breakup velocity at the base of the Hayashi track, where one is most likely to find T Tauri stars, is about 400 km/s; the process of star formation, along with any subsequent angular momentum loss during the Hayashi contraction phase, leaves young stars rotating at only a few percent of the breakup velocity.  Thus any of the mechanisms for angular momentum loss during dynamical collapse and on the Hayashi track must be very efficient.

## III ANGULAR MOMENTUM LOSS DURING AND AFTER PMS CONTRACTION

The angular momenta of young low mass stars would lead to rotation velocities on the main sequence of a few hundred km/s.  However, as reflected by the break in the Kraft curve, low mass main sequence stars rotate slowly, with velocities typically less than 10 km/s.  Therefore, they must have lost a significant fraction of their angular momentum during main sequence evolution.  This angular momentum loss is a result of magnetic braking, which can remove angular momentum quite efficiently from stars with surface convection zones.  While stars above $1.3M_\odot$ do not have significant convection zones on the main sequence, they did while they were approaching it.  Therefore, one can try to determine how much angular momentum intermediate mass stars could have lost while evolving to the main sequence.  To approach this problem quantitatively, we use lower mass stars to determine the properties of magnetic braking, and then apply the empirically derived braking rates to higher mass stars to see if magnetic braking could have removed significant quantities of angular momentum during the early lives of intermediate mass stars.

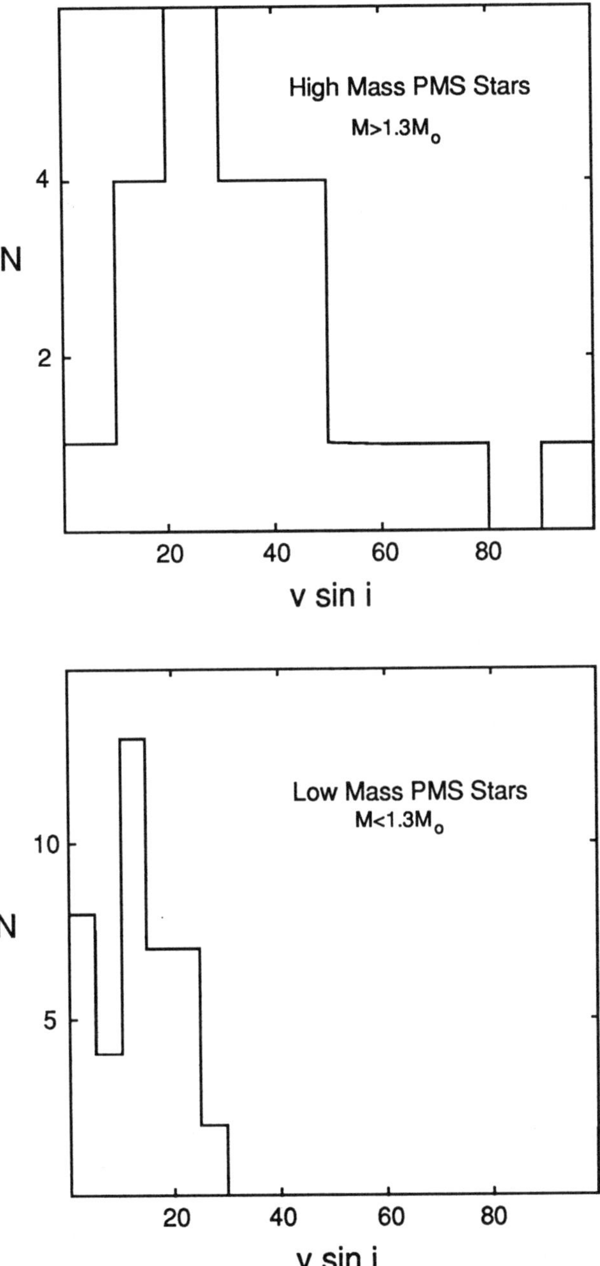
High Mass PMS Stars
M>1.3M$_o$
N
4
2
20
40
60
80
v sin i
Low Mass PMS Stars
M<1.3M$_o$
N
10
5
20
40
60
80
v sin i

## A. The Braking Rate in Low Mass Stars

As the radiative core grows in evolving pre-main sequence stars, the contracting core spins up with respect to the convective envelope, and differential rotation develops. When this contrast in rotation rate between the convective envelope and radiative core develops, dynamo generated magnetic field modulation begins, modifying (and organizing) the primordial magnetic field. This intimate relationship between rotation and magnetic fields extends out into the stellar wind; the global magnetic field constrains the wind to co-rotate with the star out to very large distances. This greatly enhances the efficiency of the wind in removing angular momentum (as initially suggested by Schatzman 1962).

With some simple assumptions about the velocity structure of the stellar wind, and using the observed relationship between the rotation rate and global magnetic field strength, Kawaler (1988) rewrote the Mestel (1984) prescription for angular momentum loss via a magnetic stellar wind as:

$$\frac{dJ}{dt} = K \, \Omega^3 \left[ \frac{R}{R_\odot} \right]^{\frac{1}{2}} \tag{3}$$

for stars of one solar mass. Skumanich (1972) and Soderblom (1983) determined the rotation rate for solar mass stars in clusters of known age; their empirical determinations of the spin down rate for solar mass stars show that rotation velocity decreases as $t^{-1/2}$. Rotation velocities of sub-solar mass stars follow this same trend, as demonstrated by Kawaler (1989) using the Hyades as an example. On the main sequence, where the stellar radius and moment of inertia change very slowly with time, the above angular momentum loss law can be integrated to yield

$$v_{eq} \propto t^{-1/2} \tag{4}$$

which matches the observed spin-down rate for low mass stars older than about $10^8$ years.

This model of angular momentum loss extends backwards in time to allow us to trace the early angular momentum history of the stars. Converting the loss rate to a time scale for angular momentum loss, one finds

$$\tau_J \approx 2.3 \times 10^{10} \left[ \frac{I}{I_\odot} \right] \left[ \frac{\Omega}{\Omega_\odot} \right]^{-2} \text{ years.} \tag{5}$$

It is easy to see that angular momentum loss occurs only when I, the stellar moment of inertia, is small and $\Omega$ is large. In the T Tauri phase, the evolutionary time scale is about $10^6$ to $10^7$ years, and $\Omega$ and I are larger, by factors of about 4 , than on the main sequence. Thus the angular momentum loss time scale of several x $10^9$ years is far longer than the evolutionary time scale in the T Tauri phase, and pre main sequence stars continue to spin up as they contract. On the approach to the main sequence, the evolutionary time scale increases and the braking time scale shortens (since $\Omega$ increases and I decreases) . These factors result in the increased importance of angular momentum loss for low mass stars when they reach the main sequence. For the Sun as an example, to spin down to the current rotation rate at the current age, its rotation rate must have been about 20 to 30 times the current rate when it first arrived on the main sequence.

The dependence of the spin-down rate on stellar mass and time can be tested by observations of rotation velocities in young clusters, as demonstrated by Stauffer and Hartmann (1986, and references therein). The $\alpha$ Perseus cluster (age $\approx$ 3-5x$10^7$ yr) shows a wide distribution of rotation velocities for all spectral types, and contains many rapidly rotating (v sin i $\approx$ 100 km/s) G, K, and M stars. In the Pleiades (age $\approx$ 7x$10^7$ years) the G stars have all spun down to below 10 km/s, while there are still rapidly rotating K and M stars. At the age of the Hyades (5x$10^8$ years), F and early G rotation velocities remain relatively unchanged, but K and M stars rotate very slowly.

As suggested by Stauffer and Hartmann (1986), this trend is clear evidence that the convective envelope spins down first in low mass stars. G stars have relatively thin convective envelopes, and therefore the envelope contains only a few percent of the total stellar angular momentum. Since the wind acts primarily on the convective envelope, G stars will spin down relatively quickly. K and M stars have thick envelopes which contain a large fraction of the total stellar angular momentum. Therefore the magnetic wind needs to work for a long time to reduce the surface rotation velocity of lower mass stars. This accounts quite naturally for the observed rapid spin down of G stars compared to K stars in young clusters. By the age of the Hyades, however, the data are most consistent with solid body rotation (Kawaler 1988). The spin down of the convective envelope causes a large shear to develop at the base of the envelope. Eventually , this shear becomes hydrodnamically unstable, and drives angular momentum transport between the envelope and the interior. This coupling of the core and envelope therefore occurs over a time scale similar to the age of the Hyades. Solar models that include redistribution of angular momentum show exactly this behavior (Pinsonneault *et al.*1989).

## B.Angular Momentum Loss in Intermediate Mass Stars?

We can use this successful model of magnetic braking to examine angular momentum loss in stars that contain substantial convection zones. Magnetic braking is only effective when a star retains a convective envelope deep enough to develop a dynamo-modulated magnetic field and vigorous enough to pump out a stellar wind. The surface convection zone in stars more massive than about $1.3M_\odot$ disappears on the way to the main sequence. Therefore magnetic braking causes angular momentum loss only during the brief pre-main sequence phases in intermediate mass stars. The following table lists the fraction of the initial angular momentum that remains with a model on the main sequence, under the assumption of solid body rotation. The braking law of Kawaler (1988) was applied to standard models of pre-main sequence stars as described in that paper, for two different initial angular momenta.

|  | $J_{final}/J_0$ | |
| --- | --- | --- |
| $M/M_\odot$ | $J_0=J_{Kraft}$ | $J_0=2.5\,J_{Kraft}$ |
| 1.00 | 0.014 | 0.008 |
| 1.25 | 0.040 | 0.014 |
| 1.50 | 0.960 | 0.810 |
| 2.00 | 0.998 | 0.990 |

It is clear that stars above $1.3M_\odot$ retain essentially all of their initial angular momentum, even when they rotate at nearly breakup velocity when reaching the zero-age main sequence.

## IV. SUMMARY

Somehow, stars solve the angular momentum problem during the dynamical collapse phase of protostellar evolution. Low mass stars ($M<1.3M_\odot$) suffer further angular momentum via magnetic braking during pre-main sequence and main sequence evolution. Their initial angular momenta, as inferred from observations of T Tauri stars, shows a peak in $J_0$ at low velocities, with no high velocity tail. On the other hand, intermediate mass stars ($6.0 < M/M_\odot < 1.5$) suffer little magnetic braking; their initial angular momentum survives through the main sequence phase. $J_0(M)$, the mean angular momentum for stars of a given mass, is proportional to $M^2$ over this range, and indicates that for some reason stars prefer to rotate with velocities that are 1/3 the equatorial breakup velocity. However, there is a wide distribution of J for a given M, as seen in both main sequence rotation velocities and in the rotation velocities of intermediate mass T Tauri stars.

While angular momentum transport within stellar interiors does happen, the surface velocities are a reliable guide to the total stellar angular momentum. The rotation of young low mass stars in clusters shows that the surface convection zone spins down first. Eventually, hydrodynamic instabilities couple the stellar interior to the surface and bring about solid body rotation in most of the outer parts of stars. This angular momentum transport can lead to some rotationally driven mixing in stellar interiors. Pinsonneault *et al.* (1990) show that the amount of mixing is directly proportional to the initial angular momentum after a few x $10^8$ years for models of around $1M_\odot$. Since lithium is a fragile element that burns at the base of the surface convection zone in low to intermediate mass stars, the surface abundance is very sensitive to any mixing that may occur. The work of Pinsonneault *et al.* (1990) suggests that the lithium abundance observed in main sequence stars is inversely proportional to the initial stellar angular momentum. Thus, with future work, it is likely that lithium abundances will give quantitative measurements of the initial angular momenta, providing yet another probe of the initial conditions for stellar rotation.

The author is grateful to Marc Pinsonneault for many illuminating discussions. This work is sponsored in part by NASA grant NAGW-1364 to Iowa State University.

REFERENCES

Bouvier, J., Bertout, C., Benz, W., and Mayor, M. 1986, *Astr. Ap.*, **165**, 110.
Fukuda, I. 1982, *Pub.A.S.P.*, **94**, 271.
Hartmann, L., Hewett, R., Stahler, S., and Mathieu, R.D. 1986, *Ap. J.*, **309**, 275.
Hartmann, L., and Stauffer, J. 1989, *Aston. J.*, **97**, 873.
Kawaler, S. 1987, *Pub.A.S.P.*, **99**, 1322.
Kawaler, S. 1988, *Ap. J.*, **333**, 236.
Kawaler, S. 1989, *Ap. J. (Lett.)*, **343**, L65.
Kraft, R. P. 1970, in *Spectroscopic Astrophysics*, ed. G.H. Herbig (Berkeley: University of California Press), p. 385.
McNamara, B. 1989, preprint.
Mestel, L. 1984, in *3rd Cambridge Workshop on Cool Stars, Stellar Systems, and the Sun*, ed. S. Baliunas and L. Hartmann, (New York: Springer), p. 49.
Pinsonneault, M., Kawaler, S., Sofia, S., and Demarque, P. 1989, *Ap. J.*, **338**, 424.

Pinsonneault, M., Kawaler, S., and Demarque, P. 1990, *Ap. J. Supp.*, in press.
Schatzman, E. 1962, *Ann. d'Ap.*, **25**, 18.
Skumanich, A. 1972, *Ap. J.*, **171**, 565.
Smith, M.A., Beckers, J.M., and Barden, S.C. 1983, *Ap. J.*, **271**, 237.
Soderblom, D.R. 1983, *Ap. J. Supp.*, **53**, 1.
Stauffer, J. and Hartmann, L. 1986, *Pub.A.S.P.*, **98**, 1233.
Tarafdar, S.P. and Vardya, M.S. 1971, *Ap. Space Sci.*, **13**, 234.
Vogel, S.N. and Kuhi, L.V. 1981, *Ap. J.*, **245**, 960.
Wolff, S.C., Edwards, S., and Preston, G.W. 1982, *Ap. J.*, **252**, 322.

# THE HERBIG Ae AND Be STARS: MASS AND ANGULAR MOMENTUM LOSSES

C. CATALA
*Département de Recherche Spatiale et URA 264, CNRS*
*Observatoire de Paris, Section de Meudon*
*92195 Meudon Principal Cedex, France*

ABSTRACT: The Herbig Ae and Be stars are pre-main sequence stars of intermediate mass $(2 \leq M/M_\odot \leq 5)$. Although they are in the radiative phase of their contraction and possess no outer convection zone, they exhibit definite signs of strong stellar winds and activity.

The relevant observational material concerning mass loss and chromospheric activity for these stars is reviewed, and current estimates for mass loss rates and radiative losses are discussed.

The relatively high mass loss and rotation rates of these stars provide a strong indication that angular momentum loss might play a major role in their evolution, and might even be the source for their paradoxical active phenomena.

## 1. Introduction

In a pioneering work almost 30 years ago, Herbig (1960) conjectured that the "Be and Ae stars associated with nebulosity" were in fact objects of intermediate mass still in their pre-main sequence (PMS) stage of evolution. He proposed a list of 26 candidates verifying four criteria, which are now commonly used as definition for the Herbig Ae/Be stars:

1. The spectral type is earlier than F0: this criterion allows us to select only massive objects.

2. They exhibit emission lines in their spectra: the presence of emission lines was known to be a characteristic of T Tauri stars, and therefore was considered as a criterion of youth.

3. They lie in an obscured region: another criterion of youth; young stars have not had enough time to escape from their parental clouds.

4. They illuminate a bright reflection nebula in their immediate vicinity: this criterion eliminates those stars that would simply be projected on dark clouds on the celestial sphere.

As Herbig himself mentioned, his original list of 26 candidate Herbig Ae/Be stars was likely to be incomplete, because it was not the result of a systematic survey. A quarter



*L. A. Willson and R. Stalio (eds.), Angular Momentum and Mass Loss for Hot Stars*, 65–83.

of a century later, Finkenzeller and Mundt (1984) extended the list to 57 field Herbig Ae/Be stars, using the same criteria as Herbig. This list is presently being extended in two directions: Hu et al. (1989) have surveyed the IRAS catalogue and selected new candidates on the basis of far infrared colors and near infrared spectra, in addition to the classical criteria. Finally, Thé et al. (1985a,b) identified some Herbig Ae/Be stars in young open clusters. The advantage of observing these stars in open clusters lies in the possibility of independent estimate of the age of the clusters.

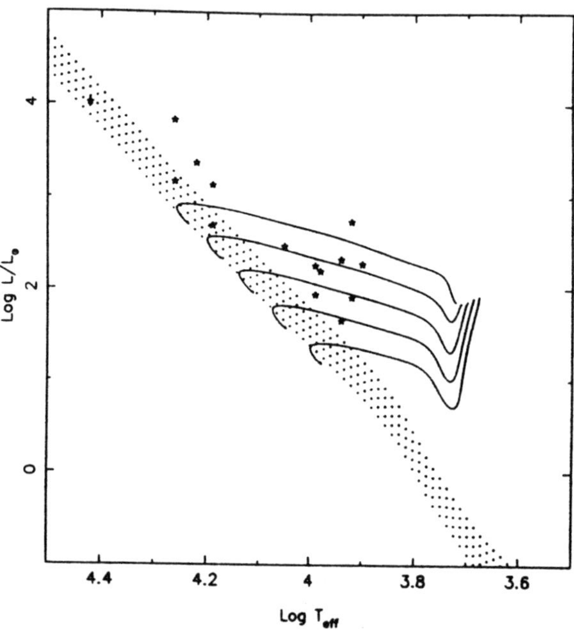

Figure 1: The Herbig Ae/Be stars in the HR diagram. Only those stars for which the effective temperature and the absolute luminosity are well known have been represented (star symbols). The thick dotted line is the main sequence. The full lines represent evolutionary tracks for 2, 2·5, 3, 4, and 5 $M_\odot$, computed with Gilliland's (1986) stellar evolution code.

Since Herbig's pioneering work, a considerable effort has been made to verify his hypothesis about the PMS nature of the Herbig Ae/Be stars. The discovery of strong infrared excesses in most of them (Mendoza, 1966, 1967; Cohen, 1973, 1975, 1980) provided a further indication that they are surrounded by dust shells, which can be considered as a sign of youth. Strom et al. (1972), then Cohen and Kuhi (1979) showed that they lie above the main sequence in the HR diagram, another crucial argument in favor of their PMS nature. The physical association of the Herbig Ae/Be stars with their associated dark clouds was checked by Finkenzeller and Jankovics (1984). Finally, the measurement of the projected rotational velocities (*vsini*) of the Herbig Ae/Be stars by Davis et al. (1983), then by Finkenzeller (1985) showed that they are intermediate rotators and excluded their interpretation as evolved objects. In summary, very little doubt is left today regarding Herbig's original conjecture that the Herbig Ae/Be stars are young objects still in a pre-main sequence stage of their evolution. This point is illustrated by Fig. 1, representing the location of the Herbig Ae/Be stars in the HR diagram.

As can be seen in Fig. 1, the Herbig Ae/Be stars are probably in the radiative phase of their contraction toward the main sequence, and are not supposed to possess outer convection zones, according to the most recent calculations of PMS evolution (e.g. Gilliland, 1986). However, as we shall see below, they exhibit signs of intense active phenomena; such phenomena, when observed in later type stars, are usually thought to be linked to the presence of thick subphotospheric convection zones. In this respect, the activity of the Herbig Ae/Be stars may be regarded as paradoxical, and its study is likely to bring us new and valuable information about this phenomenon for all types of stars.

The Herbig Ae/Be stars also possess strong stellar winds, as discussed below. None of the models already proposed to explain stellar winds for various types of stars across the HR diagram has been applied successfully to the Herbig Ae/Be stars: because of the absence of detection of hot coronae around these stars, we suspect that thermal pressure is unable to drive the observed winds; the Herbig Ae/Be stars are not luminous enough for their winds to be simply explained by the standard radiatively-driven wind theory, although the role of such agents like multiple scattering of photons, fast rotation or magnetic fields have not yet been investigated in sufficient detail in the case of these stars; finally, wave- and shock-driven wind theories have not yet been applied to such stars.

The winds and activity of the Herbig Ae/Be stars present us with a challenge: the ultimate sources of energy and momentum, as well as the mechanisms by which they are made available above the stellar photosphere, are unknown. The first step for solving this exciting problem consists of determining the main characteristics of these winds and activity, like mass loss rates, wind velocities, angular momentum losses, sizes and temperatures of the chromospheres, as well as radiative losses from the chromospheres (which provide an estimate of the amount of non-radiative energy deposited above the stellar photosphere). The present paper is an attempt to summarize our current knowledge on these characteristics of the winds and activity of the Herbig Ae/Be stars, and to indicate several possible tracks we can follow to solve the problems they raise.

## 2. Winds from Herbig Ae/Be stars

### 2.1. OBSERVATIONAL EVIDENCE FOR WINDS FROM HERBIG Ae/Be STARS

The first clear evidence for mass outflow on a statistically representative sample of Herbig Ae/Be stars came from the high resolution profiles of the H$\alpha$ line obtained by Garrison and Anderson (1977) for 14 of these stars. This type of work was extended by Finkenzeller and Mundt (1984) who published high resolution H$\alpha$ profiles for 57 candidate Herbig Ae/Be stars. The conclusion of both studies is that these stars can be divided in three subclasses according to the shape of their H$\alpha$ line: the "double-peak" emission profile subclass, the "single-peak" emission profile subclass, and the "P Cygni" profile subclass. The population of the three subclasses in the work of Finkenzeller and Mundt (1984), which is the most complete to date, are the following: double-peak: 50%, single-peak: 25%, P Cygni: 20%. Figure 2 shows an example of the three types of profiles.

The stars showing a P Cygni profile at H$\alpha$ definitely possess a stellar wind. Detailed cal-

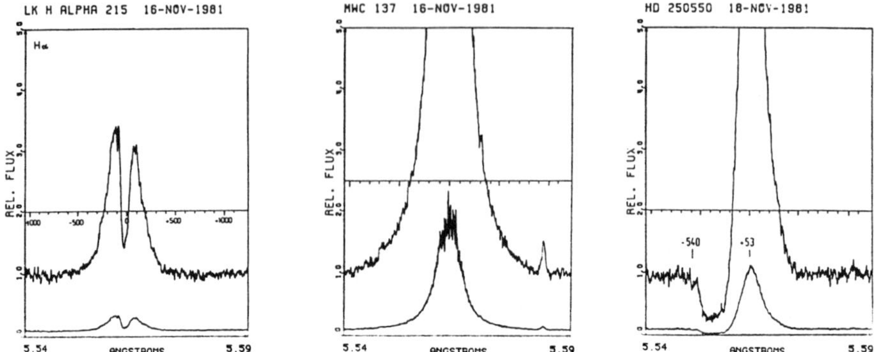



Figure 2: Examples of Hα profiles for Herbig Ae/Be stars. *Left:* double-peak emission. *Middle:* single-peak emission. *Right:* P Cygni. Adapted from Finkenzeller and Mundt (1984).

culations of the Hα profile formed in the winds of Herbig Ae/Be stars have been performed by Catala and Kunasz (1987). The interpretation of the other two types of profiles poses more serious problems. In principle, a double-peak emission profile can also be formed in a stellar wind. For instance, it has been proposed in the past that double-peak emission profiles (very similar to the type III P Cygni profiles in Beal's (1951) classification) can be interpreted as due to the combination of expansion and rotation in a stellar wind (Mihalas and Conti, 1980). In this interpretation, the emission peaks are formed in the inner parts of the wind which are corotating with the star, while the absorption component is formed in the outer regions, where expansion dominates. The relatively high rotation rates observed for the Herbig Ae/Be stars ($vsini = 100$ kms$^{-1}$; Finkenzeller, 1985) make this interpretation possible. However, the calculations of line profiles with this type of models are still preliminary (see e.g. Mazzali, 1988) and no attempt has yet been made to apply this idea to Herbig Ae/Be stars. Therefore, the interpretation of double-peak emission profiles in terms of mass outflow is still somewhat ambiguous. The same is true for single-peak emission profiles, which can also in principle be formed in stellar winds, but for which other possible interpretations exist (chromosphere with strong turbulent motions, for instance).

Adding to this confusion, some Herbig stars have their Hα profile shift from one type to another, as examplified in Fig. 3. Another example of this behavior can also be found in the Mg II resonance lines of HD 163296 (Catala et al., 1989), which changed their profile from a double-peak emission profile to a P Cygni profile in less than 15 months. This peculiar phenomenon suggests that the ultimate cause for the observed differences in the Hα profiles, whatever its nature, is not linked to the basic stellar parameters like mass or evolutionary status.

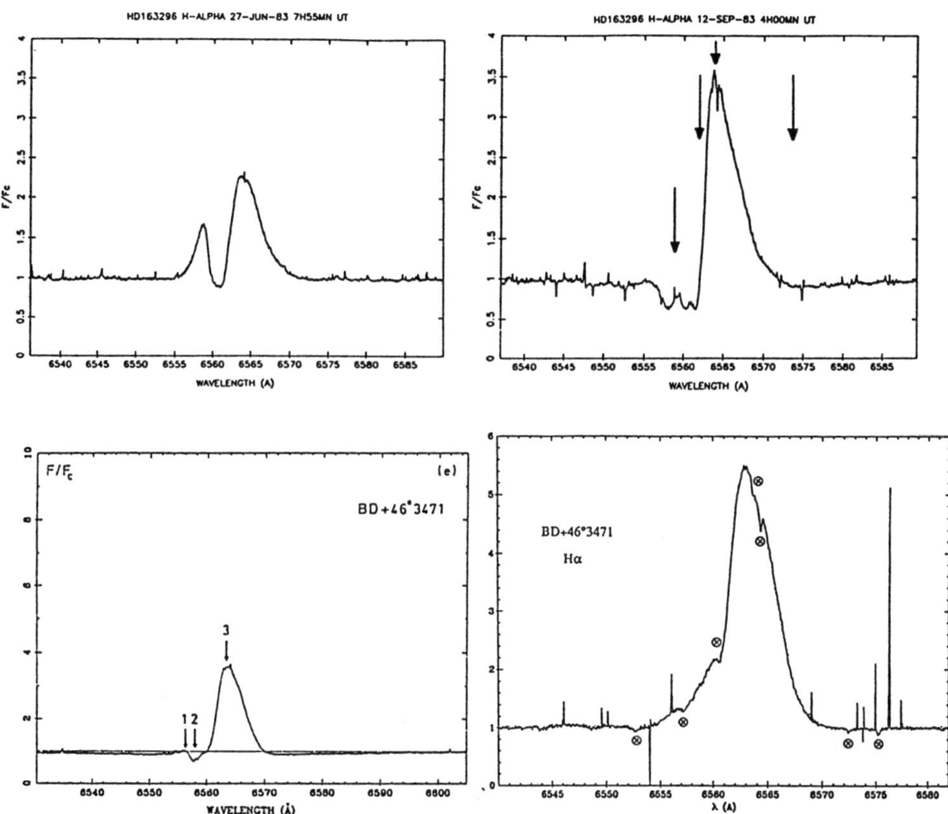



Figure 3: Two examples of Herbig stars having their Hα profile shift from one type to another. *Top:* two spectra of HD 163296 at Hα, obtained 3 months apart; the profile changes from double-peak emission to P Cygni (adapted from Thé et al.: 1985, *The Messenger* 41,8). *Bottom:* two spectra of BD +46°3471 at Hα, obtained 3 years apart; the profile changes from P Cygni to single-peak emission. The crossed circles on the bottom right spectrum indicate telluric water vapor lines (adapted from Catala et al. (1986a) and Catala et al.: 1988, *The Impact of Very High S/N Spectroscopy on Stellar Physics, G. Cayrel de Strobel and M. Spite (eds.), IAU Symp.* **132**, p. 105).

## 2.2. THE P CYGNI SUBCLASS

A considerable effort has been made in the past few years to better understand the structure of the atmosphere of the stars in the P Cygni subclass. This choice of the P Cygni subclass as the starting point for this type of detailed analysis is justified by the absence of ambiguity in the interpretation of P Cygni profiles in terms of outflows. I hope that in the not-too-distant future, the same type of work can be performed for the other two subclasses in spite of the problems posed by the ambiguity in the interpretation of their profiles.

Catala et al. (1986a) have obtained high resolution profiles for several spectral lines in stars of the P Cygni subclass: Mg II h and k lines in the UV, Ca II K line, He I 5876 A line, Na I D lines, H$\alpha$ line, Ca II IR triplet. In a complementary study, Catala and Talavera (1984) and Talavera and Catala (1989) have also secured high resolution spectra of the C IV resonance lines at 1550 A. Both studies conclude that all of these lines have a similar profile in the stars belonging to the P Cygni subclass, as can be seen on the example shown in Fig. 4. We note that the signature of the presence of winds is obvious on all of these lines, either by a P Cygni profile in the case of H$\alpha$ and the Mg II resonance lines, or by a blue-asymmetry in the case of the Ca II K line and the C IV resonance lines, which appear purely in absorption. The papers mentioned above also show that these different lines constitute probes of various regions in the winds of these stars: the H$\alpha$ line and the Mg II resonance lines are formed in a very extended region of the wind, from the photosphere up to several tens of stellar radii; the Ca II K line and the Ca II IR triplet, because of the lower abundance of calcium, are formed in a compact region at the base of the wind; the Na I D lines are formed in a remote region of the wind, because sodium is entirely ionized closer to the star. Therefore, the similar profiles observed for these lines suggest strong similarities in the structures of the winds of these stars, from the photosphere up to the distant regions where the Na I D lines are formed.

This result also allows us to concentrate on only one of these stars for a more detailed analysis, considering it as representative of the whole subclass. An obvious choice for such a detailed study is the A0V star AB Aur, the brightest Herbig Ae star in the Northern hemisphere. The quantitative interpretation of line profiles and continua in this star provided precise constraints on the structure of its atmosphere, as discussed in Section 2.3 and 3 below.

## 2.3. ESTIMATES OF MASS LOSS RATES

2.3.1. *Estimates Based On Line Profiles.* The availability of high resolution profiles for many lines in Herbig Ae/Be stars makes this method very appealing for deriving mass loss rates for these stars. The principle of the method is simple. First, one assumes a wind model (i.e. mass loss rate $\dot{M}$, laws of density, velocity and temperature in the wind), built under a set of basic assumptions. The emergent profile for a given line is then computed and compared with the observed profile, and the model is modified until a good agreement between computed and observed profiles is reached. The various applications of this method can differ by two important factors: first, the basic assumptions under which the wind model is built (for example spherical symmetry or more complex geometry, acceleration

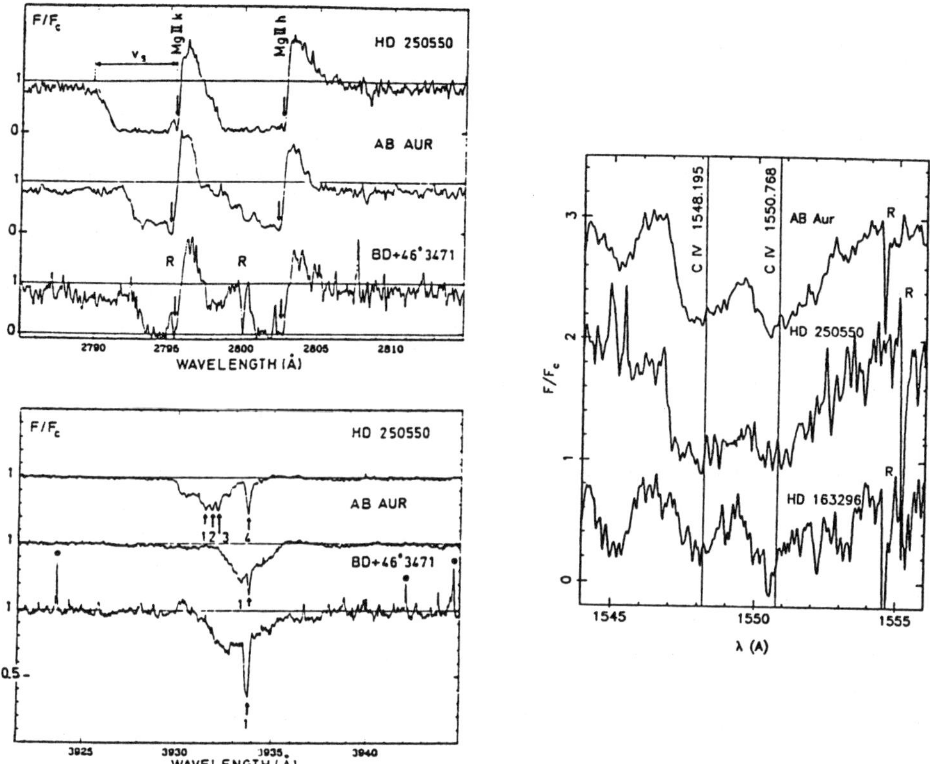



Figure 4: Spectral similarities for Herbig stars of the P Cygni subclass. *Upper left:* Mg II resonance lines for 3 Herbig Ae stars of the P Cygni subclass; note the well-marked type IV P Cygni profile for these lines in the 3 stars (from Catala et al., 1986a). *Lower left:* Ca II K line for the same 3 stars; note the similarity in the profile of this line for the 3 stars (from Catala et al., 1986a). *Right:* C IV resonance lines for 3 Herbig Ae stars of the P Cygni subclass. The 3 stars are of spectral type A0 (from Talavera and Catala, 1989).

or deceleration of the flow, etc...), and second, the method used for treating the radiative transfer problem in the lines considered.

Several difficulties are encountered when trying to derive mass loss rates by fitting line profiles. First of all, it is clear that one has to take into account non-LTE effects when solving the transfer equation in a bound-bound transition. This is not a serious problem, though, since numerical techniques exist for solving the non-LTE transfer equation accurately. Second, the presence of velocity fields in the model introduces strong Doppler effects between different regions of the wind, and can complicate seriously the treatment of the transfer problem. Again, powerful tools have been developed in the past two decades to deal with this kind of problems. The major difficulty comes from the fact that, usually, several very different models can fit reasonably well the observed profile of a given spectral line, and that there remains a strong ambiguity in the actual structure of the wind. This problem can in particular lead to enormous uncertainties on the mass loss rates derived in this way. An example of the nature of this problem can be found in Catala and Kunasz (1987), who have computed the H$\alpha$ profile for Herbig Ae stars, and find that, under certain circumstances, models differing by a factor of 10 in $\dot{M}$ lead to the same H$\alpha$ profile (see their Fig. 3). However, this problem and the related uncertainties on $\dot{M}$ can be reduced by considering many transitions, the treatment of each of them providing a set of constraints to be placed on the wind model. This process can be long and painful, but the reward is a reliable estimate of the wind parameters, including $\dot{M}$.

Kuan and Kuhi (1975) derived estimates of the mass loss rates for a set of stars showing P Cygni profiles in several hydrogen Balmer lines, including 4 Herbig Ae/Be stars. Their models assumed spherical symmetry, analytical velocity laws for accelerating or decelerating flows, and a temperature law derived from simple assumptions about heating and cooling processes in the wind. The density law in the wind was derived from the assumed mass loss rate by ensuring mass flux conservation. The lines used for this analysis were H$\beta$, H$\gamma$ and H$\delta$. Unfortunately, these lines have strong photospheric Stark wings in which the P Cygni profiles due to the wind contribution are dug. Because of the difficulty to disantangle the wind contribution from this photospheric contribution to the lines, the authors chose to use only the intensity of the emission peak with respect to the extrapolated core of the photospheric contribution. This intensity was compared to that derived from the line profile calculations, which did not include the photospheric contribution. Besides, these calculations were made under the Sobolev approximation, which assumes that the velocity gradient normalized to the intrinsic width of the line is much larger than the inverse of the photon mean-free-path. This approximation is probably not good near line center, which is precisely the part of the line considered for this study. We can therefore expect important uncertainties in the mass loss rates derived in this manner. Let us now turn to the obtained results, keeping these caveats in mind. First, the authors favor the models with decelerating flows, on the basis of the ratio between the velocity shifts of the redward and blueward edges of the lines. In their models, this ratio is primarily sensitive to the occultation effect of the star. Catala and Kunasz (1987) have argued against this argument, showing that small values for this ratio can be obtained without occultation effect. The second result of the work by Kuan and Kuhi (1975) is that, under the set of assumptions described above (including the assumption of decelerating flow), the mass loss rates for the 4 Herbig stars considered are in the range $1{\cdot}4 \leq \dot{M} \leq 3{\cdot}3 \times 10^{-7}$ M$_\odot$yr$^{-1}$. In particular, the mass loss rate for AB Aur, for which other estimates are available (see below) was found to be $3 \cdot 3 \times 10^{-7}$ $M_\odot yr^{-1}$.

A totally independent analysis was carried out by Catala et al. (1984), Catala and Kunasz (1987), and Catala (1988), who focused their efforts on AB Aur, previously shown to be a good representant of the P Cygni subclass (Catala et al., 1986a). In these models, the flow was assumed to be spherically symmetric, accelerated, and the velocity law was given as an empirical piece-wise linear function of distance to star's center. The temperature distribution was also empirically specified, and included a chromospheric region at the base of the wind, as discussed in Section 3. The wind model was adjusted at its base to match a standard photospheric Kurucz model corresponding to the star's spectral type. The mass loss rate, density law and velocity law in the wind were linked by the requirement of mass flux conservation. The transitions treated in this series of papers were the Mg II resonance lines, the hydrogen Balmer lines and continua, and the C IV resonance lines. The line transfer problem was treated without the Sobolev approximation, but by solving the non-LTE equation of transfer written in the comoving frame of the flow, following the technique described in Mihalas and Kunasz (1978) and references therein. The results are probably more reliable than those obtained by Kuan and Kuhi (1975) because a) the number of transitions used is much greater so that the ambiguity on the solution is reduced, b) the lines used are not or very mildly affected by the photospheric Stark contribution, and c) the Sobolev approximation is replaced by a more direct and appropriate transfer technique. These results differ significantly from those of Kuan and Kuhi (1975). They indeed yield a mass loss rate between $1 \cdot 0$ and $1 \cdot 8 \times 10^{-8}$ $M_\odot yr^{-1}$ for AB Aur, i.e. between 18 and 33 times less than Kuan and Kuhi's estimate. This estimate is based on the detailed interpretation of all the lines mentioned above, but is particularly sensitive to the emission component of the $H\alpha$ P Cygni profile and to the excess in the Balmer continuum mentioned in the following section. Because these features are probably variable in time, as we will see later, this estimate should also be taken with caution.

The extension of these results to the other stars of the P Cygni subclass is not straightforward, because they have different underlying photospheres that must be included for a proper calculation of the line profiles. Precise determinations of their mass loss rates must wait for a detailed interpretation of their line profiles and continua, similar to that conducted for AB Aur. However, Catala and Kunasz (1987) have explored in detail the influence of $\dot{M}$ on the $H\alpha$ profile for an A0 star with a wind. From this study and from the compilation of $H\alpha$ profiles for Herbig Ae stars published by Finkenzeller and Mundt (1984), we can tentatively conclude that those stars with spectral types around A0 have mass loss rates in the range $[10^{-8} - 10^{-7}]$ $M_\odot yr^{-1}$.


2.3.2. *Estimate Based On Balmer Excess.* Garrison (1978) has obtained spectrophotometric observations of 17 Herbig stars in the region of the Balmer discontinuity. After dereddening the energy distributions and comparing them with Kurucz models corresponding to the stars' spectral types, he found excesses in the Balmer continuum for 16 of them, between $0 \cdot 08$ and 1 magnitude. These excesses were interpreted as due to optically thin emission in the Balmer continuum by a warm expanding shell. Under the assumptions that the shell producing the excess in the Balmer continuum is geometrically thin and spherically symmetric, that its temperature is constant and can be estimated from the slope of the

Paschen continuum, that the regions of formation of the Balmer continuum excess and of the H$\alpha$ line are identical so that the velocity derived from H$\alpha$ profiles can be used, and with some assumed values for the electron density in these regions, the author was able to derive estimates for the mass loss rates. He found a wide range of values between $1\cdot2 \times 10^{-8}$ and $9\cdot3 \times 10^{-6}$ M$_\odot$yr$^{-1}$. However, these estimates of $\dot{M}$ appear to be proportional to the assumed radial velocity of the region producing the excess, which was taken as identical to that measured on the H$\alpha$ profile. Now it seems more reasonable to assume that the excess in the Balmer continuum is formed near the base of the wind, where the velocity can be lower by several orders of magnitude than that of the region of formation of H$\alpha$, which was shown to extend as far as 10 - 30 stellar radii from the photosphere by Catala and Kunasz (1987). Therefore, the mass loss rates derived by Garrison can be overestimated by several orders of magnitude.

2.3.3. *Estimates Based On Radio Continuum Observations.* The uncertainty in the estimates of $\dot{M}$ by the methods described above are basically due to our poor knowledge of the velocity law in the accelerating (or decelerating) regions near the star, which contribute significantly to the formation of most of the lines and continua in the UV-visible-IR region. The situation is very different if one considers radiation at centimetric wavelengths. Winds with mass loss rates of the order of $10^{-8}$ M$_\odot$yr$^{-1}$ are indeed optically thick up to very large distances at these wavelengths. At these distances, the flow can be reasonably assumed to have a constant radial velocity, and the radio flux can be simply related to the mass loss rate under the additional assumptions of spherical symmetry and constant temperature (Wright and Barlow, 1975). Thus, an estimate of $\dot{M}$ can be obtained from the radio flux independently from a detailed modelling of the regions closer to the photosphere.

Unfortunately, there exist very few detections of Herbig stars at centimetric wavelengths. Brown (1987) reported detection of TY CrA at 6 cm, with the VLA. He proposed a mass loss rate of $2 \times 10^{-8}$ M$_\odot$yr$^{-1}$ for this star. Güdel et al. (1989) detected AB Aur at 3.6 and 6 cm, also with the VLA. A mass loss rate of $1\cdot5 \times 10^{-8}$ M$_\odot$yr$^{-1}$ was derived from the observed flux at 3.6 cm. It is interesting to notice that this estimate is in good agreement with that derived from line profile analysis by Catala and Kunasz (1987). The flux at 6 cm suggested some deviation from a simple thermal emission mechanism. We will come back on this point in Section 3.3.

In summary, we have seen that 20% of the Herbig Ae/Be stars definitely lose mass, and that the remaining 80% also probably possess stellar winds. The estimates of the mass loss rates by methods involving line profile and/or continua analysis show important discrepancies with one another, but the most probable values for the mass loss rates in these stars are in the range $[10^{-8} - 10^{-7}]$ M$_\odot$yr$^{-1}$. These values seem to be confirmed so far by radio continuum observations.

## 3. Activity and magnetic fields

### 3.1. EXTENDED CHROMOSPHERES IN HERBIG Ae/Be STARS

The presence of a chromosphere in AB Aur was inferred by Praderie et al. (1982), from the shape of the Ca II K line, which showed intermittently a central emission component inside the absorption line. The presence of a chromosphere in this star was also confirmed by Felenbok et al. (1983), who observed the He I 5876 A line and the Ca II IR triplet, all of them in emission. Later, Catala et al. (1984) showed that an extended chromosphere was necessary to explain the profile of the Mg II resonance lines. Finally, the quantitative interpretation of the C IV resonance lines (Catala and Talavera, 1984; Catala, 1988) yielded some precise constraints on the structure of the chromosphere: its maximum temperature was shown to be between 16000 and 18000 K, and its radial size between 1 and 2.5 stellar radii. Again, because of the spectral similarities between AB Aur and the other stars of the P Cygni subclass, we can tentatively conclude that all stars of this subclass have qualitatively the same atmospheric structure, i.e. possess extended chromospheres. A more detailed analysis of each of them is still needed to derive quantitative conclusions, though.

## 3.2. SOURCE OF ENERGY FOR THE ACTIVITY OF HERBIG Ae/Be STARS

In order to maintain the extended chromospheres around Herbig stars, a certain amount of non-radiative energy must be deposited above the stellar surface. This amount of energy can be measured by the so-called radiative losses, i.e. the extra radiated energy relative to a model in radiative equilibrium. Using the semi-empirical model built for AB Aur, and the solution of the radiative transfer equation in the main transitions, Catala (1989) estimated the radiative losses due to the chromosphere of this star, and found the considerable value of $2 \times 10^{34}$ erg s$^{-1}$, that is to say 6% of the bolometric luminosity of the star! Although such a detailed computation has not been performed for the other stars of the P Cygni subclass, the spectral similarities mentioned in Section 2.2 suggest that a similar conclusion can be reached for the whole subclass. Therefore, one can conclude that the Herbig Ae/Be stars of the P Cygni subclass are *intensely active.*

This result raises the problem of the origin of the non-radiative energy necessary to compensate for the radiative losses. The general belief is that stellar activity is linked to the presence of convection in the subphotospheric layers. However, the computed evolutionary models for the masses corresponding to the Herbig Ae/Be stars (above 2 $M_\odot$) (see Iben, 1965; Gilliland, 1986) indicate that they possess at the most very thin outer convection zones, totally unable to provide the energy involved here. The source of energy responsible for the activity of the Herbig stars must be found elsewhere.

First, we may wonder if this source of energy could come from outside the star, i.e. from an accretion disk, as it is proposed by Bertout et al. (1988) for the T Tauri stars . The gravitational energy of the disk could be transformed into thermal energy in an accretion boundary layer, thus producing the observed chromosphere. Accretion rates of the order of $10^{-7}$ $M_\odot$yr$^{-1}$ would provide the necessary energy. We note that there is now a growing evidence for the presence of accretion disks around T Tauri stars. However, the usual evidences suggesting accretion disks around young stars are absent in the case of Herbig stars:

- A blueshift of forbidden emission lines is interpreted as evidence for a wind whose receding part is hidden by an opaque disk (Appenzeller et al., 1985). Such a phenomenon is often observed in the spectra of T Tauri stars (Jankovics et al., 1983). However, Finkenzeller (1985) showed that the [OI] line at 6300 A is symmetric in Herbig Ae/Be stars. This does not eliminate the idea of disks around Herbig stars, but would imply that they are optically thin at the distance of formation of forbidden lines.

- Another strong evidence for disks around T Tauri stars is the ability of disk models to explain both the IR excess by dust emission and the Balmer continuum in emission by the boundary layer at the stellar surface. In the case of Herbig Ae/Be stars, however, the Balmer continuum, although in excess relative to models in radiative equilibrium, remains in absorption (Garrison, 1978). The excess in the Balmer continuum can easily be explained by the presence of the stellar wind (Garrison, 1978; Catala and Kunasz, 1987). It can be argued that this absence of emission in the Balmer continuum can be explained by a contrast effect, even in the presence of a boundary layer, because the photosphere of A and B-type stars is bright at these wavelengths. However, the main point here is that, unlike for T Tauri stars, we do not have evidence for the presence of disks around Herbig stars.

The only piece of evidence for disks around Herbig stars comes from the maps of linear polarization which, in 3 cases, can be interpreted in terms of multiple scattering in a disk (Bastien and Ménard, 1988). We note that 2 of these 3 cases (R Mon and R CrA) correspond to very untypical Herbig stars, with an F0 spectral type for R CrA, and an unknown spectral type for R Mon. However, the survey of polarization maps in the direction of Herbig stars is certainly far from being complete, and no conclusion can be drawn at this moment.

The source of the non-radiative energy necessary to explain the chromospheres of Herbig stars can also be looked for in the interior. In the absence of outer convection zone, the only source of energy is rotation. The measurement of projected rotational velocities for Herbig stars by Finkenzeller (1985) shows that they are intermediate rotators with *vsini* of the order of 100 kms$^{-1}$. Assuming a typical stellar radius of 3 $R_\odot$ and the internal structure derived from Gilliland's (1986) calculations, and assuming furthermore that the internal rotation is constant, a total rotational energy of about $10^{48}$ ergs can be estimated. This energy, if spent on a Kelvin time scale ($10^6$ yrs) is more than sufficient to provide the necessary energy rate.

However, the mechanism by which rotational energy of the stellar interior can be made available as mechanical energy above the photosphere is still mysterious. As we will see below, Vigneron et al. (1989) suggest that the loss of angular momentum itself could play a major role in this process.

## 3.3. IS THE ACTIVITY OF THE HERBIG Ae/Be STARS MAGNETIC?

Irrespective of the ultimate source of energy responsible for stellar activity, it is usually believed that magnetic fields play a dominant role in this phenomenon, in particular as an agent for transporting energy and depositing it in the chromospheric layers. This point of view is directly derived from our knowledge of the physics of the solar atmosphere, and its extension to stars is usually referred to as "solar-stellar connection". In particular, stellar activity provides a convenient tracer of surface magnetic fields, and allows us to test current theories of their dynamo generation. In this context, it is of utmost importance to find out if the activity of Herbig stars is magnetic. The current theories of dynamo generation of magnetic fields involve the interaction of differential internal rotation with non-axisymmetric motions, usually provided by convection. The discovery of magnetic activity in stars with no convection zone would provide a very powerful constraint on such theories.

A first indirect evidence for the presence of a magnetic field at the surface of a Herbig Ae star was obtained by Praderie et al. (1986). They showed that the violet edge of the P Cygni profile of the Mg II resonance lines in AB Aur was highly variable on a short-time scale, and that these variations were periodic with a period of 45 hours. This period is close to the expected rotation period of the star and Praderie et al. interpreted their observations as rotational modulation of the lines by corotating structures in the wind, as schematically shown in Fig. 5. The flow consists of fast and slow streams alternating on the line of

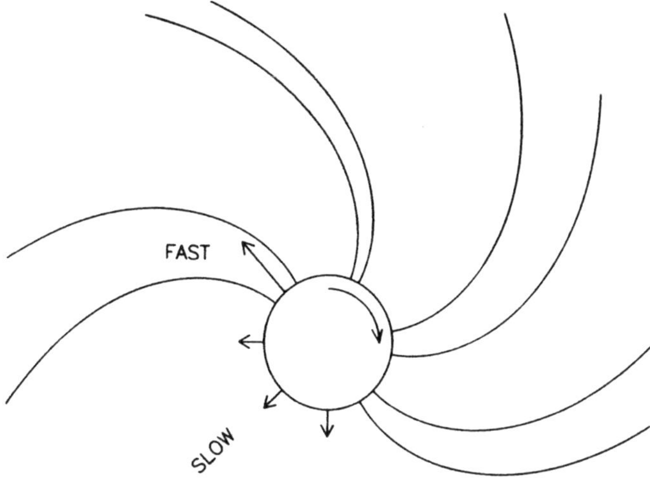


Figure 5: Schematic representation of the model of corotating structures in the winds of Herbig stars. Fast and slow streams presumably controlled by the surface magnetic field alternate on the line of sight as the star rotates. A modulation of lines formed in the wind with the rotation period is expected if one hemisphere is globally richer in fast streams than the other one, as depicted here.

sight as the star and its envelope rotate, thus producing the observed modulation of the violet edge of the line. This model is greatly inspired from the structure of the solar wind, where corotating fast and slow streams are also present. In the solar wind, the corotating structure is entirely controlled by the surface magnetic field, the fast streams originating from regions where the magnetic lines are open, and the slow streams from regions with closed magnetic lines. By analogy, Praderie et al. suggested that the corotating structure needed to explain their observations of AB Aur could also be controlled by a magnetic field at the stellar surface.

Further work in this direction showed that the Ca II K line was also modulated by the rotation in AB Aur (Catala et al., 1986b), and the same type of behavior was discovered in two other Herbig stars, HD 163296 (Catala et al., 1989), and HD 250550 (Catala et al.,

78

1990).

Although the interpretation of the periodic variations mentioned above by a model involving corotating structures is seductive, it is certainly not the only one. For instance, Baade and Stahl (1989) propose to interpret the variations they observed in the Si II $\lambda\lambda$ 6347,6371 A, Mg II $\lambda$ 4481 A, Ca II K and H$\alpha$ lines in terms of nonradial pulsations.

Therefore, there is a clear need for direct detection of magnetic fields in Herbig Ae/Be stars. Unfortunately, these stars are not suitable for any of the known spectroscopic techniques for detecting magnetic fields, because they are too faint and rotate too rapidly. However, there is a possibility that their radio continuum can be detected with the VLA. In case of the presence of a magnetic field, a non-thermal component could be expected in this radio emission, due to synchrotron radiation. As mentioned earlier, Güdel et al. (1989) have detected radio emission from AB Aur at 3.6 and 6 cm, with the VLA. The flux at 3.6cm is consistent with thermal emission from a wind with a mass loss rate of $1 \cdot 5 \times 10^{-8}$ $M_\odot \mathrm{yr}^{-1}$, but the observed spectral index differs from that expected for thermal emission in a spherically symmetric wind, and this may indicate a non-thermal component.

In summary, the Herbig Ae/Be stars of the P Cygni subclass are intensely active. The problem of the ultimate source of energy for this activity is still unsolved. There are some clues that this activity is magnetic, but no definite proof has been found so far. Magnetic activity in these stars would be somewhat paradoxical because they do not possess outer convection zones.

## 4. Loss of angular momentum

### 4.1. ROTATION RATES

Davis et al. (1983), then Finkenzeller (1985) measured the projected rotational velocity *vsini* for a statistically representative sample of Herbig Ae/Be stars, using the He I 4471 and Mg II 4481 A lines. Both studies classified these stars as intermediate rotators with values of *vsini* of the order of 100 kms$^{-1}$. A possible caveat of this type of study must be taken into account, though. The presence of strong stellar winds in these stars can possibly perturb the formation of the He I 4471 and Mg II 4481 A lines. In fact, Baade and Stahl (1989) presented a time series of high resolution profiles for these lines in the Herbig Ae star HD 163296, which showed some asymmetries and an important variability. The equivalent width of the Mg II 4481 A line, for example, varies from $0 \cdot 65$ to $0 \cdot 86$ A in the course of their observations, i.e. on a time scale of 3 days. This type of variations is reminiscent of the stream model described in Section 3.3, and shows that the broadening of the Mg II 4481 A line in HD 163296 is due not only to the rotation, but also to some contribution of the stellar wind. Therefore, the value of *vsini* derived by Finkenzeller for this star (120 kms$^{-1}$ is probably an overestimate. We may wonder if this is not also the case for the other Herbig stars. In the absence of a monitoring of the Mg II 4481 A line for a larger sample of Herbig stars, we must consider this question as open.

Because the lines formed in the winds of several Herbig Ae/Be stars are modulated by the rotation of the stars and of their extended atmospheres, rotation rates can be determined directly from observations of rotational modulation in a few cases. A period of 32 hrs was found in the modulation of the Ca II K line of AB Aur, and was interpretated as the rotation period of the photosphere (Catala et al., 1986b), whereas the 45 hr period in the Mg II data (Praderie et al., 1986) was ascribed to the rotation of the envelope further out. In the case of HD 163296, the Mg II resonance lines are modulated with a period of 50 hrs, but it is not clear if this period corresponds to the the rotation of the photosphere, or to that of the extended atmosphere farther away (Catala et al., 1989). Finally, the Ca II K line of HD 250550 was also shown to be modulated with a period of 43 hrs, probably corresponding to the rotation period at the photosphere (Catala et al., 1990). These three stars have a spectral type of A0V, from which a stellar radius of 3 $R_\odot$ can be inferred. The corresponding rotation velocities can therefore be estimated to 114, and 85 kms$^{-1}$ for AB Aur, and HD 250550, respectively, and a lower limit of 73 kms$^{-1}$ can be proposed for HD 163296. These values are significantly lower than the estimates by Finkenzeller (1985), which is understandable according to the argument given previously. However, they do not question the classification of the Herbig stars as intermediate rotators.

## 4.2. ROLE OF ANGULAR MOMENTUM LOSSES

We have seen that the Herbig Ae/Be stars of the P Cygni subclass lose mass at a rate $\dot{M} \approx 10^{-8}$ $M_\odot yr^{-1}$, and that they are intermediate rotators. Besides, we have some clues that they might have important surface magnetic fields. Therefore, these stars must lose angular momentum at a high rate, and we may wonder what role these angular momentum losses can play in their evolution toward the main sequence.

The rate of angular momentum loss $\dot{J}$ can be estimated from the mass loss rate $\dot{M}$, the rotation rate $\Omega$, and the Alfvén radius $r_a$:

$$\dot{J} \;=\; \frac{2}{3}\dot{M}r_a^2\Omega$$

In the absence of a surface magnetic field, $r_a$ must be simply be replaced by the stellar radius $R_*$ in the above expression. Now the total angular momentum $J$ of a uniformly rotating star can be estimated as:

$$J \;\approx\; kMR_*^2\Omega$$

where $M$ is the stellar mass, and $k$ a coefficient of order 0.1 measuring the moment of inertia of the star in units of $MR_*^2$.

The time scale $\tau$ for angular momentum losses is therefore:

$$\tau \;\approx\; J/\dot{J}$$
$$\approx\; \frac{3}{2}k\left(\frac{\dot{M}}{M}\right)^{-1}\left(\frac{r_a}{R_*}\right)^{-2}$$

For typical values of the mass loss rate ($10^{-8}$ $M_\odot yr^{-1}$) and of the mass (2.5 $M_\odot$), and assuming an Alfvén radius of 2 $R_*$, we find a typical time scale of $10^6$ years, which is comparable to the Kelvin time scale! In the absence of a magnetic field, the time scale for angular momentum loss is of the order of $4 \times 10^6$ years, which is still comparable to the contraction time toward the main sequence. Therefore, we reach the conclusion that, *if they retain the rate of angular momentum loss inferred from the observations, the Herbig Ae/Be stars are likely to lose most of their angular momentum before they reach the main sequence.* However, we are not able to determine the duration of the phase during which these stars lose mass. The fact that all known Herbig stars seem to possess stellar winds may well correspond to an observation bias, the most important criterion for detecting such stars being the presence of H$\alpha$ in emission. The question remains open of whether there exists a class of pre-main sequence stars of intermediate mass with no activity or mass loss. Two scenarios are therefore possible concerning the early history of angular momentum at these masses:

- If the duration of the mass loss phase is comparable to the Kelvin contraction time, then these stars must lose most of their angular momentum on their way to the main sequence. Therefore, they must reach the main sequence as slow rotators, unless the spin up due to the remaining contraction can compensate the braking due to loss of angular momentum. From their precise location in the HR diagram compared to evolutionary tracks, we know that at least some of these stars have reached a radius very close to the final one. For these stars at least, the spin up due to the remaining contraction must be negligible.

- If the duration of the mass loss phase is short compared to the Kelvin contraction time, then these stars may arrive on the main sequence as fast rotators.

Only a better understanding of the driving mechanism for the wind and its evolution with time can allow us to choose between these two possibilities and to find out of what type of main sequence stars the Herbig Ae/Be stars are the progenitors.

## 4.3. ANGULAR MOMENTUM LOSS AS A POSSIBLE SOURCE OF ACTIVITY

In Section 3.2, we mentioned the problem of the ultimate source of energy for the activity of the Herbig Ae stars. We saw that internal rotation constitutes a sufficient reservoir of energy. Now we have just seen that these stars are losing angular momentum at a high rate, and we may wonder whether these losses of angular momentum could be the engine for the activity observed in these stars.

Vigneron et al. (1989) proposed a qualitative model in which the torque exerted at the surface by the angular momentum losses produces a pseudo-convective zone in the subphotospheric layers. Their argument goes as follows: since the structure of the sub-photospheric layers is radiative, and since angular momentum transport is very unefficient in such a structure, the braking due to the angular momentum losses affects only a thin region near the stellar surface, creating very quickly a strong rotation velocity gradient in this region. Therefore, shear instabilities are likely to develop, giving rise to a turbulent layer slowly penetrating into the stellar interior. Under very simple assumptions on the structure of this layer, suggested by a comparison with geophysical and laboratory fluid experiments, Vigneron et al. were able to propose a dimensional analysis of the problem and showed that the turbulent region is likely to reach down to a significant depth in a time scale comparable to the Kelvin time scale. They suggest that this turbulent layer may play, in the case of the active phenomena observed in these stars, the role played by convection zones in the case of solar-type stars. Although this model is still qualitative, and the details of the development of the shear instabilities still need to be worked out, the idea seems definitely worth pursuing.

## 5. Conclusion

The pre-main sequence Herbig Ae/Be stars can be divided in three subclasses according to their $H\alpha$ profile. Those with a P Cygni profile at $H\alpha$ definitely possess strong stellar winds and intense activity, and the stars of the other two subclasses also probably have the same characteristics. There are some clues that the activity of the stars in the P Cygni subclass is magnetic, but no definite proof has been found so far. Using the available estimates for the mass loss rates of these stars, $\dot{M} \approx 10^{-8}$ $M_\odot yr^{-1}$, it can be shown that they are losing angular momentum at a high rate, and that their angular momentum losses may play an important role in their active phenomena during their evolution toward the main sequence.

It is clear from the present review that considerable work remains to be done, in the three usual directions of Astrophysics:

*i) Observation:*

- There is a dramatic lack of statistical quantitative information about winds and activity in these stars. Most of the results presented here deal with only a few stars. With the currently available data, we are totally unable to determine how activity and winds are related to the basic stellar parameters (mass, age, rotation). A systematic observational study of tracers of winds (Mg II resonance lines,...) and of activity (C IV or He I 10830 A) should be carried out. This study must be extended to the Herbig stars belonging to young open clusters (Thé et al., 1985a,b), because we do not have to rely on computed evolutionary tracks to assign an age to those stars.

- Spectroscopic and photometric monitoring of the Herbig stars must be continued, with the goal of determining how the phenomenon of rotational modulation depends on the same basic stellar parameters. An answer to this question would provide a clue about the origin of the hypothetical magnetic activity in these stars. Such studies are very time-consuming, but their outcome is certainly worth the effort.

- There should also be an effort to try and detect radio emission in a greater number of Herbig stars with the VLA. These detections would either allow an independent determination of the mass loss rates if the radio emission is thermal, or constitute a direct detection of magnetic fields if it is non-thermal.

*ii) Interpretation of the observations:*

- We still need to understand properly the formation of the double-peak and single-peak emission profiles observed at $H\alpha$ for most of the Herbig Ae/Be stars.

*iii) Theory:*

- The problem of the internal rotation of the Herbig stars must be addressed. We note that this problem is inserted in the more general one of the evolution of angular momentum of all types of stars, including the sun. Exciting observational results for the internal rotation of the sun (helioseismology) are to be expected for the next decade, thanks to the Soho satellite and ground-based networks. There is also a satellite project (EVRIS) for observing oscillations in some bright stars, which will provide information about the rotation of the core of these stars. In the case of the Herbig stars, we have seen that the internal rotation could be the key to the problems of stellar activity.

- We must still understand the mechanisms for mass loss in Herbig stars. They do not seem to be luminous enough for radiation pressure to play a dominant role. No evidence for coronae has been found so far, and besides it is not clear whether a coronal wind can lead to the observed mass loss rates. Effects of rotation, magnetic fields, pulsations and waves still need to be investigated.

- Finally, all of these aspects should be put together in a consistent picture of the history and the fate of the Herbig Ae/Be stars. What is the evolution with time of their mass loss rates, their activity levels, their rotation rates and their magnetic fields? Of which type of main sequence stars are they the progenitors?

## References


Appenzeller, I., Jankovics, I., Oestreicher, R.: 1985, *Astron. Astrophys.* **141**,108

Baade, D., Stahl, O.: 1989, *Astron. Astrophys.* **209**,268

Bastien, P., Ménard, F.: 1988, *Astrophys. J.* **326**,334

Beals, C.S.: 1951, *Publ. Dominion Astrophys. Obs.* **9**,1

Bertout, C., Basri, G., Bouvier, J.: 1988, *Astrophys. J.* **330**,350

Brown, A.: 1987, *Astrophys. J.* **322**,L31

Catala, C., Kunasz, P.B., Praderie, F.: 1984, *Astron. Astrophys.* **134**,402

Catala, C., Talavera, A.: 1984, *Astron. Astrophys.* **140**,421

Catala, C., Czarny, J., Felenbok, P., Praderie, F.: 1986a, *Astron. Astrophys.* **154**,103

Catala, C., Felenbok, P., Czarny, J., Talavera, A., Boesgaard, A.M.: 1986b, *Astrophys. J.* **308**,791

Catala, C., Kunasz, P.B.: 1987, *Astron. Astrophys.* **174**,158

Catala, C.: 1988, *Astron. Astrophys.* **193**,222

Catala, C.: 1989, in *Modeling the Stellar Environment: How and Why, Proc. fourth IAP Astrophysics Meeting, in honor of J.C. Pecker*, Paris, Ph. Delache, S. Laloë, C. Magnan, J. Tran Thanh Van (eds.), p. 207

Catala, C., Simon, T., Praderie, F., Talavera, A., Thé, P.S., Tjin A Djie, H.R.E.: 1989, *Astron. Astrophys.* **221**,273

Catala, C., Czarny, J., Dreux, M., Felenbok, P., Talavera, A., Thé, P.S.: 1990, in preparation

Cohen, M.: 1973, *Monthly Notices Roy. Astron. Soc.* **161**,105

Cohen, M.: 1975, *Monthly Notices Roy. Astron. Soc.* **173**,279

Cohen, M.: 1980, *Monthly Notices Roy. Astron. Soc.* **191**,499

Cohen, M., Kuhi, L.V.: 1979, *Astrophys. J. Suppl.* **41**,743

Davis, R., Strom, K.M., Strom, S.E.: 1983, *Astron. J.* **88**,1644

Felenbok, P., Praderie, F., Talavera, A.: 1983, *Astron. Astrophys.* **128**,74

Finkenzeller, U., Mundt, R.: 1984, *Astron. Astrophys. Suppl.* **55**,109

Finkenzeller, U., Jankovics, I.: 1984, *Astron. Astrophys. Suppl.* **57**,285

Finkenzeller, U.: 1985, *Astron. Astrophys.* **151**,340

Garrison, L.M., Anderson, C.M.: 1977, *Astrophys. J.* **218**,438

Garrison, L.M.: 1978, *Astrophys. J.* **224**,535

Gilliland, R.L.:1986, *Astrophys. J.* **300**,339

Güdel, M., Benz, A.O., Catala, C., Praderie, F.: 1989, *Astron. Astrophys. Letters* **217**,L9

Herbig, G.H.: 1960, *Astrophys. J. Suppl.* **4**,337

Hu, J.Y., Thé, P.S., de Winter, D.: 1989, *Astron. Astrophys.* in press

Iben, I.: 1965, *Astrophys. J.* **141**,993

Jankovics, I., Appenzeller, I., Krautter, J.: 1983, *PASP* **95**,883

Mazzali, P.A.: 1988, *A Decade of UV Astronomy with the IUE Satellite*, ESA SP-281, p. 163

Mendoza, E.E.: 1966, *Astrophys. J.* **143**,1010

Mendoza, E.E.: 1967, *Astron. J.* **72**,816

Mihalas, D., Kunasz, P.B.: 1978, *Astrophys. J.* **219**,635

Mihalas, D., Conti, P.S.: 1980, *Astrophys. J.* **235**,515

Praderie, F., Talavera, A., Felenbok, P., Czarny, J., Boesgaard, A.M.: 1982, *Astrophys. J.* **254**,658

Praderie, F., Simon, T., Catala, C., Boesgaard, A.M.: 1986, *Astrophys. J.* **303**,311

Strom, S.E., Strom, K.M., Yost, J., Carrasco, L., Grasdalen, G.L.: 1972, *Astrophys. J.* **173**,353

Talavera, A., Catala, C.: 1989, preprint

Thé, P.S., Cuypers, H., Tjin A Djie, H.R.E., Feinstein, A., Westerlund, B.E.: 1985a, *Astron. Astrophys.* **150**,345

Thé, P.S., Hageman, T., Westerlund, B.E., Tjin A Djie, H.R.E.: 1985b, *Astron. Astrophys.* **151**,391

Vigneron, C., Mangeney, A., Catala, C., Schatzman, E.: 1989, *Proc. "Inside the Sun"*, Versailles, *Solar Physics*, in press

Wright, A.E., Barlow, M.J.: 1975, *Monthly Notices Roy. Astron. Soc.* **170**,41

# ECCENTRIC SPIRAL MODES IN DISKS ASSOCIATED WITH YOUNG STELLAR OBJECTS


Fred C. Adams

*Harvard-Smithsonian Center for Astrophysics*
*60 Garden Street*
*Cambridge, MA 02318, U.S.A.*



ABSTRACT. Motivated by observations suggesting that YSO disks are moderately massive and posssess significant intrinsic luminosity, we study the growth of gravitational instabilities in star/disk systems. For spiral modes with azimuthal wavenumber $m = 1$, we find a new type of amplification mechanism. These modes can grow on nearly a dynamical timescale and may lead to disk accretion or to the formation of a binary companion.


## 1. Introduction

Circumstellar disks play an important role in the process of star formation, especially in the redistribution of angular momentum (see also the review of Kawaler [this volume]). In recent years, compelling observational evidence (cf. the reviews of Bertout 1989; Shu, Adams, and Lizano 1987) has established the presence of disks associated with young stellar objects (YSOs), although the exact properties of such disks remain controversial. The available observational evidence (see, e.g., Adams, Emerson, and Fuller 1989; Adams, Lada, and Shu 1988) indicates that YSO disks may produce significant luminosity ($L_D \sim L_* \sim 1L_\odot$) and have moderate masses ($M_D \sim M_* \sim 1M_\odot$). In this study, we consider the growth of *global* gravitational instabilities in star/disk systems; the hope is that these instabilities can lead to the observed disk luminosities. We have studied the behavior of $m = 1$ modes both numerically (Adams, Ruden, and Shu 1989) and analytically (Shu, Tremaine, Adams, and Ruden 1990); we find good agreement between the two approaches. In this present discussion, we summarize our current understanding of these modes.

## 2. The Initial (Unperturbed) State

Since the overall goal of this study is to determine the growing normal modes of a star/disk system, we must begin by specifying the basic unperturbed state. The physical system consists of a star and an accompanying gaseous disk. The growth of sprial modes is mainly determined by three elements: self-gravity, pressure, and differential rotation. The gravitational forces are determined by the potential of the star and by the disk's surface density distribution $\sigma_0(r)$, which we take to be a simple power-law in radius $r$ from the central star (the disk is also assumed to be infinitesimally thin and in centrifugal equilibrium). The pressure is determined by the temperature distribution, or equivalently, the distribution of sound speed in the disk; we take the sound speed $a(r)$ to be a power-law in radius. The

85


L. A. Willson and R. Stalio (eds.), *Angular Momentum and Mass Loss for Hot Stars*, 85–91.
© 1990 *Kluwer Academic Publishers. Printed in the Netherlands.*

86

rotation curve $\Omega(r)$ in the disk is then determined self-consistently from the potential of the star, the potential of the disk, and the pressure gradients. Since the potential well of the star dominates that of the disk everywhere except near the disk's outer edge, the rotation curve is nearly Keplerian throughout most of the disk's radial extent. (for thin gaseous disks, the pressure gradients are small compared to the gravitational forces and do not significantly affect the rotation curve). Finally, we must specify the radial extent of the disk; since observations indicate that 'typical' disk sizes are approximately 100 AU (e.g., Edwards *et al.* 1987), we consider disk sizes $R_D$ up to $\mathcal{O}(10^4)$ times the radius $R_*$ of the star (only the ratio $R_D/R_*$ enters into the calculations).

## 3. Modes with Azimuthal Wavenumber $m = 1$

Our study concentrates on modes with azimuthal wave number $m = 1$, since these modes can be global in extent and may also be the most difficult modes to suppress in unstable gaseous disks. Modes with $m = 1$ correspond to elliptic streamlines (i.e., eccentric particle orbits), a special characteristic of Keplerian potentials. Thus, for a disk with an exact Keplerian rotation curve and no interactions between particles, $m = 1$ disturbances correspond to purely kinematic modes of the system (see Figure 1); for realistic disks (with pressure), a relatively 'small' amount of self-gravity is required to 'hold the mode together' and sustain its growth.

One unique and important aspect of $m = 1$ modes is that the center of mass of the perturbation in the disk does *not* lie at the geometrical center of the system; hence, the frame of reference centered on the star is *not* an inertial reference frame. The star is actually in orbit about the center of mass (i.e., the star is accelerating) and creates an effective forcing potential – the "indirect potential." Our results show that this indirect potential is essential for the growth and maintainence of spiral modes with azimuthal wavenumber $m = 1$. In fact, the interaction of this indirect potential with the outer Lindblad resonance in the disk (see §6 below) can be the dominant amplification mechanism for these modes.

Figure 1. Schematic diagram of $m = 1$ elliptical streamlines oriented to produce a one-armed spiral. The cross denotes the position of the star.

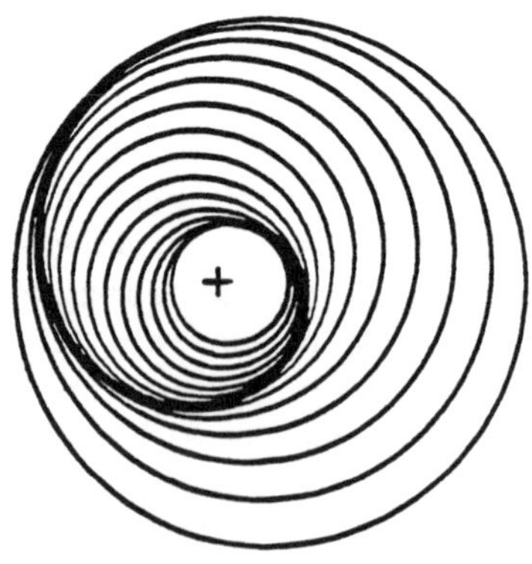

## 4. Wave Physics and Spiral Instabilities in Gaseous Disks

In the simplest description of spiral instabilities, self-excited disturbances (spiral modes) can grow through the feedback and amplification of spiral density *waves*. In this section, we quickly review the theory of spiral density waves. In the following sections, we describe the feedback cycle and the amplification mechanism for spiral modes with $m = 1$.

In the asymptotic (WKBJ) limit, the dispersion relation for spiral density waves in a gaseous disk has the form

$$k^2 a^2 - 2\pi G \sigma_0 |k| + \kappa^2 = (\omega - m\Omega)^2,$$

where $k$ is the radial wavenumber, $\kappa$ is the epicyclic freqeuncy, and where $\omega$ is the complex eigenvalue of the system (see, e.g., Lin and Lau 1979). Since the gravitational term is proportional to $|k|$, this dispersion relation has four branches,

$$k = \pm(k_0 \pm k_1),$$

where

$$k_0 \equiv \frac{\pi G \sigma_0}{a^2}, \qquad k_1 \equiv \frac{\pi G \sigma_0}{a^2}[1 - Q^2(1 - \nu^2)]^{1/2}, \qquad \text{and} \qquad \nu \equiv (\omega - m\Omega)/\kappa.$$

Here, the quantity $Q$ determines the stability of the system to *axisymmetric* disturbances ($Q > 1 \Rightarrow$ axisymmetric stability) and is defined by (Toomre 1964):

$$Q \equiv \frac{\kappa a}{\pi G \sigma_0}.$$

The overall sign of $k$ determines whether the waves are *leading* ($|k| > 0$) or *trailing* ($|k| < 0$); the inner sign determines whether the waves are *short* $[k \propto (k_0 + k_1)]$ or *long* $[k \propto (k_0 - k_1)]$.

The quantity $\nu$ is a dimensionless frequency of the spiral density waves. The radius in the disk where $\nu = 0$ (i.e., where $\Re(\omega) = m\Omega$) is known as the *corotation resonance* (CR); the energy and angular momentum of the perturbation (and the action) are positive outside the corotation radius and negative inside. Notice that for $Q > 1$, the wavenumber $k_1$ becomes imaginary for radii sufficiently close to the CR, i.e., a classical turning point exists for the density waves. The resulting "forbidden" region surrounding the CR is known as the "Q-barrier". Notice also that for long waves, $k \to 0$ at any radius where $|\nu| = 1$. The radius in the disk where $\nu = +1$ is known as the *outer Lindblad resonance* (OLR) and plays an important role in the physics of $m = 1$ modes.

## 5. Feedback Loop: The Four-Wave Cycle

We now present the feedback cycle for $m = 1$ modes in gaseous disks. One unique aspect of this feedback cycle is that all four types of waves are utilized (see Figure 2):

A. Begin (somewhat arbitrarily) with the excitation of a long trailing (LT) spiral density wave at the outer Linblad resonance (OLR) by the indirect term (see §6). The LT wave propagates inward (its group velocity is negative) until it encounters the outer edge of the $Q$-barrier.

B. At the $Q$-barrier, the LT wave refracts into a short trailing (ST) spiral density wave that propagates back outward, through the OLR to the outer disk edge.

C. The ST waves that propagate to the outer disk edge reflect there to become short leading (SL) waves. The SL waves then propagate back to the interior, through the OLR, until they encounter the outer edge of the $Q$-barrier, where they refract into long leading (LL) spiral density waves that propagate back toward the OLR.

D. At the OLR, the LL waves reflect to become LT waves. If the reflected LT wave possesses the correct phase relative to the LT wave launched from OLR by the indirect term in step A above, then we have constructive reinforcement of the entire wave cycle, and the basis for the establishment of a *resonant wave cavity*.

Using a WKBJ analysis, we have derived a quantum condition on the basis of the above four-wave cycle. This quantum condition accurately predicts the pattern speeds (i.e., the real part of the eigenfrequencies) for these modes; for strongly growing modes, the analytical and numerical results agree to within ~1 percent.

Figure 2. The Four-Wave Feedback Cycle.

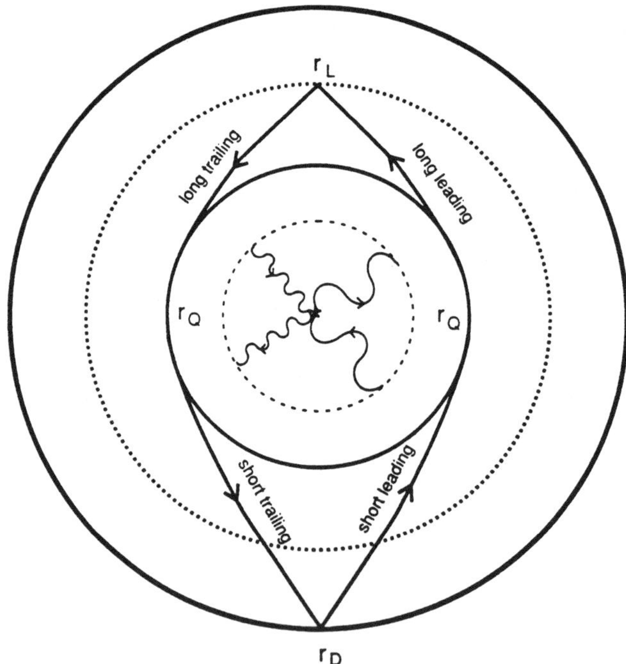

## 6. SLING Amplification

In this section, we describe the amplification mechanism for eccentric modes. Our analysis indicates that the dominant mechanism for amplification arises from the indirect potential, which provides an effective forcing term (see §3). The indirect term varies slowly with radius in the disk; since a slowly varying force can only couple to oscillatory disturbances at the disk edges or at the Lindblad resonances, the main coupling occurs at the outer Lindblad resonance for the modes considered here. Thus, this amplification mechanism differs substantially from the previously studied mechanisms, which utilize the process of *super-reflection* across the corotation resonance (super-reflection can still occur in these disks and is included in the numerical treatment, but it does not dominate the amplification). In our analytic treatment, we determine the growth rates for the modes under the assumption that *all* of the amplification arises from this coupling of the indirect term to the outer Lindblad resonance in the disk (and that the indirect term arises mostly near the outer disk edge). In other words, we conceptually regard the indirect potential as an external forcing term acting on the disk and calculate the torque exerted on the disk at the OLR. Since the long-range coupling of the star to the outer disk provides the essential forcing, we call this new instability mechanism SLING: Stimulation by the Long-range Interaction of Newtonian Gravity.

Figure 3. Contour plot for the lowest order ($n = 0$) SLING amplified mode. The two dashed circles show the location of the corotation and outer Lindblad resonances. The spiral arms trail in the sense of rotation of the material.

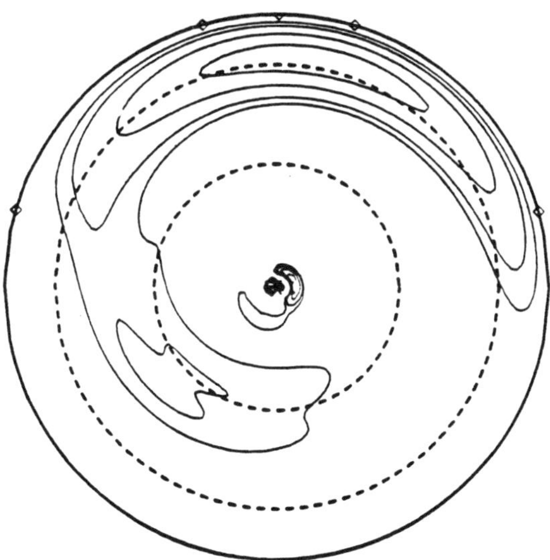

Our combined numerical and analytical treatments indicate the dependence of the growth rates (i.e., the imaginary part of the eigenfrequencies) on the parameters of the problem. Most importantly, a finite threshold exists for the SLING Amplification mechanism. When all other properties of the star/disk system are held fixed, this effect corresponds to a threshold in the ratio of disk mass $M_D$ to the total mass $M_* + M_D$. We find that the growth rates are largest for the case of equal masses $M_D = M_*$ and decrease rapidly with decreasing relative disk mass. In the optimal case, $M_D = M_*$, the grow rates can be comparable to the orbital frequency at the outer disk edge, i.e., the modes can grow on nearly a dynamical timescale. On the other hand, the presence of the finite threshold implies a critical value of the relative disk mass, i.e., the maximum value of $M_D/(M_* + M_D)$ that is stable to $m = 1$ disturbances; for the simplest case of a perfectly Keplerian disk and $Q(R_D) = 1$, this critical ratio has the value $M_D/(M_* + M_D) = 3/4\pi$.

## 7. Results and Conclusions

We have studied eccentric spiral modes in gaseous disks using both numerical and analytical treatments of the problem and have determined the (complex) eigenvalues of the system; the two approaches are in good agreement. Our analysis indicates that the basic modal mechanism involves the four-wave cycle (see §5 and Figure 2), which provides the feedback loop, and the SLING mechanism (§6), which provides the amplification. Our results indicate that a wide range of YSO disks will be unstable to the growth of eccentric distortions. When the disk mass is comparable to the stellar mass $(M_D \sim M_*)$, these distortions can grow on nearly a dynamical timescale. In addition, we find that these modes can grow when the disk is safely stable to axisymmetric disturbances (i.e., $Q$ substantially greater than 1). However, we find that our results (e.g., the exact spectrum of unstable modes) are particularly sensitive to the treatment of the outer disk edge; we are currently working toward an understanding of these effects.

The modes considered in this work may have important astrophysical applications. In the earliest stage of star formation – the protostellar phase – the mass of the disk is likely to be comparable to that of the star (see Shu, Adams, and Lizano 1987); $m = 1$ modes are thus likely to grow and may lead to mass accretion through the disk and the observed disk luminosities. Perturbations with $m = 1$ prove especially interesting because they force the star to move from the center of mass and thereby transfer angular momentum to the stellar orbit. This coupling may lead to the formation of giant planets and/or a binary companion within the disk. However, quasilinear and nonlinear analyses are needed to follow up these possibilities.

ACKNOWLEDGEMENT

The work presented at this conference was done in collaboration with Steve Ruden, Frank Shu, and Scott Tremaine.

## References


Adams, F. C., Emerson, J. E., and Fuller, G. A. 1989, "Submillimeter Photometry and Disk Masses of T Tauri Disk Systems", submitted to *The Astrophysical Journal.*

Adams, F. C., Lada, C. J., and Shu, F. H. 1988, "The Disks of T Tauri Stars with Flat Infrared Spectra", *The Astrophysical Journal,* **326**, 865.

Adams, F. C., Ruden, S. P., and Shu, F. H. 1989, "Eccentric Gravitational Instabilities in Nearly Keplerian Disks", *The Astrophysical Journal,* in press.

Bertout, C. 1989, "T Tauri Stars: Wild as Dust", *Ann. Rev. Astr. Ap.,* **27**, 351.

Edwards, S., Cabrit, S., Strom, S. E., Heyer, I., Strom, K. M., and Anderson, E. 1987, "Forbidden Line and Hα Profiles in T Tauri Spectra: A Probe of Anisotropic Mass Outflows and Circumstellar Disks", *The Astrophysical Journal,* **321**, 473.

Lin, C. C., and Lau, Y. Y. 1979, "Density Wave Theory of Spiral Structure of Galaxies", *Studies in Applied Math.,* **60**, 97.

Shu, F. H., Adams, F. C., and Lizano, S. 1987, "Star Formation in Molecular Clouds: Observation and Theory", *Ann. Rev. Astr. Ap.,* **25**, 23.

Shu, F. H., Tremaine, S., Adams, F. C., Ruden, S. P. 1990, "SLING Amplification and Eccentric Gravitational Instabilities in Gaseous Disks", submitted to *The Astrophysical Journal.*

Toomre, A. 1964, "On the Gravitational Stability of a Disk of Stars", *The Astrophysical Journal,* **139**, 1217.

# EVIDENCE THAT WOLF-RAYET STARS ARE PRE-MAIN SEQUENCE OBJECTS


ANNE B. UNDERHILL
Department of Geophysics and Astronomy
University of British Columbia
Vancouver, B.C. V6T 1W5, Canada



ABSTRACT. The evidence is reviewed that Population I Wolf-Rayet stars have solar abundances, that they are surrounded by remnant disks formed from their natal cloud, and that their rate of mass loss is moderate. These properties are consistent with Wolf-Rayet stars being young objects recently arrived on the main sequence rather than the evolved, peeled-down remnants of massive stars.


## 1. Introduction

Many people believe that Population I Wolf-Rayet stars are highly evolved massive stars which show anomalous abundances on their surfaces as a result of mass loss at a rate of the order of $5 \times 10^{-5}$ $M_\odot$ yr$^{-1}$. I shall present evidence (1) that they have solar abundances, (2) that they are surrounded by remnant disks formed from their natal cloud, and (3) that their rate of mass loss is uncertain, but possibly of the order of or smaller than $10^{-6}$ $M_\odot$ yr$^{-1}$. These properties together with the fact that Wolf-Rayet stars are associated with O and early B stars in regions which radiate interstellar CO lines points toward them being young objects with ages no greater than about $5 \times 10^6$ yrs.

The predominant emission-line spectra of Wolf-Rayet stars are suggestive of the spectra of Herbig Ae/Be stars and T Tauri stars. The major difference is that in the line-emitting regions (LERs) of Wolf-Rayet stars the electron temperature $(T_e)$ is of the order of $10^5$ K whereas in the case of the less massive pre-main sequence stars it is of the order of or less than $10^4$ K. To generate the high $T_e$'s in the LERs of Wolf-Rayet stars I postulate that mechanical energy is transformed to heat by MHD effects which may occur in a low-beta plasma. Whenever larger than normal magnetic fields are occluded as a massive star $(M_* \geq 10 M_\odot)$ is formed, a high excitation, emission line spectrum may be generated as a result of the deposit of non-radiative energy and momentum in the LER. Such spectra are the criteria by which Wolf-Rayet stars are recognized. Analysis of these spectra reveals the physical state of the LER, not of the underlying photosphere. Additional information and postulates must be made to relate the underlying stars to the theory of evolution of massive stars. These postulates and information are different for Population I than for Population II Wolf-Rayet stars.




*L. A. Willson and R. Stalio (eds.), Angular Momentum and Mass Loss for Hot Stars, 93–96.*
© 1990 *Kluwer Academic Publishers. Printed in the Netherlands.*

Study of binary Wolf-Rayet stars and of those in clusters shows that typically Population I Wolf-Rayet stars have masses of the order of $10 - 15 M_\odot$ and that their luminosities are typical of hydrogen-burning stars of these masses. Their effective temperatures estimated from integrated fluxes (Underhill 1983) and from the radiation temperatures ($T_*$) which generate their spectra (Bhatia and Underhill 1986, 1988) are appropriate for their masses, i.e. of the order of 25,000 − 30,000 K.

The Population I Wolf-Rayet stars have 6-cm fluxes of the order of or less than a mJy, and all show significant infrared excess fluxes. These properties suggest that the Population I Wolf-Rayet stars are buried in much circumstellar plasma. Many show significant polarisation, see, for example, St. Louis et al. (1987) and Drissen et al. (1987). This indicates that a Wolf-Rayet star is buried in a cloud of electrons which is not spherically symmetric. Binary Wolf-Rayet stars tend to show polarisation changes which correlate with the orbital period, see, for instance, Drissen et al. (1986), St. Louis et al. (1988), as well as Schulte-Ladbeck and van der Hucht (1989). Some Wolf-Rayet stars show small random changes in polarisation.

## 2. The Analysis of the Emission-lines of Wolf-Rayet Spectra

The first challenge is to find what range of parameters is significant for creating Wolf-Rayet type spectra. Bhatia and Underhill use the one-representative-point theory of Castor and van Blerkom (1970) for a wide range of parameters to predict the relative energies radiated in lines in the visible spectral range for comparison with observations. This theory sets up the equations of statistical equilibrium for a model atom and provides expressions for the needed radiation field in the case where a velocity gradient exists in the LER. All emission lines are assumed to be formed by photons which escape from the same body of plasma.

The "Of" emission lines are strong in Wolf-Rayet spectra. It is now known that these lines are sensitive to the selected cascade routes which may follow dielectronic recombination. However, although lines resulting from dielectronic recombination are not calculated accurately, it has been possible for Bhatia and Underhill to establish the major properties of WC and WN LERs from their studies of the statistical equilibrium of H, He, C, N, and O model atoms under a wide range of conditions and for several compositions, see BU86, BU88, and BU89.

The results are as follows:

| QUANTITY | WC LER | WN LER |
|---|---|---|
| Radiation temp., $T_*$: | 25,000 K | 25,000 K |
| Electron temp., $T_e$: | $5 \times 10^4 \leq T_e \leq 10^5$ K | $10^5 \leq T_e \leq 2 \times 10^5$ K |
| Electron density, $N_e$: | $10^{10} - 10^{11}$ cm$^{-3}$ | $10^9 - 10^{10}$ cm$^{-3}$ |
| Composition: | solar | solar |

These parameters allow one to predict successfully the observed relative intensities in many of the key lines of Wolf-Rayet spectra. See the papers by Bhatia and Underhill for details. The energy of the electrons in the LER, thus the electron temperature, is the dominant factor which causes WC spectra to be different from WN spectra. Density is a second important parameter.

## 3. Discussion

Previous conclusions that anomalous abundances are present on the surfaces of Wolf-Rayet stars are a result of the use of incomplete equations of statistical equilibrium or of the assumption of LTE and of failure to explore a wide enough range of $T_*$, $T_e$, and $N_e$ such that the true pattern of predicted line intensities as functions of the parameters can be seen. All in all, Bhatia and Underhill (1986, 1988,1989) find good agreement with key line ratios used to classify Wolf-Rayet spectra. The few discrepancies can be understood in terms of the neglect by Bhatia and Underhill of the details of dielectronic recombination, and the inadequate treatment of detailed balance in a few resonance lines. If it was all right to neglect photoionization from and radiative recombination to excited levels, the results of BU88 and BU89 would have confirmed those obtained in the NLTE studies by others. These results were not confirmed. Local thermodynamic equilibrium is not a viable hypothesis for line formation in Wolf-Rayet LERs.

BU88 find that the emissivity in He II $\lambda$5411 of typical Wolf-Rayet LERs is such that the LER has a volume of $10^{43} - 10^{44}$ cm$^3$. Because a contiguous sphere of this volume at the typical $N_e$ in an LER is nearly opaque in electron scattering, (which would imply larger apparent optical continuum radii for Wolf-Rayet stars than are estimated, see Underhill 1983), BU88 suggest the LER may be a thin ring-like disk at a radius of about $10^{15}$ cm. Evidence that line formation in a rotating, thin ring occurs is given by the observed profile of He I $\lambda$5876 in HD 191765, WN6. This line has an unchanging double-peaked shape typical of line formation in a ring-like disk where the lines are broadened by macroturbulence, see Underhill, et al. (1989).

A high value of the order of $10^5$ K requires the deposit of non-radiative energy in the LER. BU88 suggest that turbulent mechanical energy in the disk is transformed to heat by MHD effects in the presence of a small magnetic field of interstellar origin. A field of 4-8 gauss is sufficient to create a low-beta plasma which could be heated by MHD effects.

If the observed 6-cm fluxes of Wolf-Rayet stars are transformed to rates of mass loss assuming thermal bremsstrahlung in a spherical wind, there is a shortfall of radiative momentum to drive the wind by a factor of 20-50. Underhill (1984) has suggested that part of the 6-cm flux could be due to gyroresonance magnetic bremsstrahlung. Some of the 6-cm flux could also be generated in a disk-driven wind, cf. Pudritz and Norman (1986). Consequently the true $\dot{M}$ *from the star* is probably much less than $5 \times 10^{-5} M_\odot$ yr$^{-1}$. The fact that *sharp* subpeaks are seen on the profile of He II $\lambda$5411 in the spectrum of HD 191765 (Moffat et al. 1988; Underhill et al. 1989) indicates that this star is not surrounded by a dense sphere of electrons. If it were, the subpeaks would be greatly broadened by electron scattering. The rate of mass loss from a Wolf-Rayet star alone may be $\leq 10^{-6} M_\odot$ yr$^{-1}$.

The calculations of BU86, BU88, and BU89 clearly indicate that the radiation temperature $(T_*)$ of the continuous spectrum shortward of the Lyman limit is about 25,000 K. Model atmospheres for massive stars of solar composition indicate that this condition is met for $T_{eff} = 25,000 - 30,000$ K. Such effective temperatures are appropriate for stars having the masses of Population I Wolf-Rayet stars.

## 4. Conclusions

The model for Wolf-Rayet stars considered by many, see Section I, to be appropriate is supported neither by reliable analysis of the spectra of Wolf-Rayet stars nor by the properties of eclipsing spectroscopic binaries containing Wolf-Rayet stars.

It is a viable proposition to consider Wolf-Rayet stars to be massive pre-main sequence stars still surrounded by the remnant of a natal disk. Small magnetic fields appear to be present in the disk with the result that MHD effects create the high electron temperatures which Wolf-Rayet spectra imply.

The location of Wolf-Rayet stars in a galaxy indicates those regions where massive stars have formed in the presence of larger than normal interstellar magnetic fields. If these magnetic fields were not present, the newly formed stars would show spectral types in the range B1 – O8. Wolf-Rayet spectra imply high $T_e$ in the LER and moderate $T_*$. This state of affairs is a result of the deposit of non-radiative energy in the LER surrounding the star. It is not a normal result of stellar evolution.

## 5. References


Bhatia, A. K., and Underhill, A. B. 1986, *Ap. J. Suppl.*, **60**, 323. (BU86)

Bhatia, A. K., and Underhill, A. B. 1988, *Ap. J. Suppl.*, **67**, 187. (BU88)

Bhatia, A. K., and Underhill, A. B. 1989, *Ap. J.*, submitted. (BU89)

Castor, J. I., and van Blerkom, D. 1970, *Ap. J.*, **161**, 485.

Drissen, L., Moffat, A. F. J., Bastien, P., Lamontagne, R., and Tapia, S. 1986, *Ap. J.*, **306**, 215.

Drissen, L., St. Louis, N., Moffat, A. F. J., and Bastien, P. 1987, *Ap. J.*, **322**, 888.

Langer, N. 1989, *Astr. Ap.*, **210**, 93.

Maeder, A., and Meynet, G. 1987, *Astr. Ap.*, **182**, 243.

Maeder, A., and Meynet, G. 1988, *Astr. Ap. Suppl. Ser.*, **76**, 411.

Moffat, A. F. J., Drissen, L., Lamontagne, R., and Robert, C. 1988, *Ap. J.*, **334**, 1038.

Pudritz, R. E., and Norman, C. A. 1986, *Ap. J.*, **301**, 571.

Schulte-Ladbeck, R. E., and van der Hucht, K. A. 1989, *Ap. J.*, **337**, 872.

Stickland, D. J., Bromage, G. E., Budding, E., Burton, W. M., Howarth, I. D., Jameson, R., Sherrington, M. R., and Willis, A. J. 1984, *Astr. Ap.*, **134**, 45.

St. Louis, N., Drissen, L., Moffat, A. F. J., and Bastien, P. 1987, *Ap. J.*, **322**, 870.

St. Louis, N., Moffat, A. F. J., Drissen, L., Bastien, P., and Robert, C. 1988, *Ap. J.*, **330**, 286.

Underhill, A. B. 1983, *Ap. J.*, **266**, 718.

Underhill, A. B. 1984, *Ap. J.*, **276**, 583.

Underhill, A. B., Gilroy, K. K., and Hill, G. M. 1989, *Ap. J.*, **351**, March 10.

Underhill, A. B., Gilroy, K. K., Hill, G. M., and Dinshaw, N. 1989, *Ap. J.*, **351**, March 10.

# ROTATION, PULSATION AND ATMOSPHERIC PHENOMENA IN A–TYPE STARS


E. ANTONELLO
*Osservatorio Astronomico di Brera*
*Milano–Merate, Italy*



ABSTRACT. Several properties of the various classes of intermediate mass A–type stars located near the main sequence are discussed in the light of the observations made at various wavelengths from X–ray to radio. In particular, the discussion concerns the relations among rotation, chemical peculiarities and pulsation, and the possible presence of mass (and angular momentum) loss related to rapid rotation and/or pulsation in these stars. Moreover, some new results derived from the observations of the stars in the far IR are presented.


## 1. Introduction

In the present paper we have attempted to do a synthetic review of the Population I stars of spectral type from late B to early F, and of intermediate mass, $3 \geq M/M_\odot \geq 1.5$, that is stars near or on the main sequence (MS). The review is based on some observational (and theoretical) works made during the eighties. Many but not all the properties of the stars are discussed, and particular attention is devoted to those relevant to the aims of the Workshop. For a description of most of the classes of stars reported in the following, and for the results of the works up to 1982, see the comprehensive monography on A–type stars by Wolff (1983). A brief description with some references is reported also in Table 1.

In the following we will present briefly the case of pre–main sequence (PMS) intermediate mass stars in a rather young cluster, NGC 2264, just as an example, and it will be discussed in the context of a comparison with other associations and slightly older clusters. After that, we will discuss the A stars on the MS and those just evolving away from it, in the light of the main results of the observations at various wavelengths.

## 2. Young Open Clusters

### 2.1. NGC 2264

A discussion of the star formation history and evolutionary status of NGC 2264 is reported by Cohen and Kuhi (1979) and Adams et al. (1983). We remark that the cluster membership of several stars has not yet been sufficiently proven (see Sect. 2.2). In this cluster many intermediate mass stars are in the PMS stage (figure 1), and most of them do not




*L. A. Willson and R. Stalio (eds.), Angular Momentum and Mass Loss for Hot Stars, 97–121.*
© 1990 *Kluwer Academic Publishers. Printed in the Netherlands.*

TABLE 1. Population I A–type and related stars discussed in the present review

| | | Ref. |
|---|---|---|
| CP1 or Am | *typically: A1≤Sp(K)≤A6, A5≤Sp(H)≤F2, A5≤Sp(m)≤F6; classic Am: Sp(H)–Sp(K)>5; mild (proto-)Am: Sp(H)–Sp(K)<5; hot Am (e.g. Sirius): Sp(H)∼early A; evolved Am and/or δ Del (Sp(H)∼early F); binariety, slow rotation* | *1, 2, 3, 4* |
| CP2 or magnetic Ap | *cool Ap: Si, Cr, Eu, Sr (7500–12000 K); slow rot.* | |
| CP3 or Hg–Mn | *hot Ap: Si λ4200, Mn, Hg (10000–18000 K); slow rot.* | *1, 5, 6* |
| CP4 or He | *magnetic field, slow rotation; temperature sequence: (8000–30000 K) He weak Ap – He weak Bp – He strong* | *5* |
| 'sn' | *nebulous He I lines, slow rotation, B2 - A0* | *7, 8* |
| λ Boo | *early A, weak λ4481, ∼metal weak, moderately rapid rotation* | *9, 10* |
| Metal weak | *early A, ∼slow rotation (e.g. Vega)* | *11* |
| A shell | *narrow shell lines (Ti II), rapid rotation, A2 - F0* | *12, 13* |
| δ Scuti | *single-/multi–mode, radial and nonradial pulsators with 0.03<P<0.3 d; A4 – F2* | *14* |
| roAp | *rapidly oscillating cool magnetic Ap; high overtone p-modes with P<0.01 d* | *15* |
| T Tau | *irregular PMS variables; emission lines; M<∼3 M⊙* | *16, 17* |
| Orion | *photometric irregular PMS variables* | *17* |

*Remark: there is no clearcut separation among several of the various classes.*
*Ref.: 1. Wolff (1983); 2. GG (Gray and Garrison, 1987, 1989a, 1989b); 3. Jaschek and Jaschek (1987); 4. Berthet and Hauck (1989); 5. Seitter and Duerbeck (1982); 6. Faraggiana (1987); 7. Abt (1979); 8. Mermilliod (1983); 9. Gray (1988); 10. Baschek and Slettebak (1988); 11. Holweger et al. (1986); 12. Abt and Moyd (1973); 13. Jaschek et al. (1988); 14. Breger (1979); 15. Kurtz (1988); 16. Appenzeller (1982); 17. Duerbeck and Seitter (1982).*

show $H_\alpha$ emission line. Warner et al. (1977) have found infrared radiation for a significant fraction (30%) of stars later than spectral type A0. Stars earlier than A0 are on the MS, and they have rotational velocities typical of open cluster stars, while three PMS A–type stars have $v \sin i$ comparable to the mean $v \sin i$ for MS early type stars. Indeed high rotational velocities occur almost exclusively among the more luminous and massive stars; the 1.5 $M_\odot$ evolutionary track conveniently separates regimes of rapid and slow rotation (Vogel and Kuhi, 1981; hereinafter VK). X–ray observations (Simon et al., 1985) have shown that most of the X–ray sources are inconspicuous but very active stars fainter than $V \approx 14$, and only a B star (W107) and a F star (W189) were found to be bright optical counterparts. The stars W90 and W100 are the only late B stars with emission lines and/or chemical peculiarities (e.g. Perez et al., 1987); according to Rydgren and Vrba (1987), W90, which is located below the MS, is seen through a nearly edge–on circumstellar disk.

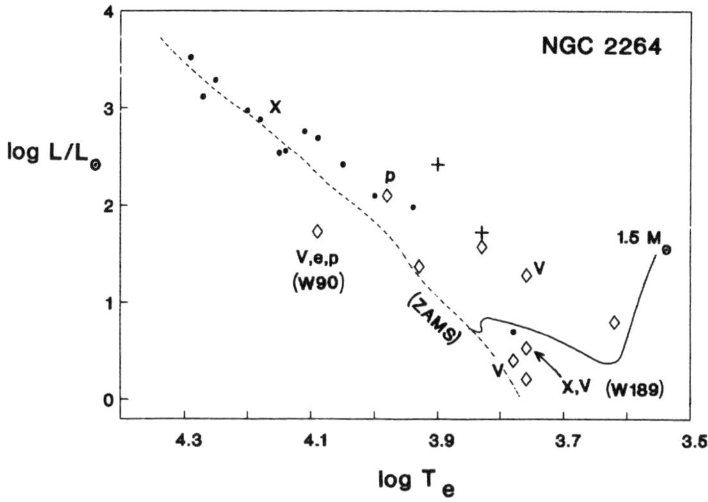



*Fig. 1. Temperature-Luminosity diagram of NGC 2264, containing some of the cluster stars. Diamonds: stars with IR radiation; plus: δ Scuti stars; dots: other MS and PMS stars; continuous line: 1.5 M⊙ evolutionary track. Characteristics: V (variability), e (emission), p (peculiarity), X (X-ray emission).*

A variability in luminosity, probably of Orion-type, has been detected in cool stars outside the instability strip, while two stars show variability of δ Scuti type (Breger, 1972a); the membership of these two stars is probable but not certain (Breger and Stockenhuber, 1984). Other possible variable stars have been reported by Sagar and Joshi (1983), and they appear to be spectroscopic binaries or variable radial velocity stars (see also Perez et al., 1989).

The observed properties of NGC 2264, along with similar observations of other open clusters and associations, allow us to begin the discussion on the interrelations between rotation, pulsation and atmospheric phenomena in the intermediate mass stars starting from about the end of their formation process.

## 2.2. ROTATION

The measured stellar rotational velocities in stellar forming regions such as Taurus-Auriga have confirmed the separation between high and low mass stars, and it should be located between 1 and 1.5 $M_\odot$. The high rotational velocity is present only in the stars with high mass (VK; Hartmann et al., 1986; Hartmann and Stauffer, 1989; see also Bouvier et al., 1986; Franchini et al., 1988). This difference has been ascribed to some difference in the efficiency of the process responsible for the angular momentum loss during the accretion and contraction phases. An early interpretation by VK was that the angular momentum is lost during the convective phase as a result of the torque exerted by stellar winds in the presence of magnetic fields, and this phase is shorter for high mass stars. Hartmann and Stauffer, looking particularly at the low mass stars, noted that rapid angular momentum transfer may be an essential component of the star formation process, that is the angular momentum loss proceeds very efficiently already in the earliest stages of star formation, and they conclude that stars older than about $10^6$ yr contract to the ZAMS at nearly constant angular momentum.

According to Bouvier et al. (1986), the massive T Tau stars studied by them should end up on the main sequence as A and F stars. Assuming the conservation of angular momentum in shells ($v \sin i \sim 1/R$) or 'homologous' contraction, and that no angular momentum is lost between T Tau and MS phases, they will arrive on the MS with rotational velocities in the range of about 40 - 150 km s$^{-1}$, in good agreement with observed rotational velocities of A and F dwarfs. The other consequence is that the MS stars should be differential rotators. A similar conclusion has been reached by Hartmann et al. (1986) for Taurus-Auriga and Orion stars. VK, on the other hand, found for their PMS stars with $1.5 \leq M/M_\odot \leq 2.5$ (almost all in NGC 2264) low projected MS velocities, and they derived the conclusion that the contraction of radiative track stars is 'nonhomologous'. In order to explain the discrepancy, Hartmann et al. (1986) suggested that some nonmembers of NGC 2264, which would be preferentially slow rotators, were included in VK sample. The questionable membership of several stars assumed as members of the cluster has been remarked also by Perez et al. (1989). From a comparison of VK sample with the observations of Perez et al., we note that one of the PMS A stars is not a member as well as the intermediate mass stars V48 and V165, while the membership of V59 and V78 is doubtful. Among the stars with similar spectral type, V48, V59 and V78 are those with the highest rotational velocity, hence there is not a strong support to Hartmann et al. conjecture. Unfortunately, there are no other F-, G-type stars in common in the samples. If we assume for the present that the estimations made by VK are correct, the different behavior of NGC 2264 could be real and related to the fact that the stars are slightly old and/or are no longer 'classic' T Tau stars.

Assuming that the behavior of NGC 2264 stars is not peculiar, but is the normal history of PMS stars, then we have to conclude that there should be some angular momentum and mass loss on the radiative tracks. VK state that no conclusion can be drawn as the existence of this mass loss in NGC 2264 intermediate mass stars; as reported in the previous section there is only indication of infrared excess in such stars. On the other hand, Levrault (1988) has found that above about 2 $M_\odot$ a clear majority of the classic T Tau and PMS stars studied by him showed incontrovertible signs of conspicuous mass loss, through either the existence of a molecular outflow or the presence of a classic P Cygni profile at H$_\alpha$; the estimated mass loss rates range from $9 \cdot 10^{-9}$ to $9 \cdot 10^{-4} M_\odot$ yr$^{-1}$, with a typical value of $3 \cdot 10^{-7} M_\odot$ yr$^{-1}$. One could speculate on the presence of angular momentum loss before the NGC 2264 evolutionary stage. A consequence is that the MS stars are not compelled to be differential rotators.

Finally, let us recall that Shu et al. (1987; see also Stahler, 1988) have stressed on the increasing evidence of circumstellar disks surrounding PMS stars, and on the need of adequate models. In fact, as discussed in Sect. 3.2, it is also evident that circumstellar disks could be a common phenomenon in nearby stars.

## 2.3. CHEMICAL PECULIARITIES

In NGC 2264 there is only a late B chemically peculiar (CP) star without emission lines. In other clusters the incidence of stars with chemical peculiarities is different. In this section we will consider mainly the magnetic Ap and the Am stars.

The results of various studies of CP stars in clusters have given two 'problems': *a*. the possible relation between the incidence of the Ap and Am phenomenon and the age of the clusters; *b*. the possible relation between the rotational velocity and the age of Ap and Am stars.

2.3.1. *Incidence of CP Phenomenon.* Relation *a*. was suggested by Hartoog (1976), and it was supported by Abt (1979) only for the Ap (and not for the Am) stars. The incidence of the chemical peculiarities seemed to increase with the age of the clusters (from Orion Nebula, $5 \cdot 10^5$ yr, to Coma, $6 \cdot 10^8$ yr), and this dependence appeared to be different according to the type of peculiarity. Moreover, Abt had the impression that the strength of the Ap(Si) phenomenon increases with time (see also Wolff, 1983). However, the photometric works in progress on a total of 56 clusters by Maitzen and coworkers (see Maitzen, 1987; Maitzen and Pavlovski, 1987) indicate that no clear correlation of the number of magnetic peculiar stars in clusters with their ages has emerged so far; on the other hand, large variations among clusters of comparable age have been found. Other series of spectroscopic works (see Glagolevskij, 1988) tend to show that the peculiarities are not related to age, but they originate before the MS phase and, on the MS, are stable. Abt and Cardona (1983) discussed a possible objection to relation *a*. Some clusters, e.g. NGC 2516 and NGC 2287, are rich in Ap stars, while other clusters of the same ages, e.g. the Pleiades, have few Ap stars. Could it be that the deficiencies in associations and very young clusters are due not to their young ages but to random differences that occur between clusters? The study of 64 visual multiple stars (Abt and Cardona, 1983) has given a further support to relation *a*., and it could be significant because the stars originated in many different clusters and associations, and therefore better represent an average effect. The limitations to the validity of the test are due to the fact that the age derived from MS primaries are only upper limits, and that the CP stars in the sample are only six. Abt and Cardona suggested, moreover, that the different incidence among coeval clusters could be due to a systematic difference in rotational velocity; as suggested by several authors, there could be different rotational velocities even in different zones of associations (e.g. Scorpio-Centaurus; Klochkova and Kopylov, 1985, hereinafter KK). On the other hand, KK, from a statistical analysis of the normal B3 - A7 stars in clusters, derived the conclusion that the differences of mean rotational velocity are not significant except for the spectral intervals B8 - A0 and A2 - A5. It will be interesting to know from the final statistical work whether the photometric observations of clusters do not support definitively relation *a*. In any case, contrarily to the visual impression of Abt, Klochkova and Kopylov, as reported by Glagolevskij (1988), have found that the degree of peculiarity is not related to age.

2.3.2. *Rotational Velocities of CP Stars.* As regards relation *b*., it can be assumed that slow rotational velocity is one of the necessary conditions for the development of the phenomenon of a peculiar star; it is unclear, however, by what cause and in what stage of evolution peculiar stars lose a large part of their angular momentum. Hartoog (1977) studied 25 Ap stars in 9 clusters and proposed that most of the angular momentum is lost in the PMS phase. Abt (1979), from the analysis of 47 Ap and 49 Am stars in 14 clusters, derived the conclusion that the rotational velocity of the peculiar stars appears to be declining with age. His interpretation was that a selection of stars of various speeds, mostly moderate ones, become abnormal and have their rotational velocities decreased through magnetic or tidal braking; the necessary requirement for the abnormality should not be an initial low rotational velocity (because not all slow rotators develop abnormality), but a magnetic field (Ap stars) or the occurrence in a close binary (Am stars). The Hg-Mn stars cannot be explained in this way because they have no magnetic fields and are not frequently found in binaries. Wolff (1981, hereinafter W81), from the study of 38 magnetic Ap stars in 15 clusters, derived the conclusion that there is a dichotomy between the hotter and the cooler magnetic stars. The former lose angular momentum primarily after they reach the ZAMS (and hence they follow relation $b$.), while the latter, which because of their lower mass require more time to contract to the ZAMS, undergo sufficient magnetic braking during PMS evolution to mask any correlation between rotation and lifetime on the MS (Wolff, 1983). The studies based on the photometric periods, which are assumed equal to the rotation periods (North, 1984, 1987; Borra et al., 1985) suggest that the magnetic Ap stars have lost their angular momentum before the MS phase, or are intrinsically slow rotators from their formation on. Moreover, the conservation of the angular momentum during the evolution of the stars on the MS is sufficient to account for the observed periods. KK, from the analysis of homogeneous data of 77 Ap stars in 9 clusters, derived the further conclusion that there is not angular momentum loss during the contraction along the radiative tracks nor on the MS. These authors conclude that the process of dissipation of angular momentum is most probably basically complete before the start of the stage of quasisteady contraction of a star.

The opposite conclusions of Abt (and W81) and of KK are rather intriguing. A comparison of the two samples of W81 and KK shows that there are 19 stars in common, and the small differences of rotational velocity (mean absolute value of 17 stars: 8 km s$^{-1}$) do not suggest responsibilities of the $v \sin i$ measurements. The strongest differences are for HD 142884 (W81: 100 km s$^{-1}$; KK: 130 km s$^{-1}$) and HD 209515 (W81: 100 km s$^{-1}$; KK: 63 km s$^{-1}$); ironically enough, the two KK values are in better agreement with relation $b$. than those of W81. There is a more serious problem: there are differences in the quantitative classification of the Ap(Si) stars, and this is a problem also for relation $a$. Probably, however, the main reason is another one. Borra et al. (1985) have discussed the observational problems related to the measurements of rotational velocities of magnetic Ap stars, in order to explain why the distribution of photometrically observed rotation periods is not related to the age as one should expect from rotational velocity data. The line broadening is related not only to $v \sin i$ but also to the magnetic field strength via Zeeman broadening. Borra (1981) has shown that the magnetic peculiar stars in Orion (and probably also Upper Scorpius) have magnetic fields stronger than the field stars, and thus one can suspect that magnetic broadening is partially responsible for the high rotational velocity values reported by Abt (1979) and W81. The fig. 4-30 in the work by Wolff (1983; or fig. 2 of W81) is suggestive of the fact that the validity of relation $b$. relies indeed essentially on the three Orion (and two Upper Scorpius) stars. It must be noted that Orion association stars were not observed by KK. Hence the opposite conclusions originate essentially from the different sample used, and in part from a different statistical treatement of data (compare figure 2 with fig. 2 of W81 and fig. 2b of KK).

The analogous case of the dependence of Am star rotational velocity from age was already discussed by Wolff (1983). Once again, Abt's results relied essentially on Orion stars, and Wolff questioned the reality of the Am nature of these stars.

2.3.3. *Conclusion.* We have to conclude that relation $b$. is not very evident; before any definitive conclusion it would be important, however, to confirm the suspicion of Borra et al. and clarify the nature of Orion CP stars. The overall impression is that, in front of the big problems *a.* and *b.*, there is a rather poor statistics and it is insufficient for giving a convincing proof of the validity of the two relations. There is a large dispersion of stellar properties within a cluster, and there could be systematic differences among the clusters. Of course, this precludes also the proof of the nonvalidity of the relations. Therefore, a different observational approach to the problem is needed.

Finally, it is disturbing to note that in several of the papers quoted in this section as well as in the recent work by Hartmann and Stauffer (1989) there are suggestions pushing to seek the climax of the angular momentum (and the peculiarity) story earlier and earlier in star life. According to Fabrika and Bychkov (1988) the reason of the slow rotation of Ap stars should be found in an early stage when the star is actively exchanging gas with interstellar medium; the most effective braking mechanism at this stage should be a magnetic stellar wind. As an example of the effectiveness of this kind of mechanism we could recall that a magnetically controlled stellar wind has been proposed to explain the variable UV doublet of C IV found by Shore and Brown (1986) in three of five magnetic He weak 'sn' stars (see the following section).

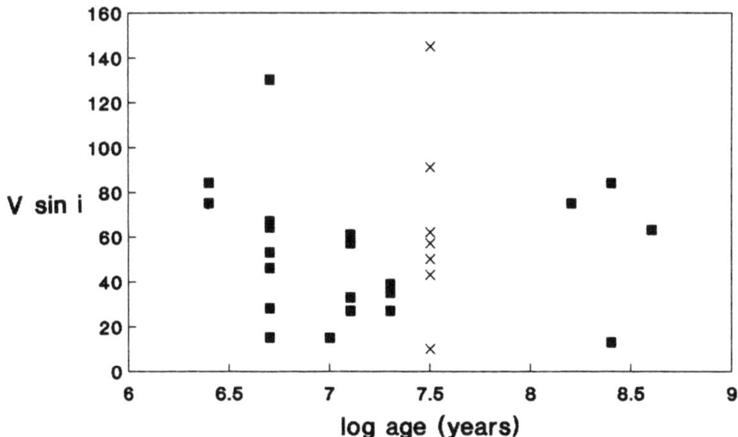


*Fig. 2. v sin i vs. age diagram for the Ap(Si) stars observed by Klochkova and Kopylov (1985, KK). The crosses indicate the Pleiades Group stars, which were not observed by Abt (1979) and Wolff (1981, W81). KK assumed slightly lower age values for young clusters than those assumed by W81. The seven stars in common with W81 are all hot Ap.*

## 2.4. X-RAY AND RADIO EMISSION

The situation found in NGC 2264 is similar to that encountered in other associations; see for example the survey of PMS stars in the Chamaeleon I forming region (Feigelson and Kriss, 1989). Recently, however, Caillault and Zoonematkermani (1989) have found twelve X-ray emitting MS B6 - A3 stars in Orion. From the sample we have to exclude the star P1956, whose spectral type is about B2 (Levato and Abt, 1976); the spectral type B8 and the low apparent luminosity given by Parenago (1954) for this star should be due to a wrong identification. The observed X–ray luminosities, from about $3 \cdot 10^{30}$ to $3 \cdot 10^{31}$ erg s$^{-1}$, are above the luminosities detected in A stars (from $3 \cdot 10^{28}$ to $10^{30}$ erg s$^{-1}$) which are thought to have X–ray emission arising from a dwarf companion. A possible explanation would be the presence of a companion of T Tau type, but there is no indication of binariety. The authors conclude that the alternative recourse is to suggest that modifications to existing theories of stellar X–ray emission are required to explain the luminosities of these stars.

Drake et al. (1987) have detected five of 34 CP stars with strong magnetic fields, observed as 6 cm continuum sources. The sources are three He strong, early Bp stars, and two Si strong, late Bp stars. Two of the stars were detected also as X–ray sources. The observations indicate that a large magnetic field is not a sufficient condition for high levels of radio continuum emission. According to the authors, the emission, which is of non-thermal origin, cannot arise from a stellar wind, but is consistent with gyrosynchrotron emission from continuously injected, mildly relativistic particles trapped in the magneto-sphere formed by the closed field regions of the surface magnetic fields. The X–ray emission could originate from a different distribution of electrons (stellar winds). Recently, another young radio and X–ray emitting magnetic B star has been detected in $\rho$ Oph molecular cloud (André et al., 1988). We will not enter into details; we just remark the fact that even for this very young object the (polarized) emission has been interpreted as due not to a strong stellar wind, but to energetic particles in a magnetosphere which features also a hot corona explaining the X–ray emission. It is interesting to note that two of the magnetic He weak 'sn' stars exhibiting stellar winds, HD 5737 and HD 21699, were also contained in the sample of Drake et al. (1987): no radio emission was detected. This could indicate a non-homogeneous behavior of the group of magnetic stars. More recently, Shore et al. (1988) have suggested that the observed UV variations in HD 5737, differently from HD 21699, are related to a magnetospheric plasma, trapped at the magnetic equator, co–rotating with the stellar photosphere.

## 2.5. PULSATION AND VARIABILITY

Pulsation of $\delta$ Scuti type is a common phenomenon in clusters. The incidence of these pulsating stars in the instability strip is estimated around 30%. The incidence in each cluster depends of course on the number of stars in the instability strip; it seems, however, that the pulsation is present in clusters of all ages, and there is a plausible decrease of the incidence for the oldest clusters owing to the depopulation of the instability strip by advancing evolution (Slovak, 1978). The situation for young clusters appears less clear because Frandsen et al. (1989) with CCD techniques have found no $\delta$ Scuti stars in Mel 105 and NGC 4755. On the other hand, other clusters with an age of less than about $10^8$ yr show a (normal) incidence of pulsation (even NGC 2264, if the membership of the variable stars will be confirmed). The discrepancy can be explained by two reasons: the detection limit, which in the CCD case was above the value of 0.01 mag typical of photoelectric photometry (Frandsen and Kjeldsen, 1988), and the stellar pulsation amplitude, which is close to 0.01 mag for stars near the MS and tends to be higher for slightly evolved stars. It was suggested that there is a difference of the mean rotational velocity between pulsating and constant stars in open clusters: it seems that, in Pleiades cluster, pulsating stars rotate more slowly than nonvariable stars, while in Hyades, Praesepe and Coma the pulsating stars rotate more rapidly (Breger, 1975). There is of course the ever present strong uncertainty due to the spread of the $v \sin i$ values and the poor statistics. The case of Hyades, Praesepe and Coma can be explained by the fact that the slowly rotating Am stars are nonvariable (see also Antonello and Raffaelli, 1983). On the other hand, the case of Pleiades is unexplained.

Despite several observational surveys, no pulsation has been detected in hot stars very beyond the blue border of the instability strip. There is no convincing evidence of pulsations in stars of luminosity class III – V and earlier than about A4, while the coolest B pulsating star is probably 27 Tau (B8 III) in Pleiades cluster (McNamara, 1987). Therefore, pulsation is absent or nondetected in the spectral range B9 – A3.

According to the opinion of both spectroscopists and photometrists, among the relatively young clusters NGC 2516 is an outstanding one. It is unique in the variety of peculiar stars, in particular the incidence of Hg–Mn stars (Abt, 1979). Guthrie (1982) remarked the unique peaked bimodal distribution of the rotational velocity of late B stars (however, see also Abt et al. 1969). The incidence of $\delta$ Scuti type pulsating stars is normal; nevertheless the cluster is unique in the high number of 'long' period variable stars (P>0.3 d) which were never seen before in other clusters and whose nature is not clear (Antonello and Mantegazza, 1986). Two surveys revealed at least eight 'long' period variable stars out of thirteen surveyed stars in the range $0.17 \leq (b - y)_o \leq 0.24$ (figure 3; Antonello and Mantegazza, 1989). Mathews et al. (1988) observed some of such stars searching for rapid oscillations in the hypothesis that they are cool Ap stars; these 'long' period variables did not show short period variations.

In the light of what reported in the Sect. 2.3, it would be important to understand the reasons of the differences among coeval clusters, and in this sense a comparison (as deep as possible) of NGC 2516 with other clusters of similar age would be very fruitful.

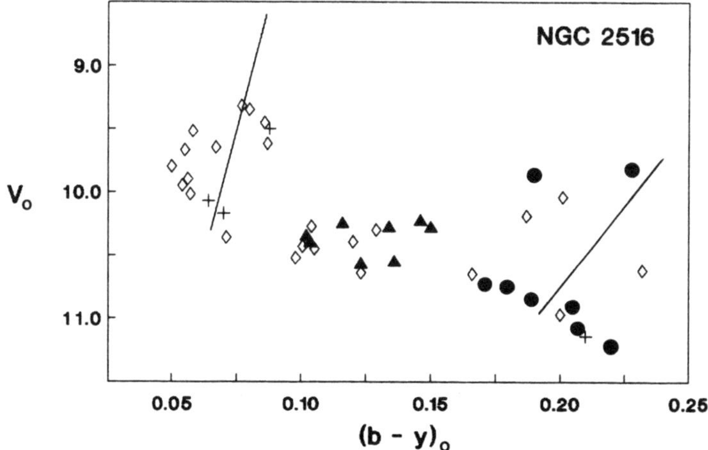


Fig. 3. *The MS stars of NGC 2516 in the region of the instability strip (continuous lines). Diamonds: nonvariable stars; triangles: short period variable ($\delta$ Scuti) stars; filled circles: 'long' period variable stars; plus: stars not surveyed for variability (Antonello and Mantegazza, 1989).*

## 3. Other Clusters and Field Stars

### 3.1. CHEMICAL PECULIARITIES, ROTATION AND PULSATION

Besides the CP stars quoted in Sect 2.3, there are other varieties found mainly among the field stars, e.g. the $\lambda$ Boo and the evolved Am stars (see Table 1). Holweger et al. (1986), Lemke and Gigas (1989) and Ramella et al. (1989), seeking the lost standard star (i.e. a substitute for Vega), have studied the sharp–lined, normal stars with spectral type near A0. The main result is that it is very difficult to find real 'normal' stars, and there are significant abundance variations from star to star of some of the elements, e.g. over– and strong under–abundance of He.

The large variety of chemical peculiarities found in A stars could be probably explained by a diffusion mechanism, which needs a sufficiently stable envelope, and hence low rotational velocity, in order to work. This is the most widely accepted mechanism proposed so far, and even if it has not been sufficiently tested to explain all the various observed features and there are several uncertainties and difficulties, yet the theoretical studies have given a proof of its capability. According to the diffusion theory (e.g. Charbonneau and Michaud, 1988, and references therein) the (classical) Am stars have a low content of helium in the envelope, which explains the fact that almost all of these stars are not pulsating. There are two exceptions: 60 Tau in Hyades (Horan, 1979) and HD 1097 (Kurtz, 1989). 60 Tau has been observed at our Observatory and its very small variability is confirmed (Poretti, 1989); we are assuming that the variability is due to the stars themselves and not to a possible companion. An interesting feature is that the two Am stars are near the cool border of the instability strip, where Cox et al. (1979; see also Valtier et al., 1979) predicted the existence of pulsating stars with chemical peculiarities and low He content in the envelope. Kurtz (1989) stressed on the fact that the pulsation of his classical Am star is excluded by Cox et al. theoretical results, which explain on the other hand the pulsation of some evolved and marginal Am stars. In our opinion, however, the importance of Cox et al. paper is just in pointing out the lack of absolute exclusion between chemical peculiarities and pulsation, and in this framework we can include also some cool classical Am stars (indeed there is no proof that they have not residual He at all). There is of course an apparent contradiction between the requirement of stability for an effective diffusion, and the velocity fields given by the pulsation. The question is: do regular pulsations produce sufficient mixing?

Some of the cool Ap stars are rapidly oscillating, and probably some could be $\delta$ Scuti variables (Weiss, 1983; see however Kreidl, 1986). Among the $\lambda$ Boo stars there are two $\delta$ Scuti variables, 29 Cyg (Gies and Percy, 1977) and HR 541 (Waelkens and Rufener, 1983). The most numerous group of pulsating peculiar stars are found among the evolved Am (and/or $\delta$ Del) stars; recently we have remarked their abnormal light curve and period ratios, which could be interpreted in the framework of the diffusion theory (Poretti and Antonello, 1988).

Ramella et al. (1989) obtained also new results on the rotational velocities of early A stars. Apparently there is an excess of slow rotators, and excluding binary stars and radial velocity variables, the distribution of rotational velocities is bimodal. This result reminds of that of B stars in open clusters; however, it is noticeable that field B stars do not show the bimodal distribution (Guthrie, 1982). Wolff et al. (1982) discussed the excess of low $v$ sin $i$ values among early B–type stars and showed that tidal and magnetic braking cannot account for it.

More rotational velocity data are needed before any attempt to give an interpretation related to the A star formation and angular momentum history. These data would be important also for updating the work by Danziger and Faber (1972), that suggested that stars on the MS or whose radii have changed by no more than a factor of two are solid body rotators, while more luminous stars are differentially rotating. The hypothesis (Abt and Levy, 1985) that a lowering of the rotational velocity due to the evolutionary expansion by a factor of 2 would change a number of normal single A stars into Am stars is consistent with the suggested solid body rotation. Presumably, the effects of the shear related to the differential rotation would prevent diffusion.

## 3.2. CIRCUMSTELLAR MATTER

In this section we will discuss three phenomena: *a.* the shells (A–shell stars), *b.* the circumstellar matter detected by IRAS, and *c.* the circumstellar disks. It is possible that they are not necessarily correlated, even if they can coexist as in $\beta$ Pic case.

*a.* The *late* A–type shell stars are rapid rotators showing some narrow lines, but differently from other stars such as 14 Com (an evolving star in Coma cluster), do not show emission lines. Jaschek et al. (1988) found variability in the shell characteristics of these stars. Slettebak and Carpenter (1983) have not found emission in UV lines, and no evidence of mass loss. We note that among these rapid rotators with variable shell but no indication of mass loss there is a $\delta$ Scuti star, 21 Vul (Garrido and Saez, 1979).

*b.* The presence of cool shells detected by IRAS around A, F and G main sequence stars is the rule, rather than the exception (Aumann, 1988), and they should be a common phenomenon also in normal B stars (Whitelock et al., 1989). A short review of the Vega phenomenon was reported by Beichmann (1987). The shells should be made of dust, but the amount of mass involved is uncertain, may be a small fraction of our planetary system. There is some debate on their origin and nature. Since small dust grains would be removed from the vicinity of the star by radiation pressure and Poynting–Robertson drag, there should be a resupply of small grains via mass loss or the grains must be large. For example, radio observations of $\alpha$ Lyr at 6 cm excluded strong winds (Hollis et al., 1985) and supported the idea of large grains. Moreover, millimetric observations failed to reveal CO and according to Walker and Wolstencroft (1988) it is one indication that these are not normal circumstellar dust shells associated with mass–loss processes.

*c.* The best studied case is that of $\beta$ Pic. The size of the circumstellar disk is of several hundreds of AU (e.g. Backman and Gillett, 1987); between the star and the disk there is a region, possibly extending out to some tens of AU, that is relatively devoided of dust grains (e.g. Smith et al., 1988).

3.2.1. *Spectroscopic Observations.* Vidal–Madjar et al. (1986) studied the gas content in the shell–disk of $\beta$ Pic, and noted that Na I and Ca II are mainly located in two different regions, the Na I in an outer region which contains nearly all the mass of the disk and in which calcium is underabundant, while most of the Ca II lies in an inner region. Slettebak and Carpenter (1983) and Lagrange et al. (1986) noted structure in UV lines profiles (e.g. Fe II, with a variable red component), suggesting velocity differences in the shell–disk. Hobbs et al. (1988) interpreted the gas component of $\beta$ Pic as due to the evaporation of comet–like bodies falling into the star. Hobbs (1986) compared the stars showing IR excess with A–type shell stars, where $\beta$ Pic is the paradigma; up to now $\beta$ Pic is the only star showing clearly the signature of both dust and gas (another star could be $\varepsilon$ Sgr, A0 II proto–shell star according to GG). There are two possible selection effects: the IRAS sensitivity, for which the IR excess cannot be detected in the other, farther shell stars, and the inclination angle of the disk, i.e. the circumstellar gas can be detected only if the disk is seen nearly edge–on. Another suggestion (e.g. Lagrange et al.) is that the shell and the disk are two phenomena not related one each other.

Owing to the common nature of the Vega phenomenon, it must be present in real slow rotators (e.g. Vega itself; see figure 4), that is non–A–shell stars. IUE observations of Mg II lines of Vega have shown a deep line core which could be explained by some superimposed absorption, probably, however, of interstellar origin (Freire–Ferrero et al., 1983); observations of other more rapidly rotating stars have suggested that some narrow absorption lines should be interstellar (e.g. Bohm–Vitense, 1981; Freire–Ferrero, 1984).

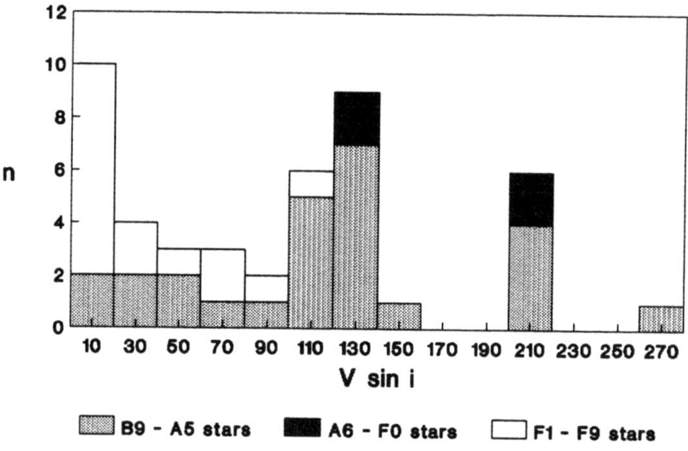


Fig. 4. The distribution of v sin i values of stars with far IR excess, for three different spectral intervals.

Molaro et al. (1984) found C IV blue shifted lines on the IUE spectra of HD 119921, a rapidly rotating star. Chiu et al. (1988) studied the IUE spectra of this and other forty-one B6–A2 stars within 200 pc; five of the program stars showed significant C IV and Si IV absorption. They are rapid rotators, and the narrow core present in Si II $\lambda1533$ indicates that HR 1147, HD 119921 and $\sigma$ Her are previously unrecognized shell stars (this feature is seen also in $\beta$ Pic). HD 119921 has variable C IV, Si IV and Si II profile, which rules out the possibility of interstellar origin. The C IV $\lambda1548$ profile of $\sigma$ Her is blue asymmetric, consistent with the existence of a strong stellar wind. The IR excess of $\sigma$ Her detected by IRAS should be due to dust and not to free–free emission as in Be stars (Coté, 1987). Another strong IR source in Chiu et al. sample, $\gamma$ Oph, does not show these UV features. Bruhweiler et al. (1989) suggest that the UV features could be explained by a heating due to shocks, related to mass outflows but not necessarily to mass loss. The mechanical energy could derive from (nonradial) oscillations, and should be dissipated through heating the outer regions of stellar photosphere or inner disk regions. However, these stars are located in a spectral range where pulsation has not yet been detected.

3.2.2. *Surveys.* Several IRAS surveys of star selections made according to various criteria have been performed; of course, there is a strong selection effect due to the brightness of the stars and their proximity (Aumann, 1985; Sadakane and Nishida, 1986; Coté, 1987; Aumann, 1988; Walker and Wolstencroft, 1988; Backman and Gillett, 1987). Among sixty surveyed bright stars with spectral types from B9 to G8 showing the far IR excess there are some $\lambda$ Boo stars ($\lambda$ Boo itself, $\pi^1$ Ori, $\theta$ Peg) and some probable metal weak stars ($\alpha$ Crv, HR 8799 and $\alpha$ Lyr); a slightly metal rich star ($\beta$ UMa; GG); a $\delta$ Scuti pulsating star ($\beta$ Cas), and other suspected pulsating stars ($\sigma$ Boo, $\gamma$ Dor, $\beta$ Leo, $\eta$ Crv and $\gamma$ Boo). A survey of CP2 – CP3 stars failed to reveal similar far IR excess (Kroll, 1987) in cool stars. Let us derive some indications from this statistics, even if it is rather biased and poor (Antonello, 1989). We can note the lack of cool Ap and classical Am stars. The lack should not be put in relation with the low rotational velocity and binariety, because other slow rotators and members of binary systems with periods similar to those of Am stars have the far IR excess. The lack should not be related to a possible low number of bright peculiar stars: there are 37 Am stars brighter than 4th mag (Curchod and Hauck, 1979), and Sirius is one of them. This impression is strengthened by considering on the other hand the high incidence of the $\lambda$ Boo stars; about twenty stars are known to be members of this class, and the brightest and/or hottest ones were detected. Among the 40 stars in the sky with $V < 3$ mag, spectral type B9 – F2 and luminosity class III – V, there are at least 10 detected stars out of 29 'normal' stars, and 1 of 12 chemically peculiar (Ap–Am) stars (1 of 13 if we include $\rho$ Pup). As a possible consequence of this indication, we recall a recent study on the accretion of interplanetary dust by Ap and Am stars made by Kumar et al. (1989). The authors discuss the shortcomings of the diffusion theory, and propose their hypothesis not just as an alternative but rather as a complement able to solve the problems. According to them, comet impacts with a rate comparable to that for the Sun and lasting $10^7$ yr should be able to produce some of the element overabundances observed in Ap and Am stars. Does the possible lack of circumstellar matter in these stars support or not the accretion hypothesis? One must assume, respectively, a short early phase of comet impact, or the lack of these events. Can this explain some of the differences between Am–Ap stars and $\lambda$ Boo–metal weak stars? In any case, the indication given by the IR excess looks of some importance for understanding the origin of these stars. It must be confirmed by further data, since there are some selection effects owing to the different criteria used by the various authors. For example, Coté excluded intentionally peculiar stars from the survey, while the other authors included them (Sadakane and Nishida report $\lambda$ Boo as A0p star); we remark that the contribution of each reference source to the sample of sixty stars is of about ten stars.

Figure 5 shows the spectral distribution of the stars. The dip for the late A type stars is a rather common feature of observed spectral distributions (see Jaschek and Jaschek, 1987). Since, however, in the corresponding range of effective temperature there is the highest incidence of Am stars (Smith, 1973), one could suspect that the ever present dip is related to this incidence. In fact Am stars are sometimes classified from Ca II K rather than from hydrogen lines, and hence they can appear with an earlier spectral type. For comparison purposes with normal stars, we should consider the hydrogen lines. The papers

110

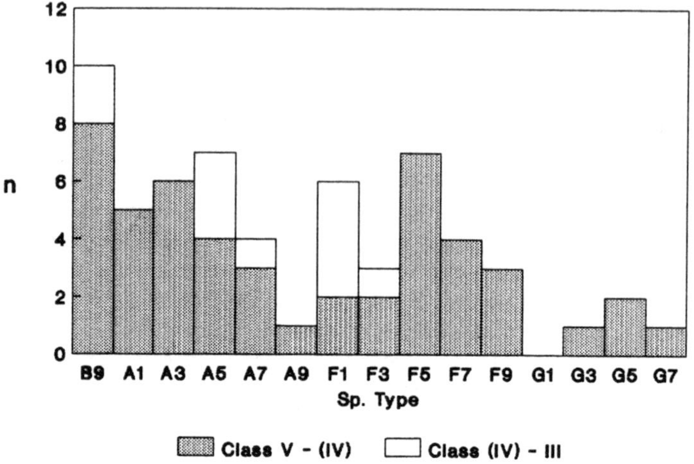



*Fig. 5. The spectral type distribution of stars showing far IR excess. Each bin contains two subclasses (e.g. B9–A0). In the text the dip at A9–F0 is discussed.*

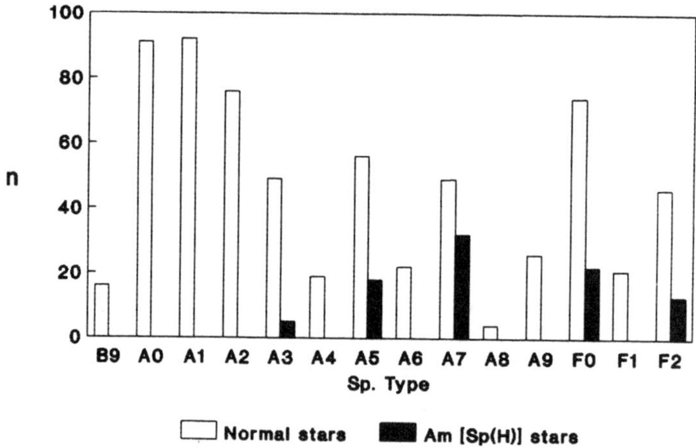



*Fig. 6. The spectral type distribution of early A, late A and early F stars with luminosity class V–III (data from Gray and Garrison, GG). The filled areas are the Am–Fm stars classified on the basis of the hydrogen lines. Note the strong variation between adjacent subclasses.*

by GG offer us the possibility of making a small test, shown in figure 6 (by the way, one should note the curious dips at A4 and A6). Here the Am–Fm stars are classified according to the hydrogen lines. The sample at our disposal shows that the dip at A8 is not affected and hence it should be intrinsic; however, owing to the curious strong variation from one subclass to the next, a test with the color index $b - y$ (or the index $\beta$) is required, taking also into account the $v \sin i$ values. According to the current interpretation (Böhm–Vitense and Canterna, 1974), a narrow gap in this temperature range is the signature of the abrupt onset of convection. Am stars could be present because, according to the diffusion theory, they must have a stable envelope; in fact, their incidence shows a rapid decline for lower $T_{\mathrm{eff}}$ (see Smith, 1973, and Wolff, 1983). This point would require a thorough treatment. In particular, we remark that the work by Bohm–Vitense and Canterna must be updated taking into account the accurate photometric studies of clusters; moreover, it is not clear whether the high incidence of Am stars is due to their intrinsically high number or to the low number of normal stars, or it is only a rotation–convection effect. Anyway, in figure 7 we show the distribution of $\beta$–index for Am–Fm stars (GG data). For normal stars, one would expect a dip (corresponding to the Böhm–Vitense–Canterna gap) at $\beta \sim 2.79$–2.80 ($\sim 7700$ K). Of course, the apparent dip at $\beta \sim 2.79$ in figure 7 can be due simply to the small number of stars; nevertheless the coincidence is very intriguing. If confirmed, it would be dangerous for the diffusion theory (the stars with $\beta < 2.79$ do not appear evolved).

As for our discussion on circumstellar matter, the conclusion is that, on the whole, the dip at A9 does not appear related to the lack of Am stars with far IR excess.

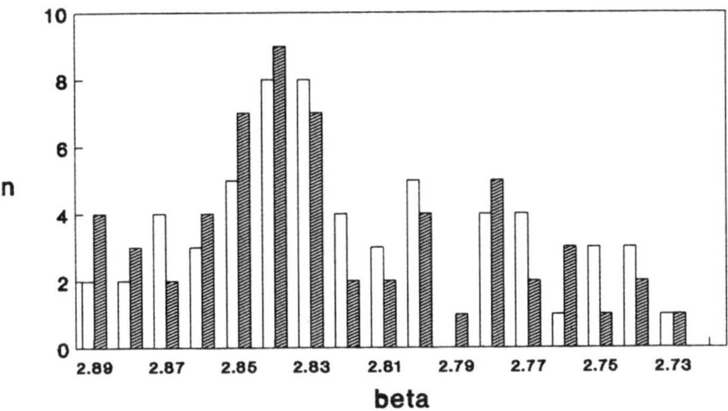

*Fig. 7. The distribution of $\beta$–index for Am–Fm stars (data from Gray and Garrison, GG), for two different bin edges (e.g. 2.810–2.800, and 2.805–2.795).*

*3.2.3. Multiple Systems.* The spectroscopic binaries showing IR excess have periods ranging from 17 d ($\alpha$ CrB) to few years ($\delta$ Gem), and the semi–major axis $a \sin i$ of the orbit is between a fraction of AU to about 5 AU. Assuming the $\beta$ Pic model, the binary system should occupy the inner region devoided of dust grains. In particular, $\alpha$ CrB is a well studied spectroscopic and eclipsing binary (Tomkin and Popper, 1986) with an inclination angle of 88°; if we assume the possible circumstellar disk co–planar with the orbit, it should be seen with a tilt angle of about 2°. One should be cautious however in the application of the model to all the stars, because the visual binaries in the sample (e.g. $\sigma^2$ UMa) have periods up to about 1000 yr, and $a \sin i$ values up to about 100 AU. As shown in Figure 8, the projected separation on the sky between the components of the multiple systems showing the IRAS excess can be of several tens or hundreds of AU, that is the size of $\beta$ Pic disk. It may be possible that the circumstellar disk or shell differs according to the stellar characteristics, e.g. the mass. The figure shows the distribution of the $a \sin i$ values of the binaries and the projected separation on the sky of the physical components of multiple systems grouped according to the spectral class (data from Hoffleit and Jaschek, 1982).

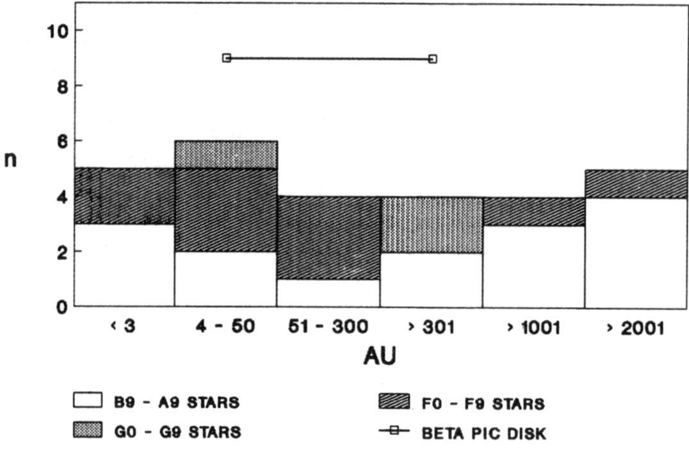


Fig. 8. *The distribution of the 'separations' of stars in multiple systems containing a bright star with far IR excess, compared with the approximate extension of $\beta$ Pic disk. The last three bins are 301–1000, 1001–2000, 2001–5000 AU.*

Of course the statistics is poor, and there is the uncertainty due to the projection factor and/or the unknown orbital characteristics (e.g. minimum and maximum value of separation). Nevertheless the case of the A stars is interesting, because there is only the star $\beta$ Leo which has a separation of some hundreds AU (about 480), and the asserted physical membership of the faint (15th mag) companion star, based on poor observations, could be questioned. The other hot stars have separations of less than 76 AU ($\kappa$ Lep) and more than 764 AU (HR 6297). At the present we are analysing all these stars in order to verify the physical reality of the systems and the possible gap for A stars corresponding to the extension of $\beta$ Pic disk. In figure 9 we show the distribution of the 'separations' for nearby (<25 pc), bright ($V$ <3), B9–A5, 'normal' and CP1–CP2 stars. Almost all the nearby stars are reported in the figure, that is only few stars are not members of well known spectroscopic/visual binaries and multiple systems (e.g. $\beta$ UMa itself; see Jaschek and Jaschek, 1987, for an interesting description of nearby Am stars). The figure suggests clearly *one of the possible reasons* for the lack of strong IR excess in nearby CP stars, that is their being members of multiple systems with separations similar to the extension of $\beta$ Pic disk. Taking into account the results of the study made by Abt and Levy (1985), it seems that such an incidence of Am stars in multiple systems is present not only in nearby stars. If we assume plausible values of the stellar masses and we change from 'separations' to periods, the distribution of our CP (Am *and* Ap) stars reminds of that of Am stars in fig. 7 (upper panel) of Abt and Levy.

Of course, since the number of stars is small, the results must be considered only as preliminary. In any case, they appear of some interest because indicate a new way in the study of the formation of CP and normal stars. Moreover, the study of multiple systems, even if with some obvious limitations, can be a test of the models of circumstellar disks.

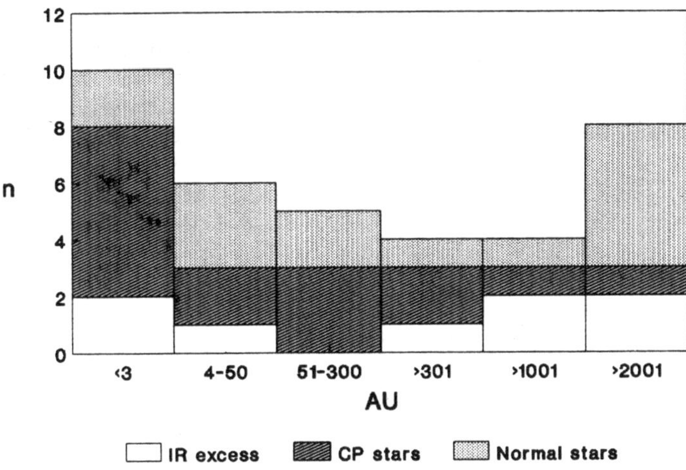


*Fig. 9. The distribution of the 'separation' of nearby bright stars in multiple systems (see figure 8). IR excess: stars with Vega–like phenomenon; CP stars: CP1–CP2 stars; normal stars: stars without strong far IR excess and without CP1–CP2 characteristics.*

## 3.3. ACTIVITY AND MASS LOSS?

The existence of chromospheres, transition regions and coronae in A–type stars, and the location along the MS where chromospheric and coronal activity begin are two problems which have been strongly debated. Wolff (1983) summarized the visual, UV and X–ray extensive data, negative as well as positive, that has been obtained in these searches. Her conclusion was that the determination of the physical conditions in the outer layers of A–type atmospheres is an exceedingly difficult observational problem; nevertheless, X–ray emission has been detected in A–type stars of all temperatures, although not in all type stars of any given temperature, and chromospheric emission has been detected in late A–type stars.

3.3.1. *UV and X-Ray Observations.* IUE observations of C II $\lambda1335$ and C IV $\lambda1549$ and ground based observations of He I $\lambda5876$ (e.g. Wolff et al., 1986) have established that virtually all of the early F dwarfs show stronger chromospheric and transition region emission than do the cooler and more deeply convective dwarf stars like the Sun (Landsman and Simon, 1988). To extend the search for UV chromospheric emission to A type stars is observationally difficult because of the strong photospheric continuum. The detection of X–ray and $Ly_\alpha$ emission from the single A star Altair suggests that stellar activity may be present in such stars. It should be absent in Vega (Praderie, 1981), and in the other early A–type stars (Freire–Ferrero and Talavera, 1984); in these stars, however, X–ray emission has been detected.

As reported by Walter et al. (1988), in the F stars the observed transition region C II and C IV fluxes increase with increasing $T_{\text{eff}}$, as the thickness of the convection envelope decreases; moreover, there is very little scatter in the fluxes among the early F dwarfs (F0 – F5), that is these stars do not show the rotation–activity correlation present in cooler stars. The C II flux drops rapidly before F0, although significant emission is visible as far as $B - V \sim 0.22$ (A7). According to these authors, the transition region emissions in A, F stars could be interpreted in terms of a basal flux component, which should be due to heating of the chromosphere by acoustic waves, and of another component due to magnetic processes similar to those observed on the Sun, since it correlates with the rotational velocity. The drop below $B - V \sim 0.30$ suggests that with the disappearance of the convective envelope, both the basal and active heating mechanisms cease to function. Conversely, if a convective envelope is sufficiently thick to produce acoustic heating, then it seems also capable of supporting dynamo activity.

The scenario resulting from the visual and UV observations can be connected with that from the X–ray data. The study of Pallavicini et al. (1981) has shown that the X–ray emission from normal stars is tied very strongly to bolometric luminosity for early–type stars and to rotation rate for late–type stars, with F stars forming a bridge between the two types of behavior. The former emission is presumably caused by some source of non–radiative heating in the extended envelopes that are produced by mass loss, while the latter is related to stellar activity. However, the most prominent exception to this clearcut rule has been the apparently anomalous behavior of A star X–ray emission, for which the large spread in X–ray luminosity showed no apparent correlation with either bolometric luminosity or stellar rotation rate. Golub et al. (1983) have shown that the low level of emission from normal (single) A stars agrees with the correlation observed for O and B stars. The high X–ray emitting A stars in general have lower mass companions, and Golub et al. suggest that the emission level is determined by the rotation rate of the secondary, according to the $L_x$ versus rotation law. If the lower mass companions are of late spectral type (dK or dM) they have presumably high rotational velocity owing to the fact that these companions of A stars are necessarily young for their spectral type. The change from early– to late–type coronal behavior is narrowed to a range between A7 and early F (i.e. Böhm-Vitense-Canterna gap).

There is some indication that apparently single Ap stars tend to emit at somewhat higher levels than Altair, but Am stars do not seem to be as strong (Golub et al., 1983). According to Cash and Snow (1983), an Am star should not be a source (at $10^{28}$ erg s$^{-1}$ level) unless it is in a spectroscopic binary with a period below about 10 days.

3.3.2. *δ Scuti Stars.* Besides in Altair (Blanco et al., 1980), $Ly_\alpha$ emission has been detected in 71 Tau and 7 And (Landsman and Simon, 1988). Moreover, Walter et al. (1988) have found C II emission in the late A stars 71 Tau and HD 90132. The star 71 Tau has raised much interest starting from the discovery that it is the most intense X-ray emitter in Hyades cluster (Stern et al., 1981). Two years ago, at the previous Trieste workshop, we stressed on the fact that it is a $\delta$ Scuti star with a high rotational velocity (resumé by Polidan, 1987). These findings open the question whether the pulsation and/or the rotation are responsible for this behavior. What do we know about the atmospheric effects of pulsation in $\delta$ Scuti stars?

Dravins et al. (1977) observed $\rho$ Pup in the visual and found a transient Ca II K emission at a phase near maximum outward acceleration. Fracassini and Pasinetti (1982) found Mg II h and k emission in $\beta$ Cas, $\rho$ Pup and in the suspected variable $\tau$ Cyg. Fracassini et al. (1983) studied a series of UV spectra of $\rho$ Pup and found that the Mg II emission is always present during a pulsation cycle, and its strength is variable, with a peak located probably near maximum light. This behavior is similar to that seen in classical Cepheids (Schmidt and Parsons, 1984) where the peak of the Mg II emission was observed during rising light. In $\beta$ Cas the chromospheric activity is modulated by the pulsation, with maximum Mg II emission found around minimum light, which is different from the Cepheids and $\rho$ Pup cases, while the mean levels of activity are comparable to other early F dwarfs (Teays et al., 1989). According to Ayres and Bennett (1986), $\beta$ Cas shows also a variable $Ly_\alpha$ emission. The chromospheric indicators and their time dependence suggest the presence of shocks. According to Garbuzov and Andrievskii (1987), the observed emission in $\rho$ Pup should be due to the radiation of gas heated by a shock wave, and its variability to the motion of the wave in an inhomogeneous medium with decreasing density.

On the whole, the weak atmospheric phenomena observed in cool $\delta$ Scuti stars can be compared with analogous, stronger phenomena in cool pulsating stars (Cepheids, RR Lyraes).

The observation of the Mg II h and k lines of hotter $\delta$ Scuti stars with high rotational velocity such as 71 Tau have not indicated the presence of emissions, but only the absorption feature probably due to the local interstellar medium. The lack of detectable emissions, however, should not be surprising owing to the smearing or washing effect of the high rotation, as pointed out by several authors (e.g. Zolcinski et al., 1982).

Micela et al. (1988) have stressed again on the outstanding character of the X-ray emission of 71 Tau. The star is a wide spectroscopic binary, with a dwarf G4 type companion, resolved by lunar occultation. As remarked by Micela et al., the companion cannot explain the strong X-ray emission. The log $(L_x/L_{bol})$ value for 71 Tau is $-4.8$, which is higher than the value found in normal early F stars (mean value: $-5.2$; Walter, 1983) and is close to that found in late F stars. We note, however, that it is also similar to the value found in the Orion B6 – A3 stars, quoted in Sect. 2.4 (the B9 – A0 stars in the sample have a value between $-4.2$ and $-5.0$); as reported above, for the present there is no explanation of such emission. Can we ascribe the X-ray emission of 71 Tau to the interaction between pulsation and rapid rotation? Micela et al. surveyed other six $\delta$ Scuti stars searching for emissions, but no one was detected; the authors do not report which stars were surveyed, hence we do not know if they are rapid rotators. Another A-type star in Pleiades is a strong emitter and has high rotational velocity, HzII 1384; Breger (1972b) found it nonvariable.

In any case, the answer to the previous question may have some importance in the light of the hypothesis on the pulsation and mass loss proposed by Willson et al. (1987), and this consideration allows us to speculate a little bit on the observational results obtained so far. According to Willson et al., the smaller $L_x/L_{bol}$ ratio of normal late A – early F stars than that of young G stars could be explained by the smaller mechanical energy fluxes disposable for coronal heating of the former stars, or by the fact that these fluxes are channelled into mass loss rather than heating and radiative losses. The second alternative should be applicable for example to the stars falling in the instability strip, with the wind driven by the coupling of pulsation and rapid rotation. According to Willson et al., the lack of observational indications of the strong wind estimated by the authors for these stars are explained by: (1) the very high temperature of the wind, where X–ray fluxes are moderate and the line signatures in visual and UV are scarce; (2) the high velocity of the wind; (3) the rapid, short–lived mass loss phase, which occurred at a very early stage (in very young clusters and associations). Assuming that in general the hypothesis is correct, the point (3), according to our opinion, would look the most plausible, because the estimated mass loss (about $10^{-9}\,M_\odot\,\mathrm{yr}^{-1}$) would give really strong observable effects which are not clearly detected in 'old' stars (unless we consider part of the observed far IR excess as due to this cause).

There are some possible observational tests, e.g. the statistics of the A–type stars in clusters, looking for the 'evolutionary' effects (gaps on the MS) of the mass loss phase, and some signatures (e.g. some X–ray emission or some small indication of mass loss in UV lines) of the proposed mechanism in the normal $\delta$ Scuti stars with rapid rotation, where it could maintain some effectiveness. The first test requires accurate observational studies of young clusters. As regards the second test, the star 71 Tau could be an important example. On the other hand, 21 Vul (Sect. 3.2) would be probably a counterexample. The star HzII 1384 in Pleiades is nonvariable, but this may be of minor importance, because one could assume a possible very low pulsation amplitude. The color index of the star, $(b - y)_o = 0.085$, corresponds to the position of the gap in NGC 2516 noted by Willson (these proceedings) and which can be probably seen also in other older clusters (remark: Böhm-Vitense-Canterna gap is located at $(b - y)_o \sim 0.15$). Is HzII 1384 losing mass? Does the relatively slow rotation of other Pleiades pulsating stars (Sect. 2.5) indicate that they already lost mass and angular momentum? It is not possible to give an answer, nor to substantiate clearly with further observational data the possibility of the mechanism. We propose the observation of another $\delta$ Scuti star in Hyades, 69 Tau, which is very similar to 71 Tau as regards period, rotational velocity and the other physical properties. Unfortunately, 69 Tau was not observed with Einstein satellite, probably because it is far from the bulk of the cluster.


## 4. Conclusion

This review is not comprehensive of all the observed properties of intermediate mass A stars. For example, among other things we have not reviewed the results of the many obervations of the surface abundances of light elements in cool stars; neither have we adequately discussed the pulsational properties and the various chemical abundances of stars.

The main conclusions which can be drawn from the discussed properties of A stars are the following ones.

*a.* Significant indications of mass loss, which could be related to some loss of angular momentum, are found only in PMS intermediate mass stars. This loss should be effective probably only for a short time interval.

*b.* The study of CP stars in young clusters and associations does not indicate a clearly evident angular momentum loss on long time scales due to magnetic and tidal braking and magnetic stellar winds. If effective, these processes should be operating on short time scales, otherwise these stars must be slow rotators since their formation.

*c.* A mass loss related to pulsation and rapid rotation has been suggested for explaining some results from UV and X–ray observations of late B – early F stars. However, this suggestion concerns very few stars and, for early A stars, is based on a hypothetical pulsation.

*d.* The diffusion processes are a significant astrophysical mechanism as proven by the widespread successful applications from the Sun to hot population II stars. Even if there are several uncertainties, most of the observed properties of the varieties of A stars are explained by them.

*e.* Circumstellar matter has been detected by IRAS in many 'normal' and metal weak (population I) stars, while it appears to be lacking in Ap and Am stars. This new finding, if confirmed, probably opens a new chapter in the study of these stars. The analysis of nearby visual multiple systems suggests one of the possible reasons for the lack: the separation between the CP star and another component of the system is similar to the extension of the circumstellar shell-disk resolved in some A-type stars. Therefore we can wonder whether, *in these cases,* multiplicity, lack of circumstellar matter and slow rotation are all related merely to the star formation process.

*f.* The Einstein and IRAS satellites have offered only a taste of A stars at X–ray and far IR wavelengths, just for making us longing for data from next satellite missions with increased sensitivity.

## 5. References


Abt H.A.: 1979, *Astrophys. J.* **230**, 485.

Abt H.A., Cardona O.: 1983, *Astrophys. J.* **272**, 182.

Abt H.A., Clements A.E., Doose L.R., Harris D.H.: 1969, *Astron. J.* **74**, 1153.

Abt H.A., Levy S.G.: 1985, *Astrophys. J. Suppl. Ser.* **59**, 229.

Abt H.A., Moyd K.I.: 1973, *Astrophys. J.* **182**, 809.

Adams M.T., Strom K.M., Strom S.E.: 1983, *Astrophys. J. Suppl. Ser.* **53**, 893.

André P., Montmerle T., Feigelson E.D., Stine P.C., Klein K.L.: 1988, *Astrophys. J.* **335**, 940.

Antonello E.: 1989, in preparation.

Antonello E., Mantegazza L.: 1986, *Astron. Astrophys.* **164**, 40.

Antonello E., Mantegazza L.: 1989, in preparation.

Antonello E., Raffaelli G.: 1983, *Publ. Astron. Soc. Pacific* **95**, 82.

Appenzeller I.: 1982, in *Landolt Börnstein*, New Series, Group VI, **2–b**, Springer–Verlag, p. 357.

Aumann H.H.: 1985, *Publ. Astron. Soc. Pacific* **97**, 885.

Aumann H.H.: 1988, *Astron. J.* **96**, 1415.

Ayres T.R., Bennett J.O.: 1986, in A.N. Cox, W.M. Sparks, S.G. Starrfield (eds.) *Stellar Pulsation*, Lect. Not. Phys., Springer–Verlag, **274**, 127.

Backman D.E., Gillett F.C.: 1987, in J.L. Linsky, R.E. Stencel (eds.) *Cool Stars, Stellar*

*Systems, and the Sun*, Lect. Not. Phys., Springer–Verlag, **291**, 340.

Baschek B., Slettebak A.: 1988, *Astron. Astrophys.* **207**, 112.

Beichman C.A.: 1987, *Annu. Rev. Astron. Astrophys.* **25**, 521.

Berthet S., Hauck B.: 1989, *The Messenger* **56**, 48.

Blanco C., Catalano S., Marilli E.: 1980, in *Second Europ. IUE Conf.*, ESA SP–157, 63.

Böhm–Vitense E.: 1981, *Astrophys. J.* **244**, 504.

Böhm–Vitense E., Canterna R.: 1974, *Astrophys. J.* **194**, 629.

Borra E.F.: 1981, *Astrophys. J.* **249**, L39.

Borra E.F., Beaulieu A., Brousseau D., Shelton I.: 1985, *Astron. Astrophys.* **149**, 266.

Bouvier J., Bertout C., Benz W., Mayor M.: 1986, *Astron. Astrophys.* **165**, 110.

Breger M.: 1972a, *Astrophys. J.* **171**, 539.

Breger M.: 1972b, *Astrophys. J.* **176**, 367.

Breger M.: 1975, in *Variable Stars and Stellar Evolution*, IAU Symp. No. 67, D. Reidel, p. 231.

Breger M.: 1979, *Publ. Astron. Soc. Pacific* **91**, 5.

Breger M., Stockenhuber H.: 1983, *Hvar Obs. Bull.* **7**, 283.

Bruhweiler F.C., Grady C.A., Chiu W.A.: 1989, *Astrophys. J.* **340**, 1038.

Caillault J.P., Zoonematkermani S.: 1989, *Astrophys. J.* **338**, L57.

Cash W., Snow T.P.: 1982, *Astrophys. J.* **263**, L59.

Charbonneau P., Michaud G.: 1988, *Astrophys. J.* **327**, 809.

Chiu W.A., Bruhweiler F.C., Grady C.A.: 1988, in *A Decade of UV Astronomy with IUE*, ESA SP–281, **1**, 397.

Cohen M., Kuhi L.V.: 1979, *Astrophys. J. Suppl. Ser.* **41**, 743.

Coté J.: 1987, *Astron. Astrophys.* **181**, 77.

Cox A.N., King D.S., Hodson S.W.: 1979, *Astrophys. J.* **231**, 798.

Curchod A., Hauck B.: 1979, *Astron. Astrophys. Suppl. Ser.* **38**, 449.

Danziger I.J., Faber S.M.: 1972, *Astron. Astrophys.* **18**, 428.

Drake S.A., Abbott D.C., Bastian T.S., Bieging J.H., Churchwell E., Dulk G., Linsky J.L.: 1987, *Astrophys. J.* **322**, 902.

Dravins D., Lind J., Sarg K.: 1977, *Astron. Astrophys.* **54**, 381.

Duerbeck H.W., Seitter W.C.: 1982, in *Landolt Börnstein*, New Series, Group VI, **2–b**, Springer–Verlag, p. 258.

Fabrika S.N., Bychkov V.D.: 1988, in Yu.V. Glagolevskij and J.M. Kopylov (eds.) *Magnetic Stars*, Leningrad Nauka, p. 241.

Faraggiana R.: 1987, *Astrophys. Sp. Sci.* **134**, 381.

Feigelson E.D., Kriss G.A.: 1989, *Astrophys. J.* **338**, 262.

Fracassini M., Pasinetti L.E.: 1982, *Astron. Astrophys.* **107**, 326.

Fracassini M., Pasinetti L.E., Castelli F., Antonello E., Pastori L.: 1983, *Astrophys. Sp. Sci.* **97**, 323.

Franchini M., Magazzu' A., Stalio R.: 1988, *Astron. Astrophys.* **189**, 132.

Frandsen S., Dreyer P., Kjeldsen H.: 1989, *Astron. Astrophys.* **215**, 287.

Frandsen S., Kjeldsen H.: 1988, in *Seismology of the Sun and Sun–like Stars*, ESA SP–286, p. 575.

Freire–Ferrero R.: 1984, in *Proc. 4th Europ. IUE Conf.*, ESA SP–218, 133.

Freire–Ferrero R., Gouttebroze P., Kondo Y.: 1983, *Astron. Astrophys.* **121**, 59.

Freire–Ferrero R., Talavera A.: 1984, in *Proc. 4th Europ. IUE Conf.*, ESA SP–218, 217.

Garbuzov G.A., Andrievskij S.M.: 1986, *Astrophysics (Astrofizika)* **25**, 251.

Garrido R., Saez M.: 1979, *Astron. Astrophys.* **79**, 347.

Gies D.R., Percy J.R.: 1977, *Astron. J.* **82**, 166.

Glagolevskij Yu. V.: 1988, in Yu.V. Glagolevskij and J.M. Kopylov (eds.) *Magnetic Stars*, Leningrad Nauka, p. 206.

Golub L., Harnden F.R., Maxson C.W., Rosner R., Vaiana G.S., Cash W., Snow T.P.: 1983, *Astrophys. J.* **271**, 264.

Gray R.O.: 1988, *Astron. J.* **95**, 220.

Gray R.O., Garrison R.F.: 1987, *Astrophys. J. Suppl. Ser.* **65**, 581.

Gray R.O., Garrison R.F.: 1989a, *Astrophys. J. Suppl. Ser.* **69**, 301.

Gray R.O., Garrison R.F.: 1989b, *Astrophys. J. Suppl. Ser.* **70**, 623.

Guthrie B.N.G.: 1982, *Mon. Not. Roy. Astron. Soc.* **198**, 795.

Hartmann L., Hewett R., Stahler S., Mathieu R.D.: 1986 *Astrophys. J.* **309**, 275.

Hartmann L., Stauffer J.R.: 1989, *Astron. J.* **97**, 873.

Hartoog M.R.: 1976, *Astrophys. J.* **205**, 807.

Hartoog M.R.: 1977, *Astrophys. J.* **212**, 723.

Hobbs L.M.: 1986, *Astrophys. J.* **308**, 854.

Hobbs L.M., Lagrange–Henri A.M., Ferlet R., Vidal–Madjar A., Welty E.: 1988, *Astrophys. J.* **334**, L41.

Hoffleit D., Jaschek C.: 1982, *The Bright Star Catalogue*, Yale Univ. Obs., New Haven.

Hollis J.M., Chin G., Brown R.L.: 1985, *Astrophys. J.* **294**, 646.

Holweger H., Steffen M., Gigas D.: 1986, *Astron. Astrophys.* **163**, 333.

Horan S.: 1979, *Astron. J.* **84**, 1770.

Jaschek C., Jaschek M.: 1987, *The Classification of Stars*, Cambridge Univ. Press, p. 225.

Jaschek M., Jaschek C., Andrillat Y.: 1988, *Astron. Astrophys. Suppl. Ser.* **72**, 505.

Klochkova V.G., Kopylov I.M.: 1985, *Sov. Astron.* **29**, 549.

Kreidl T.J.: 1986, in A.N. Cox, W.M. Sparks, S.G. Starrfield (eds.) *Stellar Pulsation*, Lect. Not. Phys., Springer–Verlag, **274**, 134.

Kroll R.: 1987, *Astron. Astrophys.* **181**, 315.

Kumar C.K., Davila J.M., Rajan R.S.: 1989, *Astrophys. J.* **337**, 414.

Kurtz D.W.: 1988, in G. Kovacs, L. Szabados, B. Szeidl (eds.) *Multimode Stellar Pulsations*, Konkoly Observatory, p. 107.

Kurtz D.W.: 1989, *Mon. Not. Roy. Astron. Soc.* **238**, 1077.

Lagrange A.M., Ferlet R., Vidal–Madjar A.: 1986, in *New Insights in Astrophysics*, ESA SP–263, 569.

Landsman W.B., Simon T.: 1988, *Bull. Am. Astron. Soc.* **20**, 697.

Lemke M., Gigas D.: 1989, *Astronomische Gesellschaft Abstract Ser.* **3**, 46.

Levato H., Abt H.A.: 1976, *Publ. Astron. Soc. Pacific* **88**, 712.

Levrault R.M.: 1988, *Astrophys. J.* **330**, 897.

Maitzen H.M.: 1987, *Hvar Obs. Bull.* **11**, 1.

Maitzen H.M., Pavlovski K.: 1987, *Astron. Astrophys.* **178**, 313.

Mathews J.M., Kreidl T.J., Wehlau W.H.: 1988, *Publ. Astron. Soc. Pacific* **100**, 255.

McNamara B.J.: 1987, *Astrophys. J.* **312**, 778.

Mermilliod J.C.: 1983, *Astron. Astrophys.* **128**, 362.

Micela G., Sciortino S., Vaiana G.S., Schmitt J.H.M.M., Stern R.A., Harnden F.R., Rosner R.: 1988, *Astrophys. J.* **325**, 798.

Molaro P., Morossi C., Ramella M., Franco M.: 1984, in *Proc. 4th Europ. IUE Conf.*, ESA SP–218, 223.

North P.: 1984, *Astron. Astrophys.* **141**, 328.

North P.: 1987, *Astron. Astrophys. Suppl. Ser.* **69**, 371.

Pallavicini R., Golub L., Rosner R., Vaiana G.S., Ayres T., Linsky J.L.: 1981, *Astrophys. J.* **248**, 279.

Parenago P.P.: 1954, *Trans. Sternberg Astr. Inst.*, Vol. 25.

Perez M.R., Thé P.S., Westerlund B.E.: 1987, *Publ. Astron. Soc. Pacific* **99**, 1050.

Perez M.R., Joner M.D., The' P.S., Westerlund B.E.: 1989, *Publ. Astron. Soc. Pacific* **101**, 195.

Polidan R.: 1987, in R. Stalio, L.A. Willson (eds.) *Pulsation and Mass Loss in Stars*, Ap. Sp. Sci. Libr. **148**, 275.

Poretti E.: 1989, private communication.

Poretti E., Antonello E.: 1988, *Astron. Astrophys.* **199**, 191.

Praderie F.: 1981, *Astron. Astrophys.* **98**, 92.

Ramella M., Gerbaldi M., Faraggiana R., Bohm C.: 1989, *Astron. Astrophys.* **209**, 233.

Rydgren A.E., Vrba F.J.: 1987, *Publ. Astron. Soc. Pacific* **99**, 482.

Sadakane K., Nishida M.: 1986, *Publ. Astron. Soc. Pacific* **98**, 685.

Sagar R., Joshi U.C.: 1983, *Mon. Not. Roy. Astron. Soc.* **205**, 747.

Schmidt E.G., Parsons S.B.: 1984, *Astrophys. J.* **279**, 202.

Seitter W.C., Duerbeck H.W.: 1982, in *Landolt Börnstein*, New Series, Group VI, **2–b**, Springer–Verlag, p. 269.

Shore S.N., Brown D.N.: 1986, in *New Insights in Astrophysics*, ESA SP–263, p. 361.

Shore S.N., Brown D:N., Sonneborn G.: 1988, in *A Decade of UV Astronomy with IUE*, ESA SP–281, p. 339.

Shu F.H., Adams F.C., Lizano S.: 1987, *Annu. Rev. Astron. Astrophys.* **25**, 23.

Simon T., Cash W., Snow T.P.: 1985, *Astrophys. J.* **293**, 542.

Slettebak A., Carpenter K.G.: 1983, *Astrophys. J. Suppl. Ser.* **53**, 869.

Slovak M.H.: 1978, *Astrophys. J.* **223**, 192.

Smith B.A., Fountain J.W., Terrile R.J.: 1988, *Bull. Am. Astron. Soc.* **20**, 875.

Smith M.A.: 1973, *Astrophys. J. Suppl. Ser.* **25**, 277.

Stahler S.W.: 1988, *Publ. Astron. Soc. Pacific* **100**, 1474.

Stern R.A., Zolcinski M., Antiochos S.K., Underwood J.H.: 1981, *Astrophys. J.* **249**, 647.

Teays T.J., Schmidt E.G., Fracassini M., Pasinetti Fracassini L.E.: 1989, *Astrophys. J.* **343**, 916.

Tomkin J., Popper D.M.: 1986, *Astron. J.* **91**, 1428.

Valtier J.C., Baglin A., Auvergne M.: 1979, *Astron. Astrophys.* **73**, 329.

Vidal–Madjar A., Hobbs L.M., Ferlet R., Gry C., Albert C.E.: 1986, *Astron. Astrophys.* **167**, 325.

Vogel S.N., Kuhi L.V.: 1981, *Astrophys. J.* **245**, 960.

Waelkens C., Rufener F.: 1983, *Hvar Obs. Bull.* **7**, 301.

Walker H.J., Wolstencroft R.D.: 1988, *Publ. Astron. Soc. Pacific* **100**, 1509.

Walter F.M.: 1983, *Astrophys. J.* **274**, 794.

Walter F.M., Schrijver C.J., Boyd W.: 1988, in *A Decade of UV Astronomy with IUE*, ESA SP–281, **1**, 323.

Warner J.H., Strom S.E., Strom K.M.: 1977, *Astrophys. J.* **213**, 427.

Weiss W.W.: 1983, *Hvar Obs. Bull.* **7**, 263.

Whitelock P.A., Feast M.W., Catchpole R.M.: 1989, *Mon. Not. Roy. Astron. Soc.* **238**, 7p.

Willson L.A., Bowen G.H., Struck–Marcell C.: 1987, *Comments Astrophys.* **12**, 17.
Wolff S.C.: 1981, *Astrophys. J.* **244**, 221.
Wolff S.C.: 1983, *The A–type Stars: Problems and Perspectives*, NASA SP–463.
Wolff S.C., Boesgaard A. M., Simon T.: 1986, *Astrophys. J.* **310**, 360.
Wolff S.C., Edwards S., Preston G.W.: 1982, *Astrophys. J.* **252**, 322.
Zolcinski M.S., Antiochos S.K., Stern R.A., Walker A.B.C.: 1982, *Astrophys. J.* **258**, 177.

# PULSATION STUDIES OF A 1.8 M$_\odot$ DELTA SCUTI MODEL


J. A. Guzik and A. N. Cox
*Los Alamos National Laboratory*
*Los Alamos, NM  87545*

D. A. Ostlie
*Department of Physics*
*Weber State College*
*Ogden, UT  84408*



ABSTRACT.  A 1.8 M$_\odot$ stellar model was evolved into the Delta Scuti instability region of the HR diagram, and used to calculate the nonadiabatic periods and growth rates of linear radial and non-radial pulsations.  Comparisons with earlier partially-analytical calculations by Lee (1985) show good agreement.  The radial modes of three 60-zone envelope models based on this same evolution model are also being studied using a nonlinear hydrodynamics code with time-dependent convection, in an attempt to reproduce the result of Stellingwerf (1979) that the amplitudes of such models grow without bound.  We find that the *linear* growth rate of the fundamental mode is very small, a few parts in a million per cycle, and that care must be taken to model the envelope to a depth that includes most of the deep radiative damping.  At present, our nonlinear calculations can detect growth rates only in excess of $10^{-4}$, but do not show a tendency for the amplitudes to grow.  Perhaps including time-dependent convection and/or modeling deeper envelopes with additional radiative damping will limit the pulsation amplitudes to observed values.


## 1. Introduction

This paper outlines work in progress to study the oscillation properties of Delta Scuti models and compare results with those of other authors.  We were motivated in part by the work of Stellingwerf (1979), who found from nonlinear hydrodynamic calculations that the amplitudes of his Delta Scuti models grow without bound (the "main-sequence catastrophe"), unless a large amount of artificial viscosity is included.  Stellingwerf suggested that some mechanism neglected in the calculations, such as mass loss or convection, limits the amplitudes to observed values.  The "main-sequence mass loss hypothesis" of Willson, Bowen and Struck-Marcell (1987) proposes substantial pulsation-driven mass loss from stars in the Delta Scuti instability region, so confirmation of Stellingwerf's result would lend plausibility to their mass loss mechanism.

## 2. Evolution

A 1.8 M$_\odot$ model was evolved using the Iben evolution code, with initial composition




*L. A. Willson and R. Stalio (eds.), Angular Momentum and Mass Loss for Hot Stars, 123–126.*
© 1990 *Kluwer Academic Publishers. Printed in the Netherlands.*

$(Y, Z) = (0.28, 0.02)$ and mixing length/pressure scale height $\alpha_{Iben} = 1.25$. The model at age 0.94 Gyr resides in the Delta Scuti instability region, in the shell hydrogen-burning phase, with $L \simeq 23\,L_\odot$ and $T_{eff} \simeq 7300$ K. The hydrogen-depleted core comprises the inner $\sim 10\%$ of the mass of the model.

## 3. Linear Non-Adiabatic Periods and Growth Rates

### 3.1. RADIAL AND NON-RADIAL MODES OF 1600-ZONE CORE + ENVELOPE MODEL

This evolution model was used to make a 1600-zone model for nonadiabatic calculations of the periods and growth rates of radial and non-radial modes in the linear approximation. Figure 1 shows the growth rates vs. $\omega$ ($2\pi/$period) calculated for the unstable radial and $l=1$ non-radial modes. Our calculations agree qualitatively with Lee's (1985) nonadiabatic numerical envelope + analytical core solutions (Fig. 2) for a 2 $M_\odot$ Delta Scuti model with $L = 27\,L_\odot$, $T_{eff} = 7080$ K, and $(Y, Z) = (0.27, 0.03)$.

### 3.2. RADIAL MODES OF THREE 60-ZONE ENVELOPE MODELS

Toward verifying Stellingwerf's results, three 60-zone envelope models based upon the structure of our evolved model were made to study their non-linear behavior. Table 1 summarizes the properties of these models. Model 1 is a deep purely-radiative envelope model, extending to where hydrogen burning has begun to affect the original composition; Model 2 is a shallow radiative model, intended to reproduce Stellingwerf's higher linear growth rates; Model 3 is a deep convective model, intended for hydrodynamic calculations with time-dependent convection (Ostlie 1989). The effective temperature of this model has been increased to lessen the amount of luminosity carried by convection, and make the hydrodynamic calculations more tractable. Table 2 lists the calculated periods and *linear* growth rates of several radial modes of each model. We note that the deep radiative model has a fundamental mode linear growth rate an order of magnitude smaller than the shallow model, due to the inclusion of deeper layers contributing to radiative damping. This extra damping may help limit the pulsation amplitudes.

## 4. Nonlinear Nonadiabatic Hydrodynamic Calculations

We have begun nonadiabatic hydrodynamic calculations using the three models above. We have taken care to limit the damping from artificial viscosity to an order of magnitude less than the driving expected from the linear calculations. At present we see no tendency for the amplitudes to grow in any of the models. Since perhaps a hundred thousand hydro cycles would be needed to see a change in amplitude if the small linear growth rates ($\sim 10^{-6}$ per period) applied, we use the Stellingwerf periodic solution method to reach limiting amplitudes after only a few dozen trial periods (see Cox 1990). At present, analysis of the Floquet matrix obtained from this method shows that we can only detect growth rates in excess of $10^{-4}$ per period; work is in progress to improve this accuracy. It is possible that including time-dependent convection and/or modeling deeper envelopes which incorporate regions of additional radiative damping will limit the pulsation amplitudes to observed values.

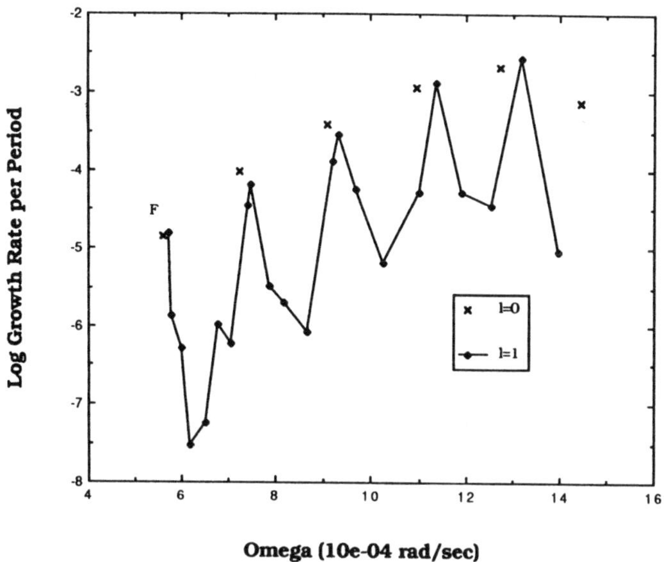



Figure 1. Calculated radial and $l=1$ non-radial growth rates for unstable modes of 1.8 M$_\odot$ model.

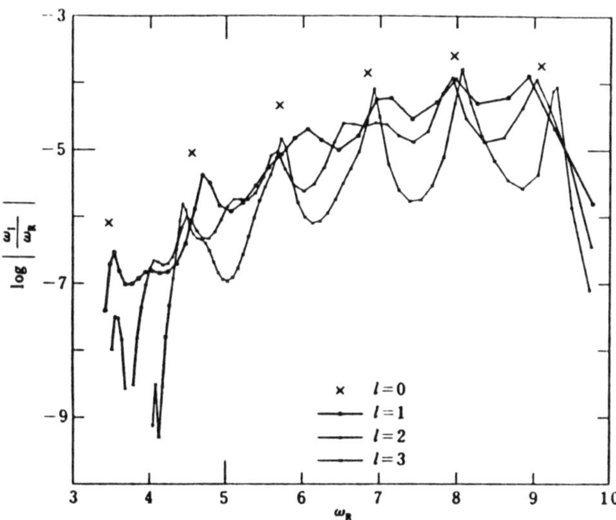



Figure 2. Lee's (1985) growth rate calculations for unstable modes of 2 M$_\odot$ model (ordinate differs by $4\pi$, and abcissa differs by $GM/R^3 = 1.38 \times 10^{-4}$ from Fig. 1).

126

TABLE 1. Model properties

| MODELS | $T_{eff}$ (K) | $T_{env.\ base}$ $(10^6$ K) | Mass$_{env.}$ $(M_\odot)$ | Radius Fraction |
|---|---|---|---|---|
| 1 Deep radiative | 7263 | 8.67 | 1.3 | 0.91 |
| 2 Shallow radiative | 7263 | 1.36 | 0.02 | 0.43 |
| 3 Deep convective | 7800 | 8.82 | 1.2 | 0.87 |

TABLE 2. Linear periods and growth rates of Models 1, 2 and 3

| | Period (days) | | | Growth Rates ($10^{-4}$) | | |
|---|---|---|---|---|---|---|
| Mode | 1 | 2 | 3 | 1 | 2 | 3 |
| Fund. | 0.130 | 0.117 | 0.105 | 0.025 | 0.13 | 0.068 |
| 1H | 0.100 | 0.087 | 0.080 | 0.47 | 2.1 | 0.97 |
| 2H | 0.082 | 0.068 | 0.065 | 2.0 | 5.6 | 5.3 |
| 3H | 0.068 | 0.056 | 0.055 | 3.5 | -15.0 | 16.7 |
| 4H | 0.059 | 0.047 | 0.047 | -3.08 | -125. | 40.2 |

## References


Cox, A. N., 'Pulsations of Delta Scuti stars', in Proceedings of Fifth Florida Workshop in Nonlinear Astronomy, Nonlinear Astrophysical Fluid Dynamics, ed. J. R. Buchler, 1990, in press.

Lee, U., 'Stability of Delta Scuti stars against nonradial oscillations with low degree $l$', P. A. S. J. **37**, 279 (1985).

Ostlie, D. A., 'Time-dependent convection in stellar pulsation', in The Numerical Modeling of Nonlinear Stellar Pulsations: Problems and Prospects, ed. J. R. Buchler, (Dordrecht: Kluwer Academic Publishers) 1989, in press.

Stellingwerf, R., 'Nonlinear Delta Scuti models: The main-sequence catastrophe?' in Lecture Notes in Physics: Nonradial and Nonlinear Stellar Pulsation, v. 125, p. 50 (1979).

Willson, L. A., Bowen, G. H. and Struck-Marcell, C., 'Mass loss on the main sequence', Comments in Astrophysics **12**, 17 (1987).

# Post-Main Sequence Evolution of Binary Am Stars


JON HAKKILA AND JAMES N. PIERCE
*Department of Mathematics, Astronomy, and Statistics*
*Mankato State University, Mankato, MN  56002*



ABSTRACT.  Similar post-main sequence evolution is suspected for most binary Am systems, based on orbits and masses of a well-studied sample.  More than 2/3 of these stars should begin Roche lobe overflow before the onset of helium burning.  It is suggested that Roche lobe overflow, followed by mass loss via a stellar wind from the evolving primary, can lead to the formation of a more widely-separated system comprised of a white dwarf and a main sequence/evolved companion.  A simple dynamical model demonstrates how many barium star orbital characteristics might result from Am binary evolution.


## 1.    Introduction

The Am stars are main sequence A stars that have earlier spectral types as obtained from calcium lines than they do from metal lines. They exhibit overabundances of heavy elements and underabundances of light elements such as Ca and Sc (Smith (1971)).  Maximum rotational velocities of 100 km/sec appear important to any explanation of their peculiar spectral properties (Abt and Moyd (1973)).

Diffusion is the most widely-accepted mechanism (Michaud (1980)) by which Am stars are believed to remove smaller cross-section elements from their surfaces and transport larger ones outward from their interiors.  Slow rotation is necessary for this, as atmospheric rotational turbulence tends to overpower weaker diffusive forces.

Most Am stars (75%) belong to binary systems, with the majority being spectroscopic binaries (Abt and Levy (1985)).  Apparently all A4 to F1 dwarfs in binary systems with orbital periods between 2.5 and 100 days show Am characteristics (Abt and Bidelman (1969)).  Close orbits and massive companions (generally $\geq$ one solar mass) indicate that rotational (tidal) braking commonly leads to slow Am rotation.

What happens to a typical Am binary system after main sequence burning? Based upon kinematic ages and binary orbital characteristics, Hakkila (1989) has suggested that evolution of some binary Am systems can lead to the formation of barium stars.  Barium stars are post-main sequence G and K stars of intermediate-disk ages exhibiting s-process




*L. A. Willson and R. Stalio (eds.), Angular Momentum and Mass Loss for Hot Stars, 127–130.*
© 1990 *Kluwer Academic Publishers. Printed in the Netherlands.*

element enhancement. They apparently all belong to binary systems of intermediate period with mass functions sharply peaked near 0.041 (McClure (1983,1989)), suggesting that the low mass companions are stars with evolutions closely linked to the barium star (some identified as white dwarfs). Mass transfer in an evolved binary system is often used to explain barium star peculiarities, as the evolved primary either filled its Roche lobe and dumped much of its mass onto the secondary, or transferred mass via a stellar wind (Boffin and Jorissen (1988)). The secondary became more massive and was enriched by s-process elements presumably formed during the more rapid primary evolution. Although the peak in the mass function supports the mass dumping model, the average orbital eccentricities of barium star systems are only slightly less than those of binary G and K giants (Webbink (1986)), which is more suggestive of stellar wind transfer. A stellar wind also appears more likely to smoothly transfer abundance peculiarities than does mass dumping.

Hakkila's suggestion of an Am star/barium star evolutionary relationship is based upon general assumptions of orbital expansion. A more thorough test of an evolutionary relationship may be made by applying results of evolutionary models to typical Am systems.

## 2.    Procedure

The binary Am stars observed by Abt and Levy (1985) comprise a sample chosen to minimize selection effects. These stars also have well-known mass functions and orbital characteristics, so 35 systems are used to study evolutionary tracks of typical binary Am stars.

Primary masses are obtained from B-V colors, as these are found to correlate well with mass for known Am eclipsing binary systems. Although some errors result from this assumption, these errors will be minimized due to the small range of Am primary masses.

Secondary masses are obtained from orbital parameters assuming $\sin(i)$ values. Many systems have either measured $\sin(i)$ values or limits. The remaining inclinations are unknown, so a mean statistical value of $32°$ is used, based upon the assumption that orbit measurements show a bias toward high $\sin(i)$ values ($i \geq 45°$).

Maximum evolutionary sizes are estimated from stellar masses using the primary's Roche lobe size at periastron. The most advanced evolutionary stage of the primary before mass dumping begins is found from stellar evolutionary models of Lattanzio (1986). Further evolution of the binary depends on separation and mass loss mechanism.

## 3.    Results

Mass transfer is apparently an important step in the evolution of a binary Am star. Some 69% of the 35 binary Am stars will apparently undergo case B mass transfer (before the onset of helium burning). Only 9% should undergo case A (main sequence) mass transfer, while 22%

will experience mass transfer after the onset of helium burning (case C). Around 60% of the binary Am stars undergoing case B mass transfer will do so after a convective envelope develops.

Evolution of stars in binary systems is not well-understood, as theoretical models do not yet explain orbits of many evolved systems. Evolutionary models of wide binaries properly predict that mass loss leads to increased orbital separations (Valls-Gabaud (1988)), whereas those of close to intermediate binaries often predict that orbital separations decrease (Thomas (1977)). This contradicts observational evidence of Oswalt and Sion (1989) showing that orbital expansion has generally occurred for intermediate binaries containing a white dwarf and a main sequence star. Resolution of this problem for binary systems rests upon better modeling and determination of binary parameters (Polidan, this conference).

Although details of this evolution are unknown, it appears that Am systems should evolve into white dwarfs orbiting main sequence/evolved companions with wider orbits than their main sequence counterparts. Roche lobe dumping is not an efficient mechanism for explaining orbital expansion, as primary mass loss during this relatively rapid phase is apparently accompanied by considerable angular momentum loss. However, as hydrostatic equilibrium is destroyed in the primary due to mass removal, initiation of a stellar wind is likely. Mass loss via a stellar wind can occur with little angular momentum loss, and can result in orbital expansion.

## 4. Discussion

Standard theoretical models of binary evolution make it difficult to identify barium star progenitors. Barium star orbital characteristics straddle the line between case B and case C mass transfer, placing in doubt a common mechanism for their formation. Rapid mass loss/dumping followed by a stellar wind could solve this problem for barium star progenitors, as all case B orbits would widen.

This would explain many properties of barium star systems. Roche lobe transfer allows some material to be exchanged, which could result in the peaked barium star mass function. A period of rapid mass loss followed by orbital expansion could decircularize the orbits (Valls-Gabaud (1988)). As pressure from helium burning dissipates the remaining atmosphere, heavy elements produced during core burning could taint the secondary's atmosphere. (Primaries discussed here all have main sequence masses high enough to favor the CNO cycle, which produces the nuclei that supply neutrons for the s-process.)

To test this theory, a simple dynamical model is used to evolve the Am star sample. Primary and secondary masses are made equal after conservative mass dumping, then a stellar wind is assumed to remove all but 25% of the initial primary mass (producing a white dwarf) with no angular momentum loss. Around 50% of the case B systems result in orbital periods (0.23 to 5.0 years), separations (0.5 to 4.5 A.U.), and mass ratios (generally close to the 3.16 average) comparable to those of barium star systems.

This theory also accounts for many differences between types of barium-like stars.  If the primary has too low an initial mass, then its main sequence burning will be via the p-p chain instead of the CNO cycle, and the secondary will receive less mass and fewer s-process elements (such as the CH stars).  More massive stars will eject larger amounts of mass, causing a wider separation with only a slight tainting of the secondary's atmosphere (such as the marginal barium stars).  If the binary components have the proper masses but are too widely separated, then a stellar wind will be unable to taint the secondary's atmosphere, and the secondary may remain a normal giant.

## 5.  Conclusions

Orbital characteristics imply that roughly 2/3 of binary Am systems will undergo case B mass transfer.  Although evolutionary details are not presently understood, it is suggested that mass transfer followed by a stellar wind could transform many Am star systems into barium star systems.  Mass exchange, orbital expansion, and stellar wind accretion would explain many barium star characteristics in terms of existing models.  Since barium stars apparently evolve from closer binaries with main sequence masses large enough to favor the CNO cycle, there should be considerable overlap between Am stars and barium star progenitors.

## 6.  References


Abt, H.A. and Bidelman, W.P. (1969), *Astrophys. J.* **158**, 1091-1098.
Abt, H.A. and Levy, S.G. (1985), *Astrophys. J. Suppl. Ser.* **59**, 229-247.
Abt, H.A. and Moyd, K.I. (1973), *Astrophys. J.* **182**, 809-816.
Boffin, H.M.J. and Jorissen, A. (1988), *Astron. Astrophys.* **205**,155-163.
Hakkila, J. (1989), *Astron. J.* **98**, 699-715.
Lattanzio, J.C. (1987), *Astrophys. J.* **311**, 708-730.
McClure, R.D. (1983), *Astrophys. J.* **268**, 264-273.
McClure, R.D. (1989), in H.R. Johnson and B. Zuckerman (eds.), *Evolution of Peculiar Red Giant Stars (I.A.U. Colloquium No. 106)*, Cambridge University, Cambridge, 196-204.
Michaud, G. (1980), *Astron. J.* **85**, 589-598.
Oswalt, T.D. and Sion, E.M. (1989) in G. Wegner (ed.), *White Dwarfs*, Springer-Verlag, Berlin, 454-457.
Smith, M.A. (1971), Astron. Astrophys. **11**, 325-344.
Thomas, H.-C. (1977), *Ann. Rev. Astron. Astrophys.* **15**, 127-151.
Valls-Gabaud, D. (1988), *Astrophys. and Space Sci.* **142**, 289-304.
Webbink, R.F. (1986), in K.-C. Leung and D.-S. Zhai (eds.), *Critical Observations versus Physical Models for Close Binary Systems*, Gordon and Breach, New York, in press.

# A STATISTICAL STUDY OF MAIN SEQUENCE A AND F STARS: TESTING THE MAIN SEQUENCE MASS LOSS HYPOTHESIS


B. M. PATTEN [1] AND L. A. WILLSON
*Astronomy Program*
*Department of Physics, Iowa State University*
*Ames, Iowa 50011*
*USA*


## 1. Introduction

The possibility that main sequence stars of spectral types A and F could lose evolution-altering amounts of mass due to a combination of pulsation and rotation has been suggested by Willson, Bowen, and Struck-Marcell (=WBS, 1987). These stars would evolve down the main sequence on a timescale that is less than or on the order of the nuclear evolution timescale. The mass loss process would end when the pulsation ends and/or when the rotation is braked.

## 2. Database Development

This work is based on a database assembled from four well known astronomical catalogs; The Bernacca-Perinotto Catalog of Stellar Rotational Velocities (Bernacca and Perinotto 1970, 1971, 1973), the Michigan Spectral Catalog (Houk 1978, 1982, Houk and Cowley 1975, Houk and Smith-Moore 1988), the Bright Star Catalog (Hoffleit 1982), and the Supplement to the Bright Star Catalog (Hoffleit et al. 1983). There are approximately 7900 total main sequence stars in the database in the spectral type range of B5 to F7. An effort has been made to separate the stars into categories consisting of main sequence single stars, main sequence binary stars (stars with close companions with orbital periods of about 50 days), and spectrally peculiar stars.

## 3. A Deficiency Of Main Sequence A Stars?

A statistically significant deficiency of stars is found in the spectral type range where a deficit is predicted by the main sequence mass loss hypothesis. This deficit is partly due to the Böhm-Vitense effect (Böhm-Vitense and Canterna 1974), but also translates into a "dip" in the mass function between 1.2 and 1.9 solar masses for standard calibrations of mass versus spectral type, although nonstandard maps of mass versus spectral type can be found that are consistent with measured masses and minimize the dip, see figure 1.

---

[1] Current address: *Institute for Astronomy, 2680 Woodlawn Drive, Honolulu, Hawaii 96832*




*L. A. Willson and R. Stalio (eds.), Angular Momentum and Mass Loss for Hot Stars, 131–134.*
© 1990 *Kluwer Academic Publishers. Printed in the Netherlands.*

132

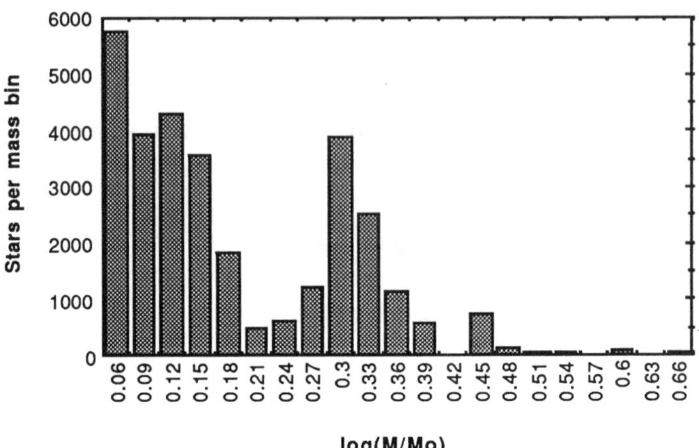

*Figure 1. Mapping spectral type into mass bins using a mass to spectral type relationship derived from eclipsing spectroscopic binaries (Habets and Heintze 1981). The dip in this plot is in the range where a deficiency of A stars is predicted by the mass loss hypothesis.*

If the mean v sin i for each mass bin is examined, see figure 2, it is found that there is a region of apparently lower than expected mean v sin i associated with the A type stars. This effect could be due to the evolution due to mass loss of the most rapidly rotating A stars from this region. Note that the spectrally peculiar stars (Am and Ap stars, for example) are not included in figure 2.

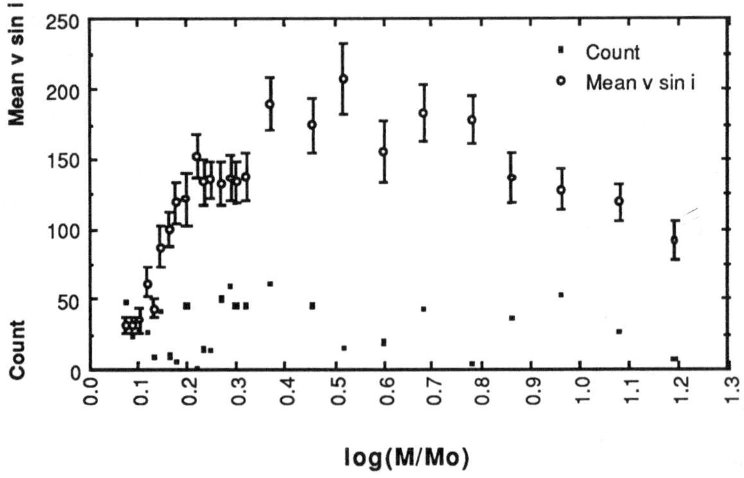

*Figure 2. Stars with measured v sin i are binned according to mass and the mean v sin i for each mass bin is determined. In approximately the same range as the dip in figure 1, there appears to be lower than expected mean v sin i for each bin. The number of stars in each bin is indicated on the plot.*

## 4. An Accumulation Of Equal Mass Binaries

Data from Popper (1980) and numerous other sources on eclipsing, double-line spectroscopic binaries show an apparent accumulation of equal mass binaries among the late F type stars. This accumulation would be expected if binary pairs initially in the instability strip are migrating towards approximately the same same final mass through mass loss.

## 5. An Infrared Survey

A survey of the IRAS Point Source Catalog using the existing database as a positional reference yielded clean IRAS flux densities for approximately 10% of the input stellar positions. Standard IRAS calibration allowed the determination of infrared magnitudes and V-[ $\lambda$] colors for each of the IRAS bands. An intrinsic relation found empirically by Waters et al. (1987) between infrared colors and B-V allowed the determination of color excesses for each of the IRAS bands for each star (though in practice only the 12, 25, and 60$\mu$m bands proved reliable enough to use). A star is defined to have a color excess if it lies $\geq$0.5mag above the intrinsic relationship for that V-[ $\lambda$] color.

It is found that 67 of the 370 (or 18%) main sequence single stars in the spectral type range of A0 to F7 with clean IRAS detections show a color excess in at least one IRAS band. 18% appears to be a large percentage of what were assumed to be otherwise normal stars. Similarly, 8 of 75 (or 11%) stars in the binaries category, 3 of 66 (or 5%) stars in the spectrally peculiar category, and 8 of 144 (or 7%) stars just above the main sequence are showing an excess in at least one IRAS band.

If these color excesses are related to mass loss, the above results are reasonable. The close binaries should experience tidal braking making the mass loss process less effective for these stars if rapid rotation is the key. The spectrally peculiar stars are for the most part all slow rotators and would not be expected to participate in evolution altering amounts of mass loss for the same reason as the close binaries.

144 of the main sequence single stars with clean IRAS detections have v sin i measurements. If these stars are binned in terms of log(v sin i) and are plotted by the fraction of stars in each bin showing a color excess versus log(v sin i) we find that there is a statistically significant, non-zero, positive slope for a simple linear regression analysis, see figure 3. This seems to indicate a correlation between rapid rotation and infrared color excess. This result is reasonable in terms of the mass loss hypothesis in that the most rapidly rotating stars should lose mass the most effectively.

## 6. Conclusions

Though there are alternate explanations for many of these phenomena, mass loss on the main sequence seems to explain each of them. At the present, the indications are that mass loss on the main sequence remains a viable hypothesis. Further work is being carried out on many of the phenomena reported here, and new avenues to test this hypothesis are being explored.

Support for this work from NASA grants NAG5-707 and NAGW 1364 is acknowledged.

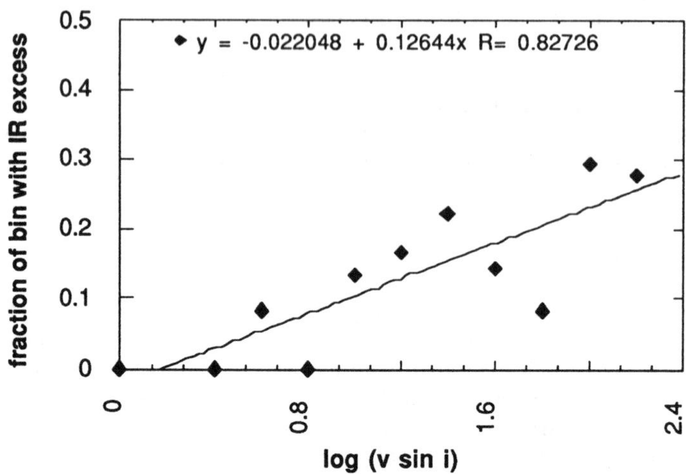


*Figure 3. Stars with measured v sin i and clean IRAS detections are binned according to log(v sin i) and are plotted against fraction of stars in each bin showing infrared color excess in at least one IRAS band. The solid line is the result of a simple linear regression.*

## 7. References

Bernacca, P. L. and Perinotto, M.   1970, Contrib. Oss. Asiago, No. 239.
     "                                  1971,            "            , No. 250.
     "                                  1973,            "            , No. 294.
Böhm-Vitense, E. and Canterna, R.   1974, Ap. J., **194**, 629.
Habets, G. M. H. J. and Heintze, J. R. W.   1981, Astron. Astrophys. Supp., **46**, 193.
Hoffleit, D.   1982, The Bright Star Catalogue, 4th ed. (New Haven:  Yale University Observatory).
Hoffleit, D., Saladyga, M., and Wlasuk, P.   1983, A Supplement to the Bright Star Catalogue (New Haven:  Yale University Observatory).
Houk, N.   1978, Michigan Spectral Catalogue, vol. 2 (Ann Arbor:  University of Michigan).
     "            1982,               "                , vol. 3 (Ann Arbor:  University of Michigan).
Houk, N. and Cowley, A. P.   1975, Michigan Spectral Catalogue, vol. 1  (Ann Arbor:  University of Michigan).
Houk, N. and Smith-Moore, M.   1988, Michigan Spectral Catalog, vol. 4 (Ann Arbor:  University of Michigan).
Popper, D. M.   1980, Ann. Rev. Astron. Astrophys., **18**, 115.
Waters, L. B. F. M., Coté, J., and Aumann, H. H.   1987, Astron. Astrophys., **172**, 225.
Willson, L. A., Bowen, G. H., and Struck-Marcell, C.   1987, Comm. Astrophys., **12**, 17.

# BASIC MAGNETIC ROTATOR THEORY WITH APPLICATION TO THE ANGULAR MOMENTUM DRIVEN WINDS OF B[e] AND WOLF RAYET STARS


J.P. CASSINELLI
Dept of Astronomy, University of Wisconsin, Madison, USA



ABSTRACT.  Some of the fundamental concepts of magnetic rotator theory for stellar winds are reviewed. The distinctions between slow magnetic rotator, fast magnetic rotator and the extreme case of centrifugal magnetic rotator winds are explained.  Special emphasis is given to the stellar properties which determine the energy deposition, mass loss rate, and the terminal wind speed.  Magnetic rotator winds are always hybrid winds, in which there is a "primary" wind mechanism that determines the mass loss in absence of rotation, and the magnetic rotator forces, which can modify the velocity structure and in the most extreme  cases, greatly increase  the mass loss rate.  Results are presented for the "Luminous Magnetic Rotator " wind model in which the primary wind mechanism is  the line radiation driving force. The results are applied to the hypergiant B[e] and the Wolf Rayet stars; two classes of stars which may have a sequential evolutionary connection. For the Wolf Rayet stars the Luminous Magnetic Rotator model may be able to explain the long standing "wind momentum problem" as well as the "spin-down problem" associated with earlier rotating wind models. Constraints on the surface magnetic field as determined from interior considerations are also discussed.


## 1.  Introduction

Strong equatorial winds can be driven from a star if the star is rotating rapidly and if the star has a sufficiently strong surface magnetic field. Elsewhere in these procedings David Friend has presented a paper on Be stars in which the winds are driven by a combination of line radiation forces, centrifugal, and magnetic forces. The model that he describes has also been applied to Wolf Rayet stars by Poe, Friend and Cassinelli (1989) and to the equatorial winds of B[e] stars by Cassinelli, Schulte-Ladbeck, Poe and Abbott(1989). So as to minimize the duplication with these other papers, I will focus on some of the fundamental physical processes that operate in angular momentum driven winds, then I will summarize some of the properties and constraints that have been imposed on magnetic rotator models for the B[e] and Wolf Rayet stars.

135


*L. A. Willson and R. Stalio (eds.), Angular Momentum and Mass Loss for Hot Stars, 135–144.*
© 1990 *Kluwer Academic Publishers. Printed in the Netherlands.*

## 2. Magnetic Rotators

It is useful to consider stars with equatorially enhanced winds as having hybrid winds. There is a "primary wind mechanism" that would drive a wind (although perhaps a very weak one) in absence of the magnetic and centrifugal forces. The magnetic rotator forces are then considered to as providing a  mechanism for advecting angular momentum from the star, for accelerating the radial flow, and for driving a higher mass loss from the star. These additional forces act primarily in the equatorial plane of the star, and the wind models that have been developed are mostly for the equatorial zone alone. I will discuss two examples of these hybrid wind models.  The first is for the case in which the primary mechanism is that of a coronal wind.  In this case it is possible to derive rather simple expressions for the mass loss rate and the terminal wind speed, in the limiting case of a very large magnetic field and rapid rotation. The second model that  will be discussed has the line radiation force as the primary mechanism, and amplified by the magnetic rotator forces.  Stars with this combination  will be referred to here as "Luminous Magnetic Rotators".

There are three subclasses of magnetic rotator winds for which it is useful to have a clear distinction: 1.) the Slow Magnetic Rotator or "SMR", 2.) the Fast Magnetic Rotator or "FMR", and finally , 3.) the extreme FMR for which a distinct name is needed, the "Centrifugal Magnetic Rotator", or "CMR". For the CMR's we will find that there is a clean separation of the dependences of mass loss rate, $\dot{M}$, and terminal velocity $V_\infty$, on the equatorial rotation rate $\Omega$, and surface magnetic field , $B_*$, respectively.

Basic magnetic rotator theory was developed by Weber and Davis (1967), who presented the equatorial wind equations as well as expressions for the conservation of wind energy and angular momentum per unit mass. They found because of the solar magnetic field the time scale at which the solar wind advects away angular momentum is comparable to the sun's age. Therefore magnetic rotator forces are quite important, even though the mass loss rate of the solar wind is extremely small by *stellar*  wind standards. Belcher and MacGregor (1976) made another major step by considering the possible time evolution of the solar rotation speed as influenced by the wind. They identified the transition from the sun's early FMR wind to the SMR wind that the sun has now. They also pointed out the significance of the "Michel Velocity", $V_M$ , that plays an important role in magnetic rotator theory for both low and high luminosity stars.

### 2.1. ENERGY DEPOSITION AND THE MICHEL VELOCITY

In a magnetic rotator wind, energy deposition required for the outflow results from a transfer of energy from the rotating magnetic field.  The flux of energy in the field is given by the Poynting vector $\mathbf{S} = c/4\pi\,(\mathbf{E} \times \mathbf{B})$, where E can be found from the frozen-in field condition.  The energy constant of the wind is

$$\mathcal{E} = \left[ \frac{1}{2} \left( V_r^2 + V_\phi^2 \right) + a^2 \ln \rho - \frac{GM}{r} \right] + \frac{S_r}{\rho\, V_r} \tag{1}$$

where $S_r$ is the radial component of the Poynting vector.  The latter term decreases with radius as energy is transferred to the gas, and at infinity

$$\mathcal{E}_{mag,\infty} = \frac{S_{r,\infty}}{\rho V_r} = \frac{\Omega^2 F_B^2}{\dot{M} V_\infty} = \frac{V_M^3}{V_\infty} \qquad (2)$$

where $F_B = r^2 B_r = $ constant, and $V_M$ is the "Michel Velocity", which can be expressed as

$$V_M^3 = R_*^4 \ B_{r,*}^2 \ \Omega_*^2 / \dot{M} \qquad (3)$$

There are three especially interesting facts concerning the Michel velocity. 1) Note from equation 3 that $V_M$ is determined by conditions at or near the base of the wind, $R_*$, $B_{r, *}$, $\Omega_*$, are the stellar radius, surface magnetic field and equatorial angular velocity. Also, the mass loss rate, $\dot{M}$, in stellar wind theory is determined by conditions near the star in the "sub-sonic" or "sub-critical point" regions. 2) From Equation 2 we can deduce that if $V_M$ is large, there will be a significant amount of energy deposited in the gas outflow. 3) In the large $V_M$ case (FMR regime), the terminal velocity of the flow is equal to the Michel velocity.

$$V_\infty \approx V_M \qquad (4)$$

As a result, the energy per gram at infinity is

$$\mathcal{E}_\infty = \frac{V_\infty^2}{2} + \mathcal{E}_{mag} = \frac{1}{2} V_\infty^2 + V_M^2 \qquad (5)$$

indicating that there remains twice as much energy in the Poynting flux as there is in the kinetic energy of the material outflow.


## 3. The Three Wind Regimes: SMR, FMR and CMR Winds


Three wind regimes can be identified in magnetic rotator theory; the Slow Magnetic Rotator, Fast Magnetic Rotator, and the extreme FMR, or "Centrifugal Magnetic Rotator" regime.

The solar wind is an example of a slow magnetic rotator. This means that it is the nonmagnetic rotator forces, or the "coronal forces", which determine the radial velocity law, $V_r(r)$. This does not mean that the magnetic rotator effects are uninteresting or unimportant. They are responsible for the azimuthal velocity structure $V_\phi(r)$ and the not insignificant angular momentum loss. The radial velocity law $V_r(r)$ appears in the azimuthal momentum equation and in the SMR case, we can take $V_r(r)$ as given by the coronal forces. Let us call the terminal velocity law from these primary forces, $V_w$.

The transition to the Fast Magnetic Rotator region occurs if the Michel Velocity $V_M$ is greater than $V_w$. In the FMR region the radial velocity law, $V_r(r)$ is dominated by the magnetic rotator forces. The condition that $V_M > V_w$ is not a particularly extreme one.

For example, if we take the sun to have its current surface magnetic field, the solar wind would be an FMR wind if the equatorial rotation speed were greater than about 100 km/sec (Belcher and MacGregor 1976); the sun probably had such a rotation speed during its T-Tauri and early main sequence phases.

Nerney (1980) derived a useful expression for the minimum surface magnetic field of a star for it to have a Fast Magnetic Rotator wind. This is found by simply equating the Michel Velocity to the observed wind speed, taking $\dot{M}$ as known observationally, and by assuming $\Omega$ is the critical (maximal value) we get

$$B_{min} = \frac{(\dot{M} V_w^3)^{1/2}}{R_*^2 \, \Omega_{max}} \tag{6}$$

For the sun $B_{min} = 1$ gauss, for Be stars the wind field is also rather small $\approx 20$ gauss while for the WR stars which have very large terminal velocities and mass loss rates $B_{min} \approx 300$ gauss. Most of the solutions described in David Friend's presentation are FMR wind models, so it is clear that the equatorial equations can be solved for the most general case. Nevertheless, I think it is useful to consider the asymptotic behaviour of FMR wind theory, which is here called CMR theory. The CMR winds allow us to isolate the effects that are responsible for the mass loss from these that determine the terminal wind speed.

For the Centrifugal Magnetic Rotator (CMR) case, there are two rather simple concepts that determine the subsonic structure. 1) The subsonic region is co-rotating as a solid body with the stellar equator, i.e. $V_\phi(r) = r\Omega_*$ , with $V_\phi$ increasing with radial distance. 2) The sonic point in CMR theory occurs where $V_\phi(r)$ equals the circular speed. So we get for the sonic point

$$V_\phi(r_s) = r_s \Omega_* = (GM_*/r_s)^{1/2} \tag{7}$$

or letting $\alpha$ be the ratio of $\Omega_* / \Omega_{max}(R_*)$ , where $\Omega_{max} = (GM_* / R_*^3)^{1/2}$ , and we get

$$r_s = 1/\alpha^{2/3} \tag{8}$$

Thus we have one of the rare occurances in stellar wind theory in which the sonic point is known from a basic model parameter, ( $\alpha$ ).

The mass loss rate from the star can now be determined from the product of the velocity, radius, and density at the sonic point, a, $r_s$, and $\rho_s$, respectively. Because of the solid body rotation, a simple expression can be given for the density distribution and we get

$$\dot{M} = 4\pi r_s^2 \rho_s a$$

$$= 4\pi r_s^2 a \rho_0 \exp\left\{ \frac{-GM}{a^2 R_*} \left[ \left(1 - \frac{R_*}{r_s}\right) - \frac{\alpha^2}{2}\left(\frac{r_s^2}{R_*^2} - 1\right)\right] - \frac{1}{2} \right\} \tag{9}$$

Note that $\dot{M}$ is a function of $\alpha$, and that it does *not* involve the stellar magnetic field B explictly, although the field must be sufficient to enforce co-rotation. Given $\dot{M}(\alpha)$ from the relation above, $V_\infty$ can be set equal to the Michel velocity, and since $\dot{M}/\Omega^2$ depends only on $\alpha$, the terminal velocity depends only on B! Thus, there is a clean separation of dependencies of $\dot{M}(\alpha)$ and $V_\infty(B)$.

Figure 1 shows a useful diagnostic diagram for hybrid winds involving magnetic rotator forces. It is a plot of log $\dot{M}$ vs. log $V_\infty$. The "primary" wind mechanism for a given star would provide a specific $\dot{M}$ and $V_\infty$, or a "point", P, on the plot. The Michel velocity relation provides lines of slope = -3 on this plot, for specific fields B (assuming $\alpha$ = 1). Hence, the B = $B_{min}$ line passes through the point P, while Michel velocity lines for larger magnetic fields lie farther to the right on the plot. Let us consider what happens as a model star is "spun up", for a case in which B > $B_{min}$. As $\alpha$ increases the winds enter into the FMR regime, (which means that the terminal speed increases), but with the mass loss rate fixed by the primary mechanism. Eventually, as we spin up the star further, the star enters the CMR regime. Now as $\alpha$ increases, the mass loss rate increases as in equation (9) above, in which it is seen that $\dot{M}$ depends only on $\alpha$. The terminal velocity is then equal to the Michel velocity, which (since $\dot{M}$ and $\alpha$ are already known) is determined only by the stellar magnetic field.

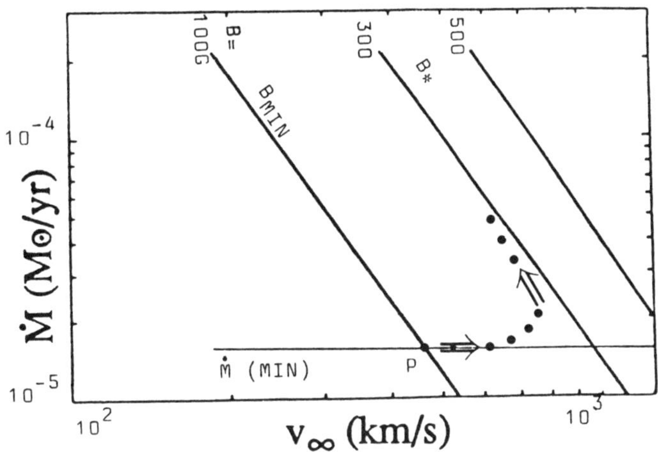


Figure 1. Plot of the logarithm of the mass loss rate versus logarithm of the wind terminal velocity. This figure illustrates the effects of magnetic fields and stellar rotation on a stellar wind. The point P represents the $(\dot{M}, V_\infty)$ that can be produced by the "Primary" wind mechanism. The diagonal lines are iso-magnetic field lines as derived from the Michel velocity relations for $\dot{M}\, V_M^3$ given in the text, in which the rotation rate has been set equal to the critical rate (i.e. $\alpha$ = 1). For stellar magnetic fields smaller than $B_{min}$, there can be negligible increase in $V_\infty$ caused by magnetic rotator effects (i.e. an SMR). However if the star has a field B* > $B_{min}$, increasing the rotation rate from $\alpha = 0$ to $\alpha = 1$, will cause the $\dot{M}$ and $V_\infty$ to change as is shown by the arrows. In the text, stars on the portion of this track moving to the right are said to have FMR winds, while the portion of the track having $\dot{M}$ increase have CMR winds. Note that the minimal mass loss rate is set by the primary wind mechanism.

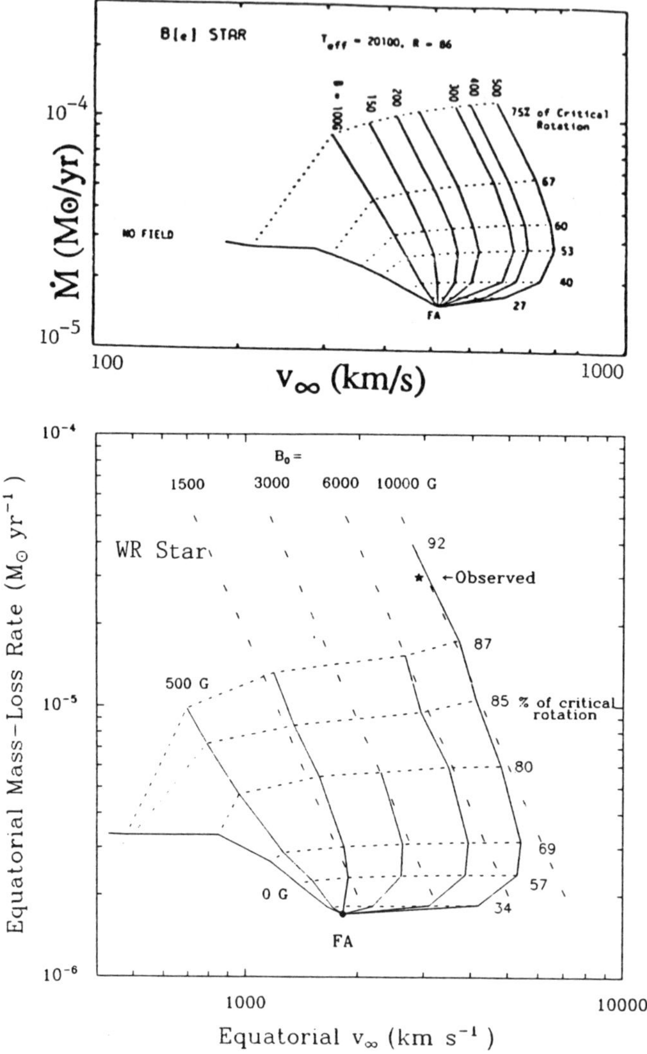



Figure 2. (a) shows an $\dot{M}$ versus terminal velocity for a B[e] star, with $T_{eff}$ = 20, 100° K, R = 86 $R_\Theta$, M = 37.3 $M_\Theta$. The point marked FA refers to the ($\dot{M}$, $V_\infty$) values for a line driven wind model for this star using the Friend and Abbott (1986) theory. The curves show the effect of increasing the rotation rate parameter $\alpha$, and results are shown for surface magnetic fields of 0, through 500 gauss. Note that in the CMR regime ($\dot{M}$ larger than the minimum value), that $\dot{M}$ depends on the rotation rate but not on the magnetic field. The maximum terminal speed on the other hand is determined by the Michel velocity relation. (b) A similar plot, but for a Wolf Rayet star, with M = 13 $M_\Theta$, R = 8.0 $R_\Theta$, and L=3 x $10^5$, from (Poe et. al. 1989). The Velocities and Magnetic fields are larger that for the B[e] star as can be seen from an application of equation 7.

## 4. Luminous Magnetic Rotators: B[e] and Wolf Rayet Stars

Now let us consider the case of a hybrid wind in which the primary mechanism is the line radiation pressure gradient. This case has been discussed for Wolf-Rayet stars by Poe, Friend and Cassinelli (1989), and for B[e] stars by Cassinelli, Schulte-Ladbeck, Poe and Abbott (1989). The results for mass loss versus terminal velocity are shown in Figures 2a and 2b. Several of the conclusions of the previous discussion are readily apparent in these figures. For example, in the CMR regime, the mass loss rate is seen to depend only on $\alpha$, the asymptotic wind speed is set by the Michel velocity relation, and the minimal mass loss rate is set by the primary wind mechanism.

Figure 2 shows that there is a large range of wind properties possible given the freedom to choose the stellar magnetic field $B_*$ and the rotation parameter $\alpha$. It is useful to consider whether some of the mass loss versus terminal velocity space can be ruled out because of physical limits to the stellar properties.

### 4.1 LIMITS ON THE SURFACE MAGNETIC FIELDS

Maheswaran and Cassinelli (1988) have shown that it is possible to limit the range of permissible magnetic fields through stellar interior considerations. This is because rapid rotation produces Eddington Sweet currents, with characteristic velocity $V_C$. If this circulation velocity is greater than the Alfvén speed in the stellar envelope, then the surface magnetic field is submerged. Therefore there is a minimum finite field that can exist at the surface. If the field is less than this minimum, there is effectively zero field in the wind. The combination of the Maheswaran limits with the wind theory $\dot{M}$ vs. $V_\infty$ plots are shown in Figure 4a and 4b. I call these "Cat" diagrams because of the feline shape of the allowed region in the $\dot{M}$ versus $V_\infty$ plots.

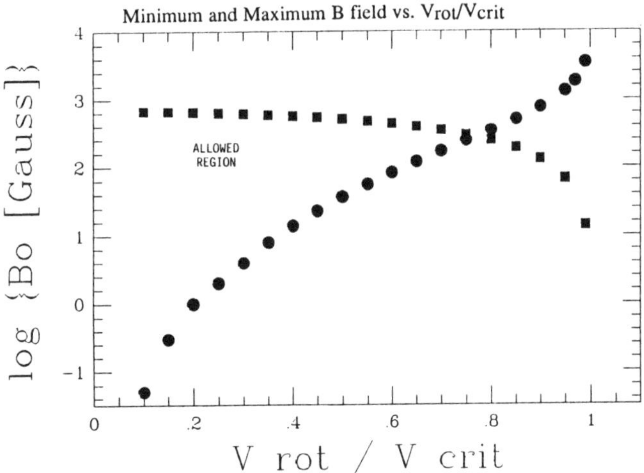


Figure 3. This figure shows the maximum and minimum finite values that are permitted by stellar interior constraints  derived by Maheswaran and Cassinelli (1988). (The results shown are for the B[e] star parameters given in the caption to Figure 2.) The rapid rotation of a star leads to circulation currents which can submerge the surface field unless the Alfvén speed is larger than the circulation speed. The upper limit is set by hydrostatic equilibrium constraints. The surface field of a star should either be approximately zero, or, lie in the allowed zone shown above.

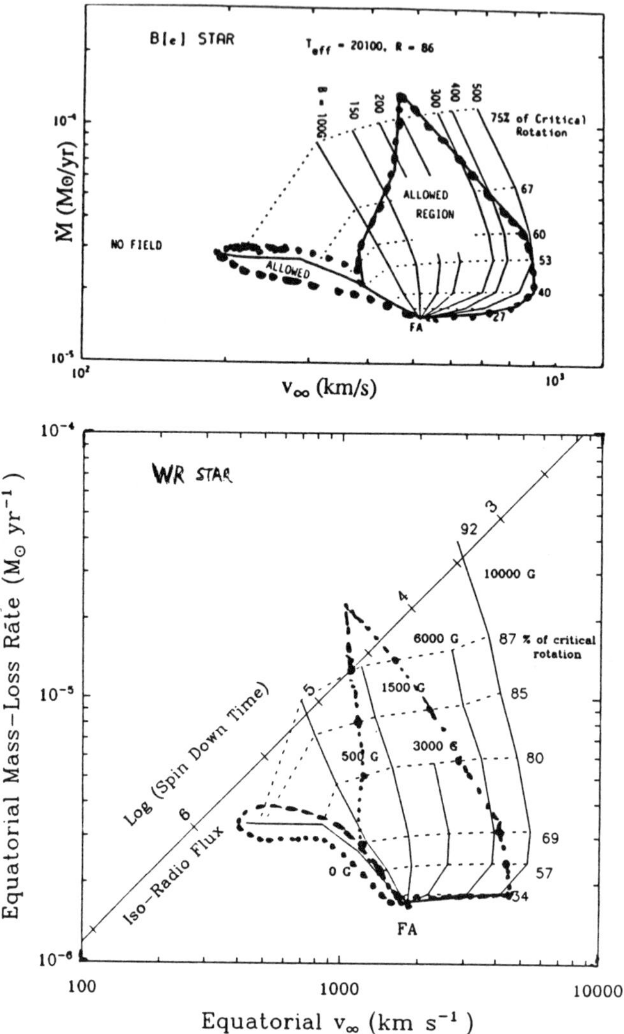



Figure 4. (a) Shows $\dot{M}$ versus $V_\infty$ plots for the B[e] star model from Figure 2 ,but now with the constraints on the field derived from Figure 3, as indicated by the dashed lines. The surface magnetic field must either be zero or in the broad region labelled "Allowed". (b) Shows a similar plot for a Wolf Rayet star, as taken from Poe et al. (1989). The curved lines show the dependence of the wind on the rotation rate and on the surface magnetic fields. The diagonal line, is one of equal radio flux for this star. Hence the large radio flux can be produced either by a star with a fast equatorial wind and a large ($\sim 10^4$ gauss) magnetic field, or by a equatorial wind with a slow speed and a relatively small field. Also indicated along this line are the logarithm of the spin down time, which are seen to range from $10^3$ to about $10^5$ years.

## 4.2 LIMITS AND CONSTRAINTS ASSOCIATED WITH WOLF RAYET WINDS

In the case of Wolf Rayet stars, Poe et al. (1989) considered other constraints on the wind models. Their goal was to determine whether magnetic rotator models could provide an explanation for the well known wind momentum problem of Wolf Rayet stars (for which wind momentum greatly exceeds the stellar photon momentum, ie. $\dot{M} V_{\infty} >> L/c$). The basic idea of their model is that the fast wind speeds, which are determined from observed UV resonance lines, comes from the polar outflow driven by line forces. On the other hand, the large mass loss rates, as inferred from radio observations, are derived from a radio flux that comes primarily from the denser equatorial outflow, which is driven by magnetic rotator forces. Figure 4b illustrates this conclusion. It shows a line labelled "iso-radio flux" on the $\dot{M}$ versus $V_{\infty}$ plot. Models which have an ( $\dot{M}$, $V_{\infty}$ ) somewhere on this line would produce the observed radio flux. A wide range of winds are thereby allowed. The high field ($10^4$ gauss) case was first suggested by Hartmann and Cassinelli (1981), and Cassinelli (1982). However, such a model produces far too rapid a spin down time ( $t < 10^4$ years). Note that spin down times are shown on the iso-radio flux line. Figure 4b shows the combination of the iso-radio flux, spin down, and Maheswaran constraints. It is seen that a WR model with B $\approx 10^3$ gauss $\alpha > .85$, can explain the dominant Wolf-Rayet problems: the momentum problem, the spin down problem, as well as the rotating interior constraints on B. Cassinelli et al. (1989) have presented results of rotational evolutionary calculations and find that the B[e] stars should naturally evolve into WR stars. They argue that it is possible that the B[e] and Wolf Rayet stars have rotation speeds near the maximal value. This is because the stars lose a significant fraction of their mass while in the Luminous Blue Variable phase, and also on their evolutionary track to the left in the HR diagram the stars are overluminous. The combination of the reduced mass and the enhanced luminosity can cause the stars to have a small maximal rotation speed. Hence as the surface of the stars increase their rotational rate as the stars evolve towards higher effective temperatures, their $\alpha$ value can become close to unity. There is observational evidence to to support this picture as is discussed by Cassinelli et al. as well as by Zickgraf elsewhere in these procedings. For example, the B[e] stars show intrinsic polarization and spectral evidence for equatorially enhanced mass loss.

## Summary

The distinctions amongst the various types of magnetic rotators are described, and for the specific case of the CMR, the mass loss rate and wind terminal velocity is shown to be related in a simple way to the rotation rate and surface magnetic field. The results are used to explain the asymptotic behaviour of the Luminous Magnetic Rotator wind models. In combination with a two component picture of the stellar winds, plausible values for the surface magnetic field and stellar rotation rate are found to be able to explain the dominant features associated with B[e] winds, as well as major problems associated with the winds of post- B[e] Wolf Rayet stars.

## References

Barnes, A. 1974, Astrophys. J., 188, 645.
Belcher, J.W. and MacGregor, K.B. 1976 Astrophys. J. 268, 498.

Cassinelli, J.P. and Hartmann, L. 1981, in <u>Wolf-Rayet Stars: Observations, Physics and Evolution,</u> IAU Symp. 99, eds. C.W>H. de Loore and A.J. Willis (Dordrecht: Reidel) p. 173.

Cassinelli, J.P., Schulte-Ladbeck, R.E., Poe, C.H. and Abbott, M. 1989, in <u>Physics of Luminous Blue Variables</u>, IAU Colloquium 113, eds. K. Davidson, H.J.G.L.M. Lamers and A.F.J. Moffat (Dordrecht: Kluwer), p. 121.

Friend, D.B. and Abbott, D.C. 1986, Astrophys. J., <u>311</u>, 701.

Hartmann, L. and Cassinelli, J.P. 1981, Bull. AAS, <u>13</u>, 785.

Maheswaran, M. and Cassinelli, J.P. 1988, Astrophys. J., <u>335</u>, 931.

Nerney, S. 1980, Astrophys. J., <u>242</u>, 723.

Poe, C.H., Friend, D.B., and Cassinelli, J.P. 1989, Astrophys. J., <u>336</u>, 888.

Weber, E.J. and Davis, L., Jr. 1967, Astrophys. J., <u>268</u>, 228.

# THE CONNECTION BETWEEN ROTATION AND THE WINDS OF BE STARS

D. B. FRIEND
Department of Physics
Weber State College
Ogden, Utah 84408
USA

ABSTRACT. It is now clear that Be stars are surrounded by at least two different types of outflowing circumstellar material: a dense, slowly expanding disk, and a fast, but rather tenuous, stellar wind. Given the fact that Be stars are rapid rotators, we should expect that rotation would play a major role in the structure and dynamics of the circumstellar material, creating an outflow which is quite different in the equatorial and polar regions. The observations can be explained, at least qualitatively, by a radiation-driven wind modified by rapid rotation and a weak magnetic field. I will describe how such a wind model is constructed, and how it leads to a wind which is slow and dense in the equatorial regions and fast and tenuous along the poles. I will also describe the limitations of the model and observational evidence which cannot yet be explained by models of this type.

## 1. Introduction

Observations of Be stars over the last several years are now leading us to the idea that their circumstellar environment is composed of two distinctly different flows. The optical emission lines (which define the Be stars as a class) show the presence of a dense region of circumstellar material, which, from the widths of these lines, seems to be confined to a disk at least partially corotating with the star (see Poeckert 1982 for a review). The real evidence that this dense material is in the form of a disk is that the emission from Be stars is polarized at a significant level (Coyne and McLean 1982). Infrared excess measured by IRAS (Waters 1986; Cote and Waters 1987) is also consistent with this dense material being in a disk, though other distributions would also be consistent with the data. The infrared measurements also suggest that this material is slowly expanding outward (at speeds of roughly 10 km/s), based on the fact that the density seems to fall off faster than $1/r^2$, indicative of an outward flow. Many Be stars have now been observed in the ultraviolet with the IUE satellite (see, for example, Grady *et al.* 1987, 1989), and it appears that there is a second type of outflow associated with Be stars. The UV line profiles of highly ionized species show the presence of a more tenuous, but much faster, stellar wind, with velocities on the order of 1000 km/s. This flow is also highly variable, and in many stars disappears completely at certain times (see Barker and Marlborough 1985).

The fact that there are two different types of circumstellar flows surrounding Be stars doesn't immediately tell us that this has anything to do with the rapid rotation of the Be stars. However, much recent evidence suggests that this is indeed the case. Dachs *et al.* (1986) have shown that the strength of the Balmer lines in Be star spectra are correlated with the observed rotational velocity of the star (that is, with v sin i). Briot (1986) has shown that this



*L. A. Willson and R. Stalio (eds.), Angular Momentum and Mass Loss for Hot Stars, 145–157.*

correlation exists with other optical emission lines too.  Waters (1986) has shown that the
infrared excess is also correlated with v sin i.  Vardya (1985) and Nieuwenhuijzen and de Jager
(1988) have tabulated mass loss rates, (based on both infrared and ultraviolet data), that
suggest that the mass loss rate of a Be star wind is correlated with its rotational velocity.
Finally, Grady et al. (1987, 1989) have found that the strengths of the UV P Cygni lines are
correlated with v sin i.  So the observational evidence that the properties of the outflows are
related to the rapid rotation of the Be stars is well established.

In the past, empirical models have been made which try to intrepret the data on Be stars in
terms of a slowly expanding disk of material.  Poeckert and Marlborough (1978) made the first
detailed model based on these ideas.  More recently, Marlborough and Zamir (1984) and
Marlborough (1987) have also tried to explain the existence of the high velocity material in
terms of a faster, hotter, polar flow.  Though these models do a good job in explaining the
observed widths of the Balmer lines, the infrared excess, and the polarization, they do not
address the question of the physical mechanism behind the flow (or flows).  In this paper I
will present a possible physical mechanism for the two different flows, based on the idea of a
rotationally-enhanced radiation-driven stellar wind.


## 2. Radiation-Driven Winds Modified by Rotation and Magnetic Fields

The line radiation-driven wind theory of Castor, Abbott, and Klein (1975; hereafter CAK), as
recently modified by Friend and Abbott (1986) and Pauldrach et al. (1986), has been very
successful in explaining the winds from hot luminous stars.  Main sequence B stars, having
much lower luminosites, are probably not able to initiate such winds, but, as Abbott (1982) has
shown, a radiation-driven wind could be maintained in such stars if some other mechanism
exists to initiate the flow.  Could rapid rotation play such a role in the Be stars?  And how would
it affect the wind once a wind was established?

Friend and Abbott (1986) considered the effect of rotation in a radiation-driven wind from
an O star, and found that the wind is changed in two ways: 1) The mass loss rate *increases* as
the rotational velocity increases, because the centrifugal force (in the frame of the rotating
star) is the dominant outward force near the base of the wind when the rotational velocity is
large.  2) The terminal velocity *decreases* as the rotational velocity increases, partly because
the amount of mass that is being accelerated is now larger, and also because the terminal
velocity in radiation-driven winds scales with the escape velocity, which is reduced by rapid
rotation.  Since the centrifugal force depends on latitude, rotation also introduces a
latitude-dependence in the properties of the stellar wind.  The polar wind should be basically
unaffected by rotation, while the equatorial wind will feel the maximum effect.

If a rapidly rotating star also has open magnetic field lines in the equatorial plane, there
could be another large force acting to accelerate the wind and change its properties: the
Lorentz force that the magnetic field exerts on the ionized, flowing gas.  Friend and MacGregor
(1984) considered the effect of such a  magnetic field on the wind from an O star, and Poe and
Friend (1986) applied the model to Be stars, incorporating the improvements of Friend and
Abbott (1986).  Friend and MacGregor found that a large magnetic field had little effect on the
mass loss rate in the wind, but could substantially enhance the terminal velocity.  In other
words, the magnetic force acts like an 'afterburner' for the stellar wind velocity.  They also
found that the azimuthal velocity of the wind could be greatly enhanced over what would be
expected from conservation of angular momentum.  This effect could be very important for Be
stars, in which the equatorial disk appears to be, at least to some extent, corotating with the
star.

The Friend and MacGregor model treats a rotating, magnetic, radiation-driven wind as a
combination of the Weber and Davis (1967) model for a rotating, magnetic, solar wind and the

CAK model for a radiation-driven wind from a hot, luminous star. The specific features of this model are as follows: 1) The flow is considered in the equatorial plane only, so that meridional flow is ignored. 2) Axial symmetry is assumed, so that the problem is one-dimensional. The velocity field has radial and azimuthal components, but each is a function of radius only. 3) The magnetic field lines are assumed to be open spirals in the equatorial plane. The magnetic field, like the velocity field, has radial and azimuthal components which are functions of just the radius. 4) The wind flow has *two* CAK-type critical points which the soluton must pass through to be a complete wind model. The reader is referred to Friend and MacGregor for the equations of the wind model.

In the next section I will describe how the Friend and MacGregor model can be applied to a Be star wind, and what it implies about the two types of flows that are observed. Most of these results are taken directly from Poe and Friend (1986), though some more recent results will also be described.


## 3. Application of the Wind Model to Be Stars

### 3.1. MODEL RESULTS

To see the effects of adding rapid rotation and open magnetic field lines to a radiation-driven wind model for a Be star, we had to choose a set of stellar parameters representative of a Be star. We chose the parameters of 59 Cygni, since it is a well-studied and apparently typical Be star. It should be emphasized, however, that we are not attemting to model the actual wind of 59 Cygni, since it is extremely variable and our model cannot begin to address such variability (see Doazan *et al.* 1989). The stellar parameters we chose are from Barker (1979): mass = 13 solar masses, luminosity = 7500 solar luminosities, and radius = 6 solar radii. In our model we varied the equatorial rotational velocity from 0 to 540 km/s, and the surface magnetic field strength between 0 and 400 gauss. Note that the critical (or 'break-up') rotational velocity for these stellar parameters is 640 km/s.

Figures 1-5 show the dependence of the mass loss rate, the terminal velocity, the radial velocity law, and the azimuthal velocity law on the rotational velocity and the magnetic field strength. Figure 1 shows the effect of increasing rotational velocity on the mass loss rate. The mass loss rate rises sharply with increasing rotational velocity, as found by Friend and MacGregor (1984) and Friend and Abbott (1986) for O stars. The increase appears to be larger for Be stars, probably because we were able to make models with rotational velocities closer to the critical value. When Poe and Friend (1986) initially made this study, they thought that the mass loss rate might increase without bound based on this figure. However, further analysis of rapidly rotating models without magnetic fields (Poe 1987) showed that there is a theoretical limit to the mass loss rate based on the nature of the critical point of the flow. Whether this result also applies to magnetic models is not yet known. As we see from the curves for different magnetic field strengths, the magnetic field has a much smaller effect on the mass loss rate, as was shown by Friend and MacGregor for an O star model.

Figure 2 is a plot of the terminal velocity of the wind vs. the rotational velocity, with each curve again representing a different magnetic field strength. How the rotational velocity affects the terminal velocity now depends on whether the magnetic field is large or small. For small field strengths, the terminal velocity goes down with increasing rotation rate, but for large field strengths, it incresases with increasing rotation rate. The physical reason for this different behavior will be described later when we discuss figures 6 and 7.

Radial velocity laws ($v_r$ vs. r) are plotted in figures 3 and 4. Figure 3 shows how the radial velocity varies with rotational velocity for a fixed field strength. We see that the entire velocity law is shifted downward as the rotational velocity is increased. In other words, the

148

velocity law becomes *shallower* when the rotation rate goes up. In figure 4 we see the effect of changing the magnetic field strength with a fixed rotational velocity. The magnetic field only affects the velocity at large radii, and has no impact on the velocity near the star. This is the 'afterburner' effect mentioned earlier.

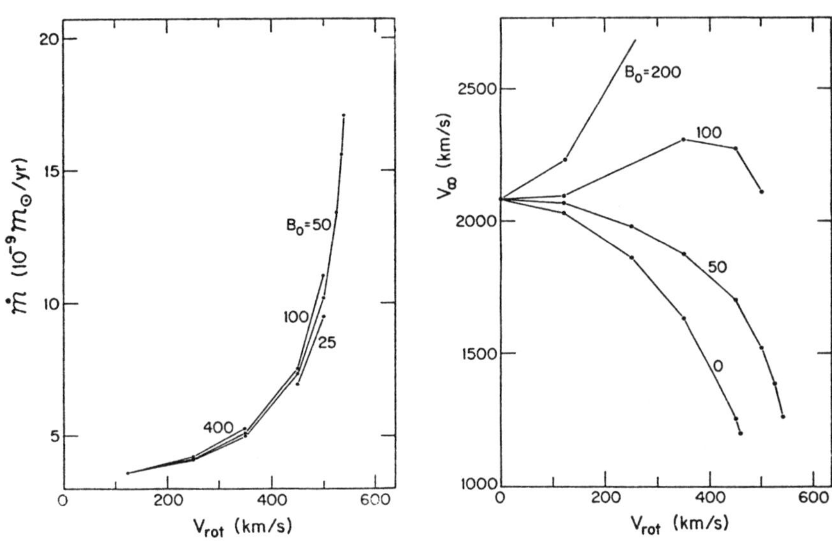


Figure 1. Mass loss rate as a function of rotational velocity. The four curves represent different values of the magnetic field strength.

Figure 2. Terminal velocity as a function of rotational velocity. As in figure 1, the different curves are for different magnetic field strengths.

Figure 5 shows how the azimuthal velocity varies with distance from the star for different values of rotational velocity and magnetic field strength. We now see the main reason why a magnetic field might have interesting consequences for a Be star wind. For low values of rotation rate and magnetic field strength the azimuthal velocity is only slightly enhanced over what would be expected from angular momentum conservation. But for the case of a large rotational velocity (450 km/s) and a large magnetic field (400 gauss) the azimuthal velocity approaches that of a solid body rigidly corotating with the star. As mentioned before, empirical modeling of the optical emission lines suggests that such a corotation may be taking place in the dense equatorial flow around a Be star.

We can understand the above dependences of the stellar wind properties by looking at the forces in the wind. In figures 6 and 7 are plotted the forces in the wind vs. radius: figure 6 is for the case of a small magnetic field strength, and figure 7 is for the case of a dynamically important one (these two models were actually made with O star parameters). Both of these models are for a rapid rotation rate of 400 km/s. In both cases we see that centrifugal force

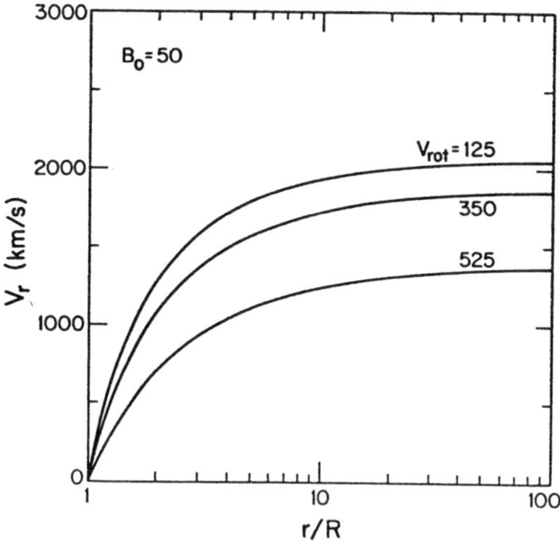

Figure 3. Radial velocity laws for different rotational velocities and a fixed magnetic field strength.

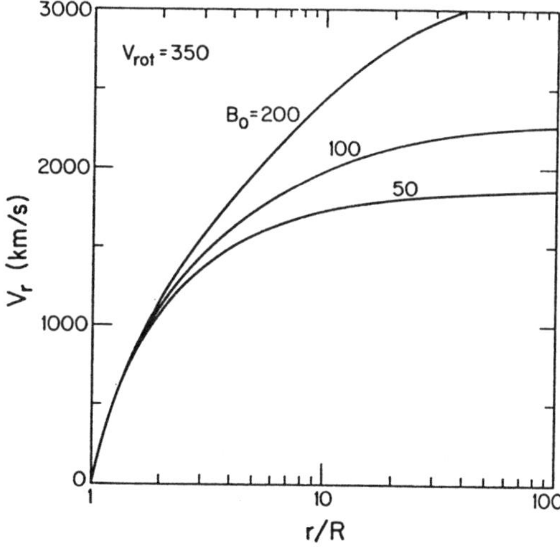

Figure 4. Radial velocity laws for different magnetic field strengths and a fixed rotational velocity.

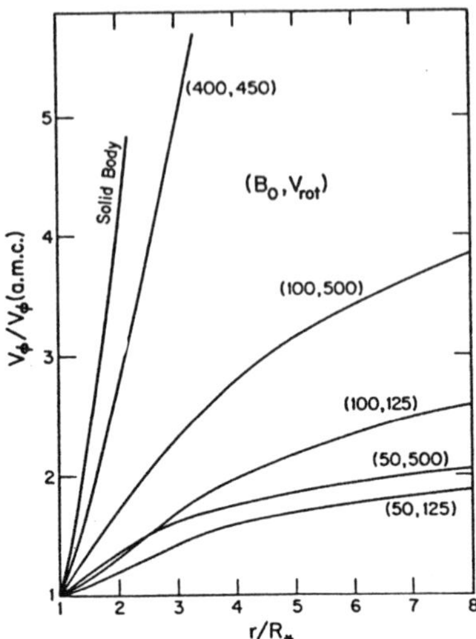



Figure 5. Azimuthal velocity laws (normalized to that expected from angular momentum conservation) for different values of the rotational velocity and magnetic field strength. The azimuthal velocity law for a solid body is shown for comparison.

and the thermal pressure gradient are the largest forces very near the star. Also in both cases, the line radiation force quickly becomes large as we move out from the stellar surface, and dominates the force balance beyond 1.5 stellar radii. There are three things that the large magnetic field does that makes figure 6 and figure 7 very different. First of all, it introduces a large Lorentz force, which is largest at roughly 3-5 stellar radii, boosting the terminal velocity to higher values. It also increases the line radiation force, since any force which increases the velocity gradient also increases the line force (see Abbott 1980). A third thing which the large magnetic field does is to increase the *azimuthal* velocity, which makes the centrifugal force fall off much less rapidly than it does in figure 6. This enhanced centrifugal force can also boost the terminal velocity, which explains why for large field strengths the terminal velocity increases with increasing rotational velocity, as seen in figure 2.

## 3.2. DISCUSSION

We can now see how a radiation-driven wind model enhanced by rapid rotation and open magnetic field lines might explain the properties of the two types of outflows from Be stars. A model with a small magnetic field strength and a large rotation rate has a greatly *enhanced* mass loss rate and a greatly *reduced* terminal velocity, which also means that its equatorial wind will be much *denser* than its polar wind. Since the polar wind should be largely unaffected by rotation and the magnetic field, we can represent the polar wind (to first order) by a nonrotating wind model with the same stellar parameters. We can then compare the

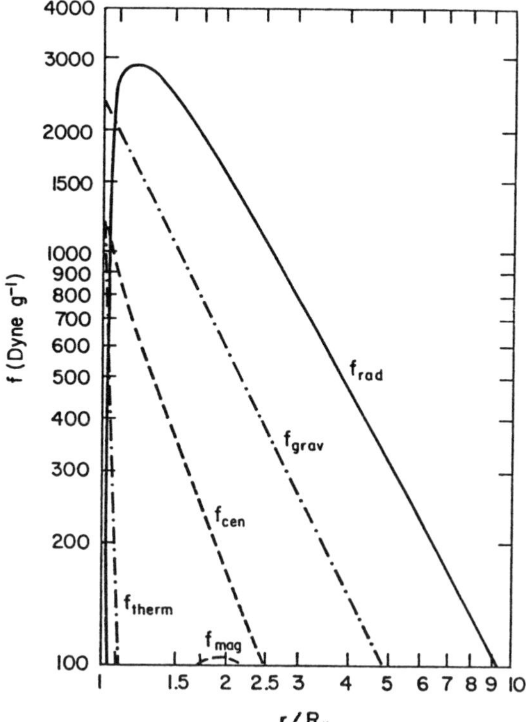



Figure 6. Forces in the wind for an O star wind model with a rapid rotation rate and a dynamically unimportant magnetic field. $f_{rad}$ is the line radiation force, $f_{grav}$ is the gravitational force, $f_{cen}$ is the centrifugal force, $f_{mag}$ is the magnetic force, and $f_{therm}$ is the thermal pressure gradient force.

density in the equatorial wind to the density in the polar wind at a given radius. This comparison is shown in figure 8. Because of the shallower velocity law in the equatorial wind, the density enhancement can be *enormous* near the star. At large radii the density contrast is smaller, approaching a value of about five for this model, as r goes to infinity.

This wind model, with its very different properties in the equatorial plane and along the poles, can explain the two types of outflows seen in Be stars, at least qualitatively. The optical emission lines and the infrared excess come from the dense equatorial flow, while the high velocity UV lines originate in the polar flow. The large density contrast near the star could also explain the polarization seen in Be stars.

There are a few problems with the model that prevent it from agreeing quantitatively with the observations. The first is that even though the velocity in the equatorial wind is lower

152

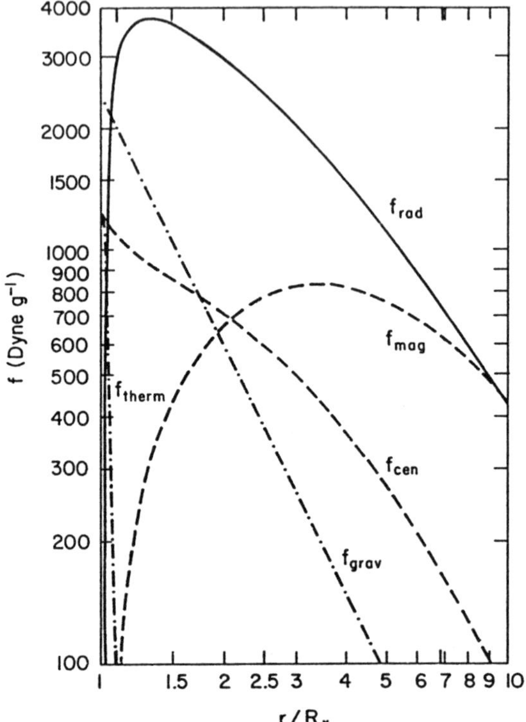

Figure 7. Forces in the wind for the same O star model, but with a dynamically important magnetic field. The forces are labelled as in figure 6.

than that in the polar wind, is is still much too high to agree with the velocities implied by the optical and infrared measurements. The lowest velocity is produced by a model with a large rotation rate and a small magnetic field, so that the Lorentz force doesn't boost the velocity at large radii. Even in this case, the radiation force accelerates the wind material to velocities on the order of 100 km/s. Either the Be stars must be rotating even closer to their critical rotation rates, which isn't supported by the observations, or there must be some other force that supplies much of the mass loss in the equatorial plane or somehow keeps the material from flowing outward. Some speculations for possible resolutions of this problem will be offered in section 4.

A second problem is that a small magnetic field strength is required to keep the radial velocity low in the equatorial plane, but a large magnetic field strength is required to make the "disk" corotate with the star and produce optical emission lines of the right width. It may be that no single value of magnetic field strength can satisfy both observational constraints.

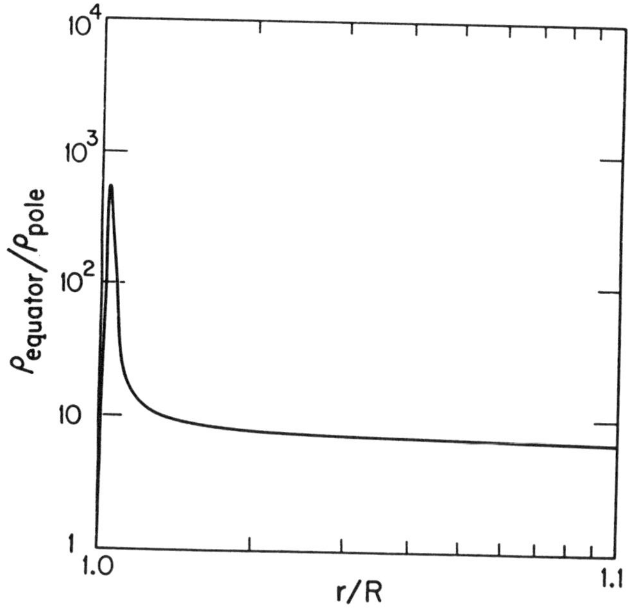

Figure 8. Density contrast between the equator and the pole as a function of radius.

In the future we plan to model the optical emission line profiles to see exactly what values of magnetic field strength in this model can explain their widths and shapes.

A third problem involves the interpretation of a zero rotation rate model with the polar flow. More generally, the problem lies in the assumption that we can treat the equatorial and polar wind flows as one dimensional, with no meridional flow. It would be much better to make a two-dimensional wind model, which treats the entire flow in a meridional plane (still assuming axial symmetry). Poe (1987) has made a preliminary attempt at such a two-dimensional model. The model is fairly simple, in that it treats the star as a uniform sphere, even though it is rapidly rotating, and it does not include magnetic fields. But it does include the finite disk correction in the line radiation force and it can find all three velocity components at any radius and polar angle. Some results of this model are shown in figures 9 and 10. Figure 9 shows the mass flux as a function of polar angle, for a model O star rotating at 300 km/s. The four curves are for four different distances from the star. We see that the mass flux is enhanced along the equator, as expected from our one-dimensional model. But we also see the presence of meridional flow, in that the enhancement in the equator *grows larger* with increasing radius. This increases the density enhancement between the equatorial and polar flows. Figure 10 is a plot of the density enhancement at large radii (compared to a zero rotation rate model) as a function of polar angle, for several different rotation rates. We see that the density is reduced along the pole and enhanced along the equator, as the rotation rate is increased. The two-dimensional model thus justifies our conclusions based on a simpler one-dimensional model. In fact, the two-dimensional model actually *increases* the effects found with the one-dimensional model, because of the meridional flow from the pole to the equator.

154

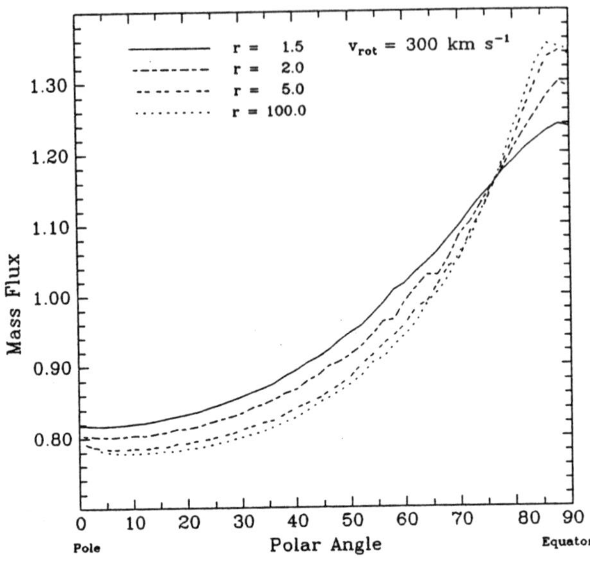


Figure 9. Mass flux as a function of polar angle in the two-dimenional wind model. The different curves are for different distances from the star.


## 4. Conclusions and Speculations

A radiation-driven wind enhanced by rapid rotation and open magnetic field lines on the equator can qualitatively explain the properties of Be star winds. A dense, slow equatorial flow is produced because of the enhanced centrifugal force and reduced effective gravity in the equatorial plane. A fast, but more tenuous, polar flow is produced because the line radiation force there is not affected by the rapid rotation or the magnetic field. A crude two-dimensional model with flow in the meridional plane verifies that these conclusions should also hold in a more accurate description of a wind from a rapidly rotating star.

The main problem with this specific model is that the flow velocity in the equatorial plane is still too high, perhaps by as much as an order of magnitude. A much more serious problem with the model, mentioned briefly in section 3, is that of the extreme variability of Be stars

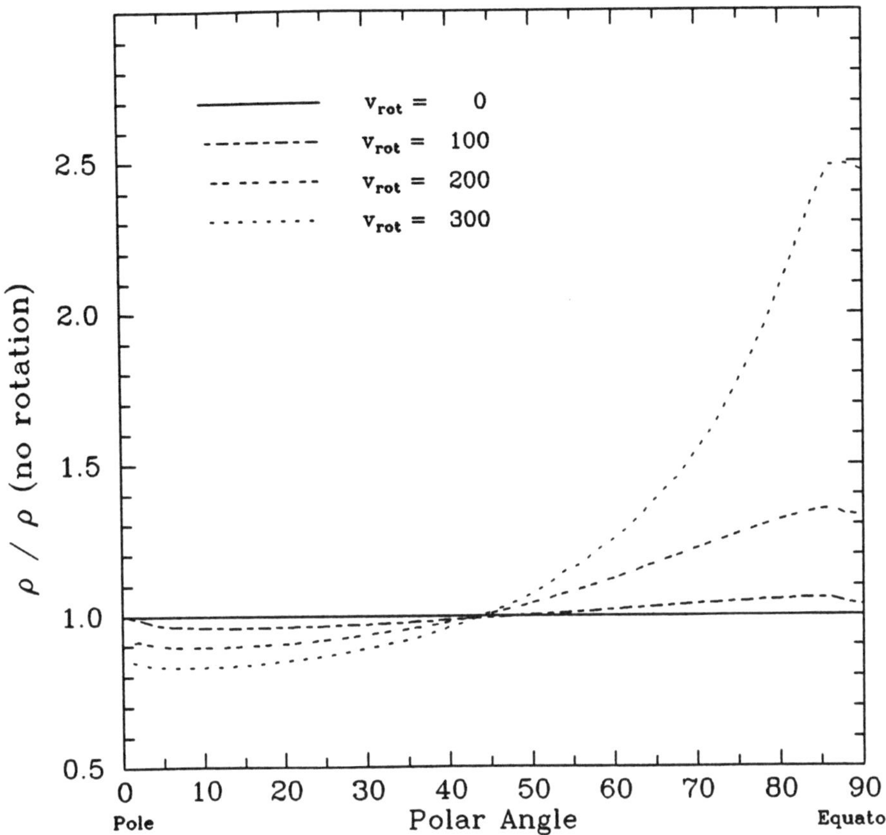



Figure 10. The density of the two-dimensional wind model at large radii, compared to the density of a non-rotating model. The different curves are for different rotational velocities.

(see Barker and Marlborough 1985). The dense material in the equatorial plane appears episodically, and can be entirely missing for long periods of time. A steady-state model, such as described here, cannot possibly address the question of variability. However, there is a possible mechanism for the variability that could also explain the low velocities in the equatorial plane. This mechanism is non-radial pulsations in the atmospheres of Be stars, the evidence for which is described by Baade (1987). Non-radial pulsations have the following relevant properties: they are time-dependent, they are coupled to the star's rotation rate, and they could possibly provide greater amounts of mass loss on the equator if the energy in the pulsation could somehow be transferred to the outflow. If non-radial pulsations could provide extra mass loss in the equator, this material would not be accelerated to as high a velocity because the radiation force would not be sufficient to accelerate the larger mass to the same velocity. A model which incorporates the time-dependent effects of non-radial pulsations at the base of a wind has yet to be made, and may not be made for a long time. It should also be noted that some the people who were supporting the idea of non-radial pulsations in Be stars as a major factor influencing their outflows are now skeptical of the idea (see, for example, Balona and Cuypers 1990 and Bolton and Stefl 1990, both in this volume).

Another possibility for keeping the velocity low in the equatorial plane would be to have a *closed* magnetic field in the equatorial regions, which would prevent outflow through the closed field lines. Pneuman and Kopp (1971) made a model for the solar wind with a magnetic field that started out as a dipole. They then solved for the wind flow and the magnetic field lines self-consistently. The wind distorted the field lines so that they were open on the poles and closed in a small region around the equator. This model was intended to explain both polar coronal holes and equatorial streamers. Such a model, applied to a Be star, could possibly explain the fast polar wind and the dense equatorial "disk" in the same manner. When the line radiation force is included in a model like this, it can no longer be solved using the approach of Pneuman and Kopp. So it is not yet clear how to apply this idea to Be star winds.

## 5. Acknowledgements

The model described in this work was done in collaboration with Clint Poe, and I thank him for permitting me to quote freely from our results. I would also like to thank Clint for permission to describe many of the results of his doctoral thesis (Poe 1987) prior to publication. I also thank Joe Cassinelli, Keith MacGregor, and Mike Marlborough for many useful discussions about rotating stellar winds. I would like to thank the Physics Department at Weber State College for graciously allowing me to miss a week of classes to attend this meeting.

## 6. References

Abbott, D. C. (1980) 'The Theory of Radiatively Driven Stellar Winds. I. A Physical Interpretation', *Ap. J.*, **242**, 1183.

Abbott, D. C. (1982) 'The Theory of Radiatively Driven Stellar Winds. II. The Line Acceleration', *Ap. J.*, **259**, 282.

Baade, D. (1987) 'Be Stars as Nonradial Pulsators', in Slettebak, A., and Snow, T. P. (eds.), *IAU Colloquium 92, Physics of Be Stars*, Cambridge University Press, Cambridge, p. 361.

Balona, L., and Cuypers, J. (1990) 'Intensive Photometric Campaigns on Be Stars: Behaviour of Short-Term Periodic Variation and its Relation to Pulsation and Mass Loss', in L. A. Willson and R. Stalio (eds.), *Angular Momentum and Mass Loss for Hot Stars*, Kluwer Academic Publishers, Dordrecht, this volume.

Barker, P. K. (1979) 'The Shell Episode of 59 Cygni (1974-1975); Rotating Magnetic Winds?', Ph. D. thesis, University of Colorado.

Barker, P. K., and Marlborough, J. M. (1985) 'Carbon IV Absorption Troughs in the Ultraviolet Spectra of Be Stars: Gone with the Wind?', *Ap. J.*, **288**, 329.

Bolton, C. T., and Stefl, S. (1990) 'An Investigation of the Correlation between Pulsation Amplitude and Shell Activity in the Be Star Lambda Eridani', in L. A. Willson and R. Stalio (eds.), *Angular Momentum and Mass Loss for Hot Stars*, Kluwer Academic Publishers, Dordrecht, this volume.

Briot, D. (1986) 'Rotational Velocity of Be Stars Correlated with Emission Characteristics', *Astr. Ap.*, **163**, 67.

Castor, J. I., Abbott, D. C., and Klein, R. I. (1975) 'Radiation-Driven Winds in Of Stars', *Ap. J.*, **195**, 157 (CAK).

Cote, J., and Waters, L. B. F. M. (1987) 'IRAS Observations of Be Stars. I. Statistical Study of the IR Excess of 101 Be Stars', *Astr. Ap.*, **176**, 93.

Coyne, G. V., and McLean, I. S. (1982) 'Polarimetry and Physics of Be Star Envelopes', in M. Jaschek and H.-G. Groth (eds.), *IAU Symposium 98, Be Stars*, Reidel, Dordrecht, p. 77.

Dachs, J., Hanuschik, R., Kaiser, D., and Rohe, D. (1986) 'Geometry of Rotating Envelopes around Be Stars Derived from Comparative Analysis of H-alpha Emission Line Profiles', *Astr. Ap.*, **159**, 276.

Doazan, V., Barylak, M., Rusconi, L., Sedmak, G., Thomas, R. N., and Bourdonneau, B. (1989) 'The First Decade of Envelope Formation of 59 Cygni in the Far UV and Optical Regions. II.', *Astr. Ap.*, **210**, 249.

Friend, D. B., and Abbott, D. C. (1986) 'The Theory of Radiatively Driven Stellar Winds. III. Wind Models with Finite Disk Correction and Rotation', *Ap. J.*, **311**, 701.

Friend, D. B., and MacGregor, K. B. (1984) 'Winds from Rotating, Magnetic, Hot Stars. I. General Model Results', *Ap. J.*, **282**, 591.

Grady, C. A., Bjorkman, K. S., and Snow, T. P. (1987) 'Highly ionized Stellar Winds in Be Stars: The Evidence for Aspect Dependence', *Ap. J.*, **320**, 376.

Grady, C. A., Bjorkman, K. S., Snow, T. P., Sonneborn, G., Shore, S. N., and Barker, P. K. (1989) 'Highly Ionized Stellar Winds in Be Stars. II. Winds in B6-B9.5e Stars', *Ap. J.*, **339**, 403.

Marlborough, J. M. (1987) 'Rotationally-Enhanced Stellar Winds', in Slettebak, A., and Snow, T. P. (eds.), *IAU Colloquium 92, Physics of Be Stars*, Cambridge University Press, Cambridge, p. 316.

Marlborough, J. M., and Zamir, M. (1984) 'Some Effects of Rotation on the Structure and Dynamics of a Radiation Driven Wind from a Hot Star', *Ap. J.*, **276**, 706.

Nieuwenhuijzen, H., and de Jager, C. (1988) 'The Relation between Rotational Velocity and Mass-Loss for Hot Stars', *Astr. Ap.*, **203**, 355.

Pauldrach, A., Puls, J., and Kudritzki, R. P. (1986) 'Radiation Driven Winds of Hot Luminous Stars. Improvements of the Theory and First Results', *Astr. Ap.*, **164**, 86.

Pneuman, G. W., and Kopp, R. A. (1971) 'Gas-Magnetic Field Interactions in the Solar Corona', *Solar Phys.*, **18**, 258.

Poe, C. H. (1987) 'The Effects of Rotation on the Winds from Hot Stars', Ph. D. thesis, University of Wisconsin.

Poe, C. H., and Friend, D. B. (1986) 'A Rotating, Magnetic, Radiation-Driven Wind Model Applied to Be Stars', *Ap. J.*, **311**, 317.

Poeckert, R. (1982) 'Model Atmospheres of Be Stars', in M. Jaschek and H.-G. Groth (eds.), *IAU Symposium 98, Be Stars*, Reidel, Dordrecht, p. 453.

Poeckert, R., and Marlborough, J. M. (1978) 'Be Star Models: Observable Effects of Model Parameters', *Ap. J. Suppl.*, **38**, 229.

Vardya, M. S. (1985) 'Stellar Rotation and Mass Loss in O and B Stars', *Ap. J.*, **299**, 255.

Waters, L. B. F. M. (1986a) 'The Correlation between Rotation and the IR Color Excess for B-type Dwarfs', *Astr. Ap.*, **159**, L1.

Waters, L. B. F. M. (1986b) 'The Density Structure of Discs around Be Stars Derived from IRAS Observations', *Astr. Ap.*, **162**, 121.

Weber, E. J., and Davis, L., Jr. (1967) 'The Angular Momentum of the Solar Wind', *Ap. J.*, **148**, 217.

UV GLIMPSE OF OB STARS

R. STALIO [1]
Dipartimento di Astronomia
Universita' di Trieste
Via G.B. Tiepolo, 11
34131, Trieste, Italy

R. S. POLIDAN
NASA/Goddard
  Space Flight Center
Code 681
Greenbelt MD 20771 USA

ABSTRACT: We discuss some observational aspects of FUV research which could be relevant for this meeting emphasizing at the same time the role of multiwavelength, coordinated observations of variable stars. We also summarize the characteristics of the Santa Maria experiment which will be devoted to a long term, multiwavelength monitoring program of variable sources.

## 1. INTRODUCTION.

One of the trends of current research concerns the extension of wavelength coverage to the FUV (912 - ~1250 Å) and to the EUV (<912 Å). These regions are essentially unexplored at medium or high resolution; they are being studied by the UV spectrometers on the Voyager 1 and 2 spacecrafts at low resolution.
  There are several classes of astrophysical sources which emit a relevant fraction of their radiation in the EUV and FUV. Studying these regions will give us clues towards understanding important physical mechanisms as it is discussed in the Magellan report released by ESA (1982) and in NASA FUSE report (1983). More recently Jenkins et al. (1988) have re-stated the scientific importance of these regions. Instrumental projects for the FUV and EUV are ready to fly or are in the planning phases: EUVE, Santa Maria, Lyman, and others. Technical problems, such as the low reflectivity of the optics and the lack of transmitting materials, are not anymore question of concern. There are several modern solutions which alleviate these problems; for example, the EUV and FUV spectrometers of Santa Maria (an experiment which will be shortly described in this paper) will use SiC coatings and

---

[1] also CARSO (Center for Advanced Research in Space Optics), Area di Ricerca, Padriciano 99, Trieste, Italy.



*L. A. Willson and R. Stalio (eds.), Angular Momentum and Mass Loss for Hot Stars, 159–170.*
© 1990 *Kluwer Academic Publishers. Printed in the Netherlands.*

intensified CCD (photon counting) detectors operating with vacuum sealed doors (Stalio *et al.* 1990).

This paper is split into two sections. In the first we discuss some observational aspects of the FUV research which could be relevant for this meeting emphasizing at the same time the role of the multiwavelength, coordinated observations. We discuss topics as (unprojected) rotational velocities ($v_{rot}$), critical $v_{rot}$, luminosities and effective temperatures, FUV variability. In the second part we summarize the characteristics of the Santa Maria experiment which will be mostly devoted to a long term monitoring program of observations of variable sources over an extended wavelength range: 200 - 7000 Å.

2. ASPECTS OF FUV (AND MULTIWAVELENGTH) RESEARCH FOR HOT STARS.

The important observables for hot stars are the intrinsic spectral energy distributions, the photospheric line profiles and strengths, the P Cygni profiles, and the flux and spectral variability. Several ground based and orbiting telescopes are providing us with these data, at fairly high quality, for a number of multiwavelength observing programs which we are carrying on. One of these programs, which is in progress, aims at determining $v_{rot}$ (the unprojected rotational velocity) of T Tauri stars (cool stars!) on  the basis of (1) the measurement of the projected rotational velocity from high resolution spectroscopy and (2) the determination of the rotational modulation in the photometric light curves due to photospheric spots.  Similar programs are in course for the Be stars $\lambda$ Eri (in collaboration with M. Smith *et al.*) and P Car (in collaboration with A. Porri *et al.*)  which are suggested targets for studying hot star activity and other phenomena affecting and being affected by the mass loss. These last two programs are presently in the "data acquisition" phase.

An example of the procedures used for the T Tauri stars, where $v_{rot}$ is considered to be an important parameter for both testing evolutionary theories and mechanisms and deriving useful correlations with chromospheric activity, is based on the following steps:

A. We determine the quantity a = $v_{rot}sini$ from about 30 photospheric line profiles using the original Gray (1976) method. The data are taken at the 3.6 m ESO telescope in La Silla with CASPEC at a resolving power of approximately 25000, a signal to noise ratio of the order of 130-300 and a spectral range coverage from 5800 to 6800 Å (Franchini *et al.* 1989).

B. From photometric monitoring of the target stars we derive a characteristic period P which we assume is indicating the existence of co-rotating spotted regions located at the star equator (Covino *et al.* 1990). The period, $P = 2\pi r_{star} sini/a$ allow us to derive $r_{star} sini$.

    C. From the same photometric data (UBVRI colors), IUE and IRAS data we derive the bolometric luminosity; thus we get $r_{star}^2 T_{eff}^4$.

    D. An automatic curve of growth method, fully tested with the Sun and atmospheric models, allows us to determine the effective temperature (Castelli *et al.* 1990) and to derive $r_{star}$ and *sini*.

The proposed procedures for studying activity in the hot stars are similar and require (i) high resolution and  high signal-to-noise UV and optical profiles for measuring $v_{rot} sini$, $v_{term}$, profile shapes, occurrence of transients and for determining $v_{rot}$, mass loss rates, the pulsation modes, etc.; (ii) UV and FUV monitoring at selected wavelength bands, using photometrically stable instruments for enhancing the detectability of light variations; (iii) the determination of intrinsic spectral energy distributions for the effective temperatures, gravities and radii.

A few hints on what we shall expect from the observations in the FUV spectral range are provided by the following short review of some of the stellar programs which are carried on from the Voyager's UV spectrometers.

*The UVS on the Voyagers.*  The UV spectrometers on the Voyager 1 and 2 spacecrafts operate in the spectral region between 500 and 1700 Å at a spectral resolution of approximately 15 Å. They are objective grating instruments (Broadfoot *et al.* 1977) and their photometric errors are typically 3% for <1200 Å and 8% for >1200 Å.  The calibration and the related uncertainties are discussed in Holberg *et al.* (1988); the procedures for spectra acquisition, background subtraction, and a discussion on the effects of systematic errors due to uncertainties of the star position on the slit are in Polidan and Plavec (1984) and Reichert *et al.* (1988).

*$T_{eff}$ 's and luminosities of OB stars.*  The effective temperature range of the B stars makes their FUV spectra very sensitive to temperature changes because we observe at the peak of the flux distribution, at shorter wavelengths  then is possible with the IUE. As the temperature decreases the peak moves to longer wavelengths and induces strong changes in appropriately defined color indices. After correcting for the interstellar reddening with their extinction curve, Longo *et al.* (1989) have measured the $[m(1059)-m(1400)]_0$ color index of a group of B stars and found that it ranges from −approximately −1.00 at B0 (−0.99 for $\tau$ Sco) to 2.25 at B8 ($\beta$ Ori). The FUV wavelengths give also sensitive criteria for the stellar luminosity. Both $T_{eff}$ and luminosity effects are illustrated in Figure 1 (Longo *et al.* 1989).

The opposite is true for the O stars. Here the peak maximum is shortward of 912 Å and we cannot observe at those wavelengths. The FUV data of Longo *et al.* confirm Massa and Savage (1985) results that the O main sequence stars do not have UV colors which are significantly bluer than a normal B0V stars and that the spectral distinction among supergiants and the other O stars is only present at the later O spectral types.

*Enhanced detectability of light variations for B stars.* To make the point we summarize some of the results of a recent study by Porri *et al.* (1990) where both Voyager and IUE data of the $\beta$ Cephei star $\nu$ Eri have been analyzed and UV light curves, equivalent width curves and intrinsic spectral energy distributions at different phases have been derived. Assuming that the spectral energy distribution of $\nu$ Eri can be mimicked at any phase by a proper standard LTE model, Porri *et al.* have compared the best UVS spectrum at minimum and the best at maximum with the prediction of Kurucz (1979) models. The fitting has been obtained by means of an automatic least square procedure (Morossi and Malagnini 1985) that allowed to determine the values of the star effective temperature, gravity and angular diameter at the maximum and minimum phases. The derived values indicate that the gravity (log g = 3.55 ± 0.20) and the angular diameters ($\phi$ = 0.27 ± 0.03 milliarcsec) remain constant (within the errors) and that the $T_{eff}$ varies from 20120 (± 680) K at minimum to 22340 (±850) K at maximum. An extensive discussion of the errors and the quality of the fits is given in Porri *et al.*'s paper.

Figure 2 compares the magnitude difference between the theoretical spectra at maximum and minimum in the wavelength range 912 - 1500 Å with the magnitude difference of the UVS spectra, corrected for extinction, at maximum and minimum. The observed 0.659 mag light amplitude at 1050 Å is more than 5 times larger than the V amplitude of 0.114 mag reported by Walker (1952). The acceptable fitting between computation and data suggests that the strong magnitude effect is essentially due to temperature difference.

The same paper by Porri *et al.* reports the observation of micro-variations of the FUV maximum flux occurring at different maxima in $\nu$ Eri and $\beta$ Cep itself; it is not likely that these variations could be observed from optical instruments at intensities predicted by scaling the FUV fluxes. In $\nu$ Eri they appear, likely temporarily, as a nice, systematic increase of the maximum flux with time at approximately 0.01 mag/cycle (Figure 3); in $\beta$ Cep the micro variability of the maxima is irregular.

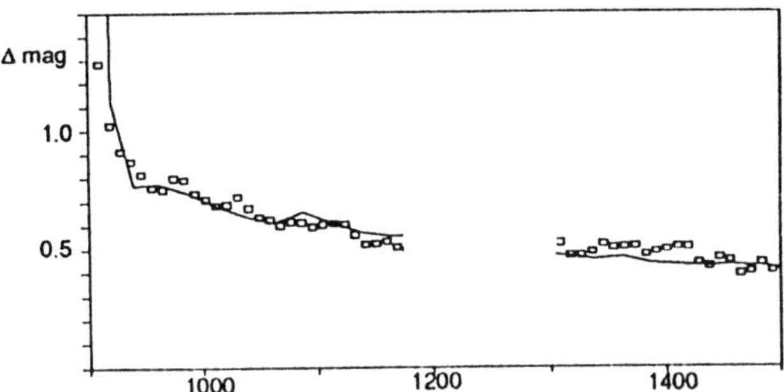

FIGURE 2:  Computed and observed magnitude differences between the maximum and minimum FUV spectra of ν Eri.

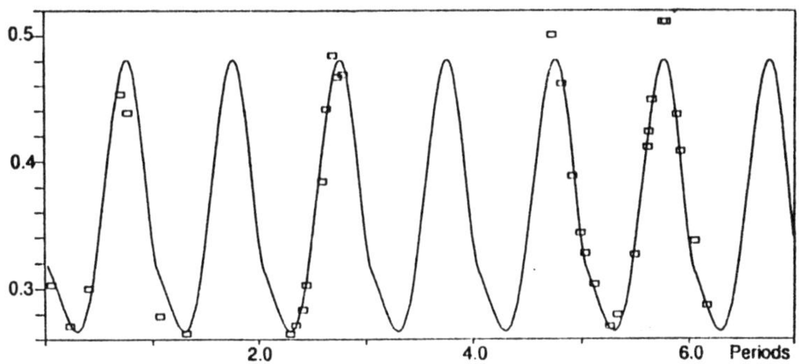

FIGURE 3:  $F_{1050}$ fluxes (integrated flux average for the spectral region between 950 and 1150 Å versus time.  The solid line represents a sequence of 6th order polynomial fits.

---

*Critical rotation in Be stars; effects on the spectral energy distributions.* Polidan et al. (1990) are studying a number of Voyager spectra of low reddened B and Be stars to determine to what extent the gravity darkening, i.e. the equatorial distension of the atmosphere of a star rotating at critical velocity, is present in the spectral energy distribution of Be stars. The method is based on  a comparison of the spectral energy distributions of low reddening B and Be stars of equal spectral type *and* $(B-V)_o$ color indices and on the use of Collins and Sonneborn (1984, CS) models computed using the method discussed in Collins and Sonneborn (1977).

CS models predict that given two equal mass stars, one, X, rotating slowly and the other, Y, rotating at critical velocity,

164

the effect of critical rotation makes (1) the star Y appearing of
later visual spectral type and later $(B-V)_0$ than the star X, and
(2) the star Y having spectral type not consistent with any
standard spectrum shortward of the light maximum (FUV). Viceversa
if the two other stars XX  and YY have the same spectral type and
$(B-V)_0$ and the star YY rotates at critical velocity, the effect of
critical rotation makes the star YY appearing with UV excess.

As an example of the analysis made by Polidan *et al.* we
consider two of their program stars: the Be star $\eta$ Cen  [B2IVe,
$(B-V)_0$ = -0.24, $v_{rot}sini$ = 380 km/s] and the slowly rotating
standard counterpart $\zeta$ Cas [B2IV, $(B-V)_0$ = -0.25, $v_{rot}sini$ < 10
km/s]. To ensure that the reference stars are free from the effects
of gravity darkening, i.e. that they are not rapid rotators seen
pole on, we have further compared their continua with model
atmospheres from Kurucz (1979) and zero inclination CS models. In
Figure 4 we display the composite flux distribution of $\eta$ Cen (upper
panel) formed using the Voyager data, TD1 data (Jamar *et al.* 1976)
and 13 color photometry of Johnson and Mitchell (1975). In the
lower panel we have plotted the difference between these
observations and similar data for $\zeta$ Cas. The match is reasonably
good over the whole spectral range, suggesting that there is no
evidence of gravity darkening in the Be star. The observed $v_{rot}sini$
for $\eta$ Cen (380 km/s) combined with the CS critical velocity for a
B2 main sequence star (472 km/s) imply that $\omega/\omega_{crit}$ > ~ 0.80 and
poses limits on the inclination angle that can only be contained
between 90 and 53 degrees. The reasonable good fitting of $\eta$ Cen
with the CS B2 model (considering the observational uncertainties)
suggests additionally that the observations are consistent with a
model for an equator-on star rotating at 75% of its rotational
velocity. Similar results are obtained for several of the other
program stars the, suggesting that critical rotation does not play
a key role for the Be phenomenon.


*Angular Momentum and Mass Loss in Binary Stars*   The study of
angular momentum and mass loss in binary systems has been rather
neglected in recent years with respect to that of single stars.
Since binary stars are a rather large and diverse group of stars we
have restricted the discussion below to a small subgroup: early-
type (O, B, and A), semi-detached (one component in contact with
the Roche surface) systems.  The exclusion of other classes of
binaries (e.g. contact systems) is not meant to argue that angular
momentum and mass loss are unimportant in these binaries, rather it
simply reflects the interests of the authors.  In the brief
discussion below we present an overview of the problems concerning
angular momentum and mass loss in these interacting binary systems.

Unfortunately, very few solutions are presented to these problems: the status of the field is such that we are quite aware of the problems but are only beginning to be aware of the solutions. The current state of the field can be found in the proceeding of the recent Algol colloquium (Batten 1989).

Mass transfer in binary systems is the "driver" of the angular momentum and mass loss. Mass transfer begins when the initially more massive star of the binary comes in contact with its Roche surface and begins transferring matter to the less massive component. Quickly ($10^4$ or $10^5$ years) matter is transferred, transforming the system from one in which the mass losing star has, typically, 55 to 70% of the total mass to one in which it has less than 25% of the total mass of the system. This process is not conservative, that is the total binary system mass and angular momentum are not conserved. Significant amounts of mass and angular momentum are lost in a very short, astronomically speaking, time. As an aside it is important to mention that the discussion below is derived primarily from studies of eclipsing binary systems. But not all binary systems are eclipsing (for typical semi-detached systems the ratio of non-eclipsing to eclipsing systems is ~8) and contrary to popular belief it is not always easy to establish if a given star is single or double.

In depth studies of specific systems looking for evidence of mass and angular  momentum loss (e.g. V356 Sgr, Polidan 1989) are quite useful in examining the details of the loss mechanisms and in quantifying the amount of mass and angular momentum being lost at the present time.  They, however, cannot address the broader question of the importance of mass and angular momentum loss to binaries as a class.  Unfortunately, the majority of statistical studies have often worked with data bases of uneven quality containing (observationally induced) systematic errors.  Despite these difficulties when the results from statistical (and, also, detailed studies of specific systems) are compared with theory the result is the same:  agreement is achieved only if on the order of 50% (with considerable uncertainty) of the binary system mass and angular momentum is lost during the mass transfer process.  How is this extensive mass and angular momentum loss accomplished?  As yet we do not have an answer (or answers), but significant progress is being made. Studies of mass transfer binary systems with IUE have shown that virtually all systems show high ionization UV resonance line emission (see discussions and references cited in McCluskey and Sahade 1987 and the papers contained in Batten 1989).  Detailed studies of some systems (Polidan 1989) have shown that these lines originate in an extensive, probably outflowing, plasma that pervades the system.  But how are these lines produced?  Radiative processes appear insufficient and while many systems contain accretion disks that could drive the outflow, many do not.  A more

"general" loss mechanism must exist if we are to to explain the
"common" evidence of mass and angular momentum loss in binary
systems.  Rapid rotation, produced when the mass gaining star is
spun up the accretion of high angular momentum material, and
magnetic fields have been discussed as important factors in
understanding binary star mass and angular momentum loss but
detailed modeling must first be done before they can be considered
key factors in the process.

## 3. SANTA MARIA.

The FUV topics described above are only a fraction of the large
research programs which are planned with the next generation of
instruments operating in the EUV and FUV bands. In addition, one
must realize that today the astrophysics research is becoming more
and more concerned with phenomena which require a wide spectral
coverage, X-rays, UV, visual, IR and radio, in order to be
explained and understood. One of the important areas of the
multiwavelength research is the study of variability occurring at
the different wavelength bands over a wide range of time scales and
time lags. Variable phenomena are present at all astrophysical
scales, ranging from the nearby flare stars to the QSO's, and are
fundamental for our basic understanding of these types of objects.
The importance of studying active objects simultaneously at many
wavelengths and for long periods of time is illustrated by the fact
attempts at coordinated observations with other observatories were
made in a large fraction of the observing programs carried on from
Exosat.
     For this reason we are performing an implementation study
(phase A) for an international orbiting observatory, Santa Maria,
that has the capability of making multispectral observations of
astrophysical targets including extragalactic, galactic and solar
system objects. This mission will be primarily devoted to
monitoring temporal and/or spectral variations through the spectral
range from 200 Å to 7000 Å and secondarily to outgrowing the
current, highly successful, program which is presently carried on
from the International Ultraviolet Explorer (IUE). The ability of
Santa Maria to obtain simultaneous spectrophotometric and
photometric observations of a wide range of targets will add a new
dimension to the study of active astrophysical and solar system
phenomena such as active galactic nuclei, outbursts in dwarf novae,
flares and related activity in late type stars, the Io torus and
planetary atmospheres. Two additional complementary packages are
proposed: a terrestrial plasmasphere experiment and an absolute
solar flux experiment. The instrumentation will be divided in three
parts:

1) The primary instrument (Table 1) consists of a pointed array of
5 co-aligned imaging spectrometers, operating simultaneously and
covering the ranges from 200 - 600 Å (EUV1), 400 - 850 Å (EUV2),
900 - 1300 Å (FUV), 1150 - 3200 Å (UV) and 3000 - 7000 Å (Vis)
respectively. The FUV spectrometer will have the largest aperture
and will consist of a telescope and of a Rowland grating
spectrograph and two intensified CCD (photon counting) detectors.
It will provide 1 Å resolution and  angular resolution will of
about 5 arcsec. The apertures of the other channels will be scaled
to the FUV channel in order to obtain comparable sensitivities and
resolving power. The two EUV channels will consist of grazing
incidence telescopes and gratings.

Table 1 -- INSTRUMENT PARAMETERS

| Channel | Configuration range (Å) | Wavelength (Å) | (cm$^2$) | Resol. ($^a$) | Apert | Sens. |
|---|---|---|---|---|---|---|
| EUV2 | 200-600 | 10 | 4X12 | ~6 | Grazing; ICCD; windowless intensifier | |
| EUVE1 | 550-850 | 5 | 4X12 | ~7 | Grazing; ICCD; windowless intensifier | |
| FUV | 800-1300 | 1-2 | 25X36 | ~8-4 | Telesc. + Rowland grating; ICCD; windowless intensifier | |
| UV | 1150-3200 | 5 | 10.5X5 | 0.4 | Telesc. + Rowland grating; ICCD; windowless intensifier | |
| Vis. | 3000-7000 | 7.5 | 10.5X15 | 0.1 | Telesc. + Rowland grating; ICCD; windowless intensifier | |

$^a$ Point source continuum flux, S/N = 10, 1200s, units $10^{-13}$ ergs
s$^{-1}$ cm$^{-2}$ Å$^{-1}$.

2) The plasmasphere instruments consists of several miniaturized
monochromatic imagers and of an energetic neutral particle imager.
3) The solar instrument is formed by a miniature solar spectrograph
and a number of EUV cameras for measuring the absolute solar flux.
   The largest fraction of the experimental payload will be for
the primary instrument. The earth plasma and solar instruments are
smaller experiments integrated into the spacecraft and will use
only a small portion of the spacecraft resources. The design of the
instruments will draw heavily on space proved intensified charge
coupled devices (ICCD) technology and matching optics to provide
maximum capability for minimum weight and spacecraft complexity.

A key aspect of the program is its international scope. The Santa Maria mission is a collaborative program between Spain, Italy and the United States. The experiment consists of an integrated package, spacecraft and scientific instruments, for a San Marco Scout mission.

AKNOWLEDGEMENTS: The authors would like to thank all the collaborators in the different programs mentioned in the text

REFERENCES:


Batten, A. H., ed.: 1989 *Algols* Proceedings of IAU Colloquium no. 107, Space Sci. Rev. **50**, Numbers 1-2.
Broadfoot, L.A., *et al.*: 1977, Space Sci. Rev., **21**, 183.
Castelli, F., Franchini, M. and Stalio, R.: 1990, *in preparation.*
Collins, G. W. II, Sonneborn, G. H.: 1977, Astrophys. J. Suppl. **34**, 41.
Collins, G. W. II, Sonneborn, G. H.: 1984, *private communication.*
Covino, E., Franchini, M., Stalio, R., Chavarria, C. and Terranegra, L.: 1990, in preparation.
Franchini, M., Magazzu', A and Stalio, R.: 1988, Astron. Astrophys. **132**, 189.
Gray, D. F.: 1976, *The Observation and Analysis of Stellar Photospheres,* J. Wiley & Sons.
Holberg, J.B., Forrester, W.T., Shemansky, D.E. and Barry, D.C.: 1982, Astrophys. J., **257**, 656.
Jenkins, E. B. *et al.*: 1988, Proc. Soc. Photo-opt. Instr. Eng., **923**, 213.
Johnson, H. L. and Mitchell, R. I.: 1975, Rev. Mexicana Astron. Astrop. **2**, 299.
Kurucz, R.L., 1979, Astrophys. J. Suppl. **40**, 1.
Longo, R., Stalio, R., Polidan, R.S. and Rossi, L.: 1989, Astrophys. J., **339**, 474.
Morossi, C. and Malagnini, M. L. : 1985, Astron. Astrophys. Suppl. Ser. **60**, 365.
Massa, D. and Savage, B.D.: 1985, Astrophys. J., **299**, 905.
McCluskey, G. E. Jr. and Sahade, J. 1987, in *"Exploring the Universe with the IUE Satellite",* p. 427.
Polidan, R. S. 1989, Space Sci. Rev., **50**, 85.
Polidan, R.S. and Plavec, M.J.: 1984, Astron. J., **89**, 1721.
Polidan, R.S., Stalio, R. and Peters, G.: 1990, *in preparation.*
Porri, A., Stalio, R., Morossi, C., Babar, A. and Polidan, R.S.: 1990, *in preparation.*
Reichert, G.A., Polidan, R.S. and Carone, T.E.: 1988, Astrophys. J., **325**, 721.
Stalio, R. *et al.*: 1990, "Berkeley Conference on EUV Astronomy", *in press.*

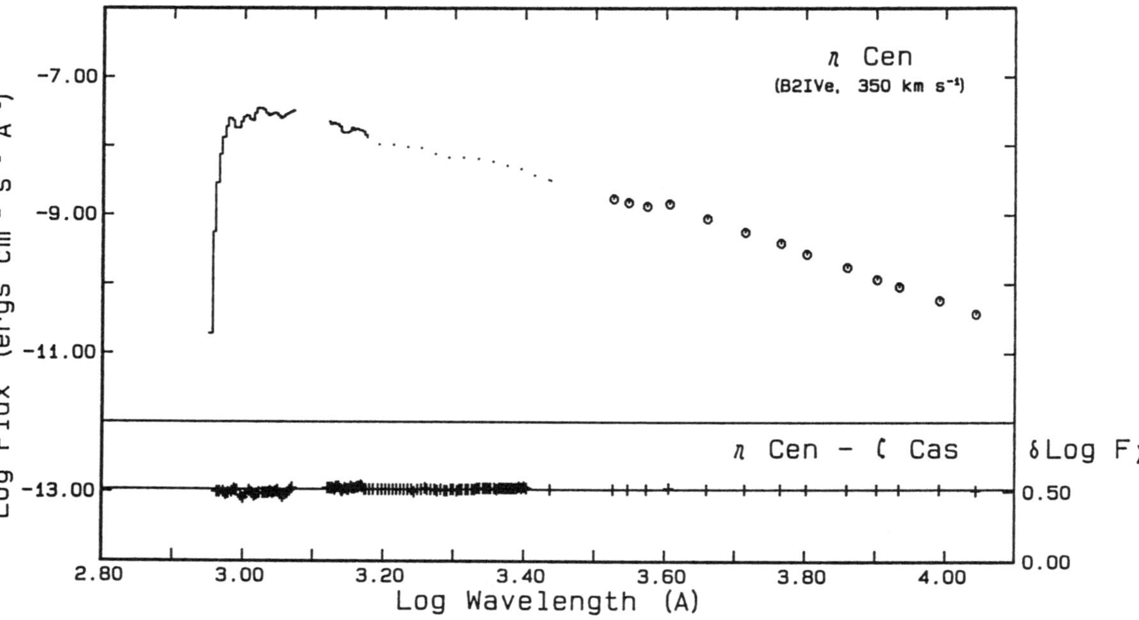


FIGURE 4: Composite flux distribution of η Cen (Voyager, TD1, 13 color photometry data) and difference spectrum between η Cen and ζ Cas.

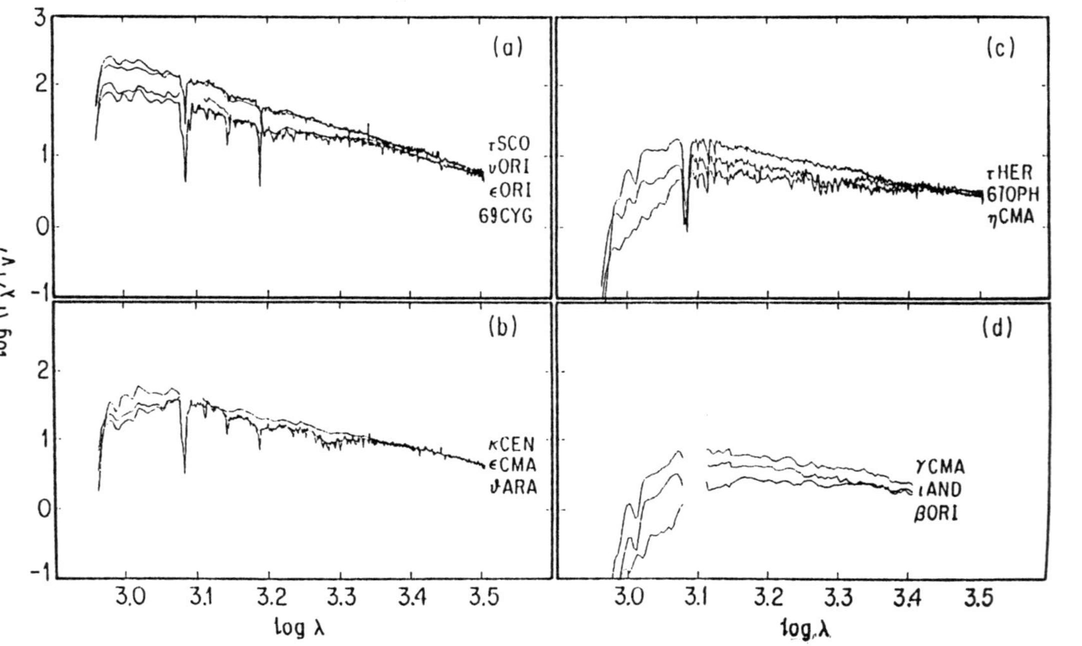

FIGURE 1: Dereddened spectral energy distributions, normalized to V of B stars with different luminosity classes: [a] B0 stars ($\tau$ Sco and $\upsilon$ Ori, luminosity class V; $\epsilon$ Ori, Ia; 69 Cyg, Ib); [b] B2 stars ($\tau$ Her, IV; 67 Oph, Ib; $\eta$ CMa, Ia); [3] B5 stars ($\kappa$ Cen, IV; $\epsilon$ CMa; II, $\theta$ Ara, Ib) and [d] B8 stars ($\gamma$ CMa; II, $\iota$ And, V; $\beta$ Ori, Ia).

# NONSPHERICAL RADIATION DRIVEN WIND MODELS APPLIED TO BE STARS

FRANCISCO X. ARAÚJO
Departamento de Astronomia
CNPq – Observatório Nacional
Rua Gal. José Cristino, 77
CEP 20921 – Rio de Janeiro – RJ
BRAZIL

ABSTRACT. In this work we present a model for the structure    of     a
radiatively driven wind in the meridional plane of a hot star. Rotation
effects and simulation of viscous forces were included in    the   motion
equations. The line radiation force is considered with the inclusion of
the finite disk correction in self-consistent computations which    also
contains gravity darkening as well as distortion of the star      by
rotation.
An application to a typical B1V star leads to      mass-flux
ratios between equator and pole of the order of 10 and mass loss rates
in the range $5.10^{-8}$ to $10^{-8}$ M⊙/yr. Our envelope models are      flattened
towards the equator and the wind terminal velocities in that region are
rather high (1000 Km/s). However, in the region near the star      the
equatorial velocity field is dominated by rotation.

## 1. INTRODUCTION

In the last years some attempts have been done in order to     explain
the winds of Be stars within the context of radiation driven    winds
(see the review by Marlborough, 1987). From an analytical study of the
conditions in the critical point, Marlborough and Zamir (1984)
concluded that the mass-loss rate is only slightly altered by rotation.
Poe and Friend (1986) developed a model which includes the effects   of
a magnetic field and the finite disc correction (Pauldrach et al. 1986;
Friend and Abott 1986). Their model however is spherically symmetric.
Therefore, in order to be able to compare their results with      the
observations they have to let the high rotation models simulate    the
equatorial regions, while the low rotation models would describe    the
polar regions. In this situation they have reached some good
qualitative results.
In this work we present the results of our exploratory
analysis of a non-spherical (axy-symmetric) model for the envelope of a
Be star. Besides incorporating rotation we work with the three
equations of motion and we simulate the effects of viscosity.    We

171

*L. A. Willson and R. Stalio (eds.), Angular Momentum and Mass Loss for Hot Stars, 171–176.*
© 1990 *Kluwer Academic Publishers. Printed in the Netherlands.*

172

consider also the distortion of the star due to rotation and the
variation of the photospheric temperature as a consequence of  the
gravity darkening. The mass flux and the velocity field were calculated
at different latitudes of the star, in particular, at the equator  and
poles. This approach enable us to make more direct comparisons  between
theory and observations. Our model does not take into account   time
variability, which is likely to be related with nonradial pulsations
(see the review by Baade, 1987). However, we believe that there is  a
continuous background component even for stars with highly variable
winds and to this steady component we adress our work.

## 2. THE HYDRODINAMIC EQUATIONS

We consider the equations for the conservation of mass and momentum  of
a fluid subject to: gravity, gas pressure, radiation force       from
continuum and lines and centrifugal acceleration. In addition,      we
simulate a viscous force that makes the wind to deviate from   angular
momentum conservation. The assumptions adopted are: steady    state ,
azimuthal symmetry and meridional flows not significant. We     assume
also the equation of state of a perfect gas:

$$p = a^2 \rho \tag{1}$$

Within these approximations the mass conservation can be  immediately
integrated and we obtain

$$\Phi(\theta) = r^2 \rho v_r \tag{2}$$

where $\Phi(\theta)$ is the mass flux per unit of solid angle. The total    mass
loss rate is given by the relation

$$\dot{M} = 2\pi \int \Phi(\theta) \sin\theta \ d\theta \tag{3}$$

For the radial component of the velocity we can write

$$v_r \frac{\partial v_r}{\partial r} - \frac{v_\phi^2}{r} + \frac{1}{\rho} \frac{\partial p}{\partial r} + \frac{GM(1-\Gamma)}{r^2} - \frac{1}{\rho} F^\ell = 0 \tag{4}$$

The second term represents the centifugal force, the fourth is  the
effective gravitational acceleration and the last term is the radiative
force due to line opacity. The parameter $\Gamma$ is given by the expression

$$\Gamma = \frac{\sigma_e L}{4\pi GMC} \tag{5}$$

All the symbols have their usual meaning.
The equation for the azimuthal component reduces to

$$v_r \frac{\partial v_\phi}{\partial r} + \frac{v_r v_\phi}{r} = f \tag{6}$$

where f is an unknown viscous force per unit mass. We assume as a solution of this equation

$$v_\phi(r,\theta) = \chi \left[\frac{GM(1-\Gamma)}{R}\right]^{1/2} \sin\theta \left(\frac{R}{r}\right)^\beta , \quad -1 \leqslant \beta \leqslant 1 \tag{7}$$

where $\chi$ is the ratio between the centrifugal acceleration at the equator and the effective gravity and $\beta$ is an adjustable viscosity parameter. Finally, the equation for the meridional component of the velocity leads to the relation

$$-v_\phi{}^2 \cot\theta + \frac{1}{\rho} \frac{\partial p}{\partial\theta} = 0 \tag{8}$$

We bypass the energy equation assuming a variation of the temperature throughout the envelope of the form

$$T(r,\theta) = t_1(r) \cdot t_2(\theta) \tag{9}$$

which, if combined with (1), (2), (7) and (8) requires a more restrictive functional relation

$$T(r,\theta) = t(R,\theta)(R/r)^{2\beta} \tag{10}$$

This last relation together equation (7) tell us that a non-viscous flow ($\beta = 1$) and an isothermal envelope ($\beta = 0$) are inconsistent assumptions.
The line radiation force used assumes Sobolev approximation (for discussion see Pauldrach et al. 1986), single scattering (see Puls 1987) and the correction for the finite size of the source of radiation. Therefore the expression is

$$\frac{F^\ell}{\rho} = \frac{\sigma_e L}{4\pi c r^2} \frac{K}{(\sigma_e \rho v_{th})^\alpha} \left(\frac{\partial v_r}{\partial r}\right)^\alpha \frac{\{1 - [1-(R/r)^2+(R/r)^2\frac{2v/r}{d_v/d_r}]^{1+\alpha}\}}{(1+\alpha)(R/r)^2 \frac{[1-v/r]}{d_v/d_r}} \tag{11}$$

where K and are the radiative parameters.
If now we combine equations (1), (2), (4), (7) and (11) we obtain the basic equation that describes the wind

$$\left[v_r - \frac{a_o{}^2}{v_r} \left(\frac{R}{r}\right)^{2\beta}\right] \frac{dv_r}{d_r} + \frac{GM(1-\Gamma)}{r^2} \left[1-\chi^2 \sin^2\theta \left(\frac{R}{r}\right)^{2\beta-1}\right] = \tag{12}$$

$$= \frac{2a_o{}^2}{R} \left(\frac{R}{r}\right)^{2\beta+1}(1+\beta) + f(r,v,\frac{dv}{dr}) \frac{C}{r^2} \left(r^2 v_r \frac{dv_r}{d_r}\right)^\alpha; \quad C = \frac{\Gamma GMK}{[\sigma_e v_{th} \Phi(\theta)]^\alpha}$$

$$\left\{ \frac{1 - \left[ 1-(R/r)^2 + (R/r)^2 \frac{v/r}{dv/dr} \right]^{1+\alpha}}{(1+\alpha)(R/r)^2 \left[ 1 - \frac{v/r}{dv/dr} \right]} \right\} = f\left(r, v, \frac{dv}{dr}\right) \qquad (12)$$

This equation is solved using the additional boundary condition $\tau(R) = 2/3$.

## 3. PARAMETERS AND RESULTS

### 3.1 Stellar and wind parameters

In order to model a typical B1 star we adopted $M = M_\odot$ and $L = 9700\ L_\odot$ (Stettebak et al. 1980). The corresponding radius of a non-rotating star would be $R = 5.3\ R_\odot$, while the effective temperature would be $T_{eff} = 25000$ K. As we have mentioned we considered the effects of rotation in the star itself obtaining a functional dependence of temperature and radius on latitude. We had also to establish an adequate range of variation for the parameters $\beta$ and $\chi$ . For the viscosity parameter we have taken values in the interval 0.00 to 0.49. For the rotation rate $\chi$, in view of the uncertainties related to this quantity, we decided to use four different values: 0.5, 0.66, 0.75 and 0.9. Finally, the line radiation parameters K and $\alpha$ were taken from the work of Abbott (1982). In our calculations we have used $\alpha=0.5$ and $K = 0.5$ as representative of the conditions prevailing in an early B star envelope.

### 3.2 Results and discussion

Figures 1 to 4 summarize the main results of our computations. In order to emphasize the contrast between equatorial and polar regions we restrict our analysis to these directions. From figure 1 we obtain a ratio between equatorial and polar mass flux of about 10 ($\chi = 0.9$ model) and a global mass loss in the range 10 M$\odot$/yr to 5.10 M$\odot$/yr. Figure 2 shows the dependence of the terminal velocity with rotation and viscosity. In summary we could say that polar terminal velocities of the order of 2000 Km/s are obtained while the equatorial velocity decreases from 2000 Km/s to about 1000 Km/s as the rotation rate increases from 0.5 to 0.9 . These results agree well with the velocities derived from UV lines but are in conflict with the usual belief of low velocities in the equator. However, in figure 3 we see that the velocity field in this region is dominated by rotation at the inner parts (r<3R) and by expansion at the outer parts. This could possibly be important to understand the extended wings of some H$\alpha$ profiles that reveal velocities larger than 1000 Km/s (Andrillat 1983). Finally, figure 4 shows the isodensities curves which characterize the envelope. We can see that it possesses an asymmetric (concentrated towards the equator) structure.

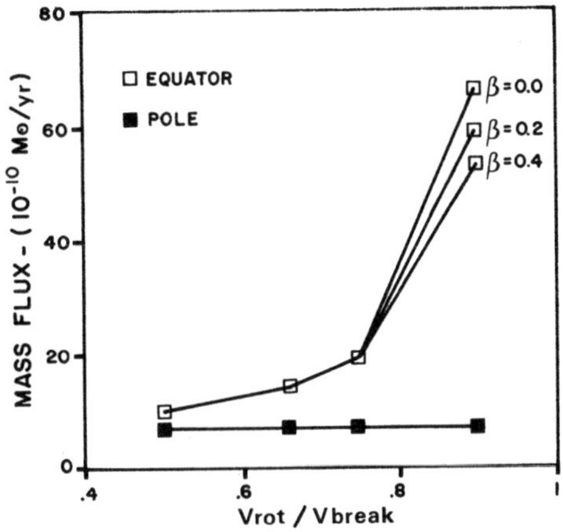

FIG. I – EQUATORIAL AND POLAR MASS FLUX AS A FUNCTION OF ROTATION AND VISCOSITY ($\beta$).

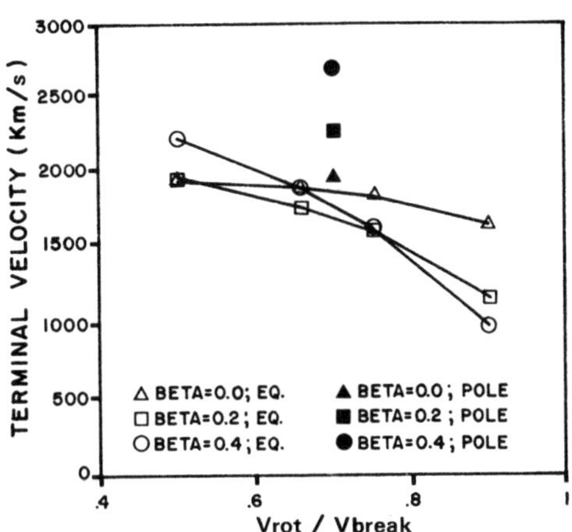

FIG. 2 – EQUATORIAL AND POLAR TERMINAL VELOCITY AS A FUNCTION OF ROTATION AND VISCOSITY ($\beta$).

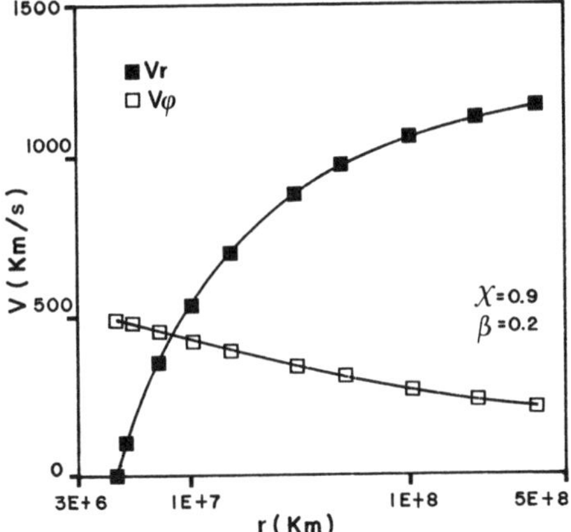

FIG. 3 – EQUATORIAL VELOCITY PROFILES FOR THE RADIAL AND FOR THE TANGENTIAL COMPONENT.

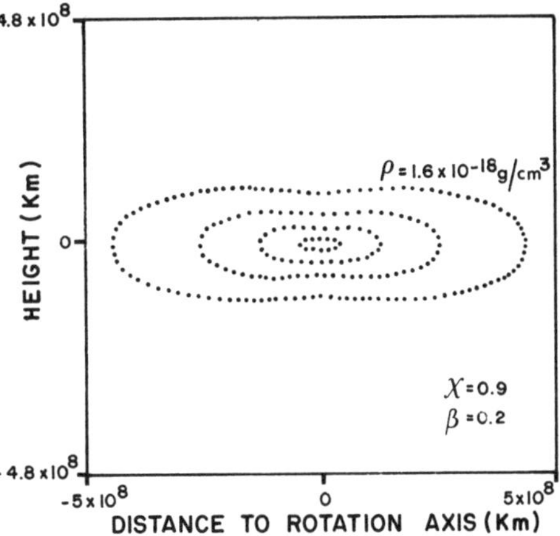

FIG. 4 – MERIDIONAL ISODENSITY CURVES.

## 4. CONCLUSIONS

Some of our results encourage us to make a comparison with Be stars
observations. For example, our model seems to support the usually
assumed ad hoc density enhancement at the equator. In addition, the
mass loss predicted are in good agreement with the rates obtained from
the IR excess and a discrepancy between equatorial and polar mass flux
arises naturally. On the other hand the radial velocity laws are rather
problematic, particularly the equatorial one. In fact, we do not claim
that our model includes all the physics necessary to explain the Be
phenomena. Nonradial pulsations for instance, which are likely to be
linked with the observed time variability and may be responsible for
periods of enhanced mass loss. The inclusion of such a driving
mechanism, among others, would probably greatly improve the present
work.

## 5. REFERENCES


Abbott, D.C. 1982, Astrophys. J., 259,282.
Andrillat, Y. 1983, Astron. Ap. Suppl., 53, 319.
Baade, D. 1987, in IAU Colloquium n 98, Physics of Be stars, A.
    Slettebak and T.P. Snow (eds.), Cambridge University Press, p. 361.
Friend, D.B. and Abbott, D.C. 1986, Astrophys. J., 311, 701.
Marlborough, J.M. 1987, in IAU Colloquium n 98, Physics of Be stars,
    A. Slettebak and T.P. Snow (eds.), Cambridge University Press, p.
    316.
Marlborough, J.M. and Zamir, M. 1984, Astrophys. J., 276, 706.
Pauldrach, A., Puls, J. and Kudritzki, R.P. 1986, Astron. Ap., 164, 86.
Poe, C.H. and Friend, D.B. 1986, Astrophys. J., 206, 182.
Puls, J. 1987, Astron. Ap., 184, 227
Slettebak, A., Kuzma, T.J. and Collins, G.N. II 1980, Astrophys. J.,
    242, 171.

# A SIMPLE CRITERION TO IDENTIFY RAPIDLY ROTATING STARS VIEWED AT SMALL TO INTERMEDIATE INCLINATION ANGLES[*]


Dietrich Baade
European Southern Observatory
Karl–Schwarzschild–Str. 2
D–8046 Garching, W. Germany



ABSTRACT.   Weak, roughly central quasi-emission bumps have been discovered in various absorption lines of a few B-type stars. Various explanations are discussed. Since the features have nearly zero velocity, they are probably evidence of differences between the line spectra of polar and equatorial regions. Rotationally induced temperature and gravity gradients or polar spots are the simplest explanation at the moment; a reliable discrimination requires observations of a larger number of ions. Regardless of the true physical cause, geometric projection effects and limb darkening let smooth equator-to-pole variations have a noticeable effect on observed line profiles only if the pole is not too far from the center of the visible stellar disk. Therefore, the phenomenon provides an easy-to-use criterion for the identification of (rapidly rotating) stars seen at small to intermediate inclination angles.


## 1. INTRODUCTION

The significance of observational studies of rotational effects on stellar structure and evolution depends largely on the extent to which the equatorial velocity, $v_{eq}$, and the inclination angle, $i$, of the rotational axis can be determined separately. 'Simple' methods include restrictions to

- o statistical deconvolutions, assuming that $v_{eq}$ or $i$ or both are randomly distributed;
- o double-lined binaries, assuming alignment of orbital and rotational axis;
- o a particular type of stars (e.g., Be stars) and assumption of a model (e.g., critical rotation);
- o stars with circumstellar envelopes, assuming a model for the envelope.

The problems inherent to such simplistic approaches are obvious; most importantly, these methods are not applicable to individual, 'normal' stars.

Observed spectra represent an integral over the visible stellar hemisphere. Any rotation-induced variations of $T_{eff}$, $g$, etc. with stellar latitude will affect observed line profiles as a function of both, $v_{eq}$ and $i$. That this can be exploited to separate the two quantities, was already pointed out (e.g., Hutchings and Stoeckley 1977) when the S/N of standard spectra still limited the application of this idea to a comparison of line widths which may even yield ambiguous results (cf. Carpenter et al. 1984).

Only few problems require the separation of $v_{eq}$ and $i$ for *any* star, i.e. a full analysis. E.g., it often is sufficient just to identify intrinsically rapidly rotating stars seen under small inclination angles. This is the subject of the present study.

---


[*]Based on observations obtained at the European Southern Observatory, La Silla, Chile





*L. A. Willson and R. Stalio (eds.), Angular Momentum and Mass Loss for Hot Stars, 177–180.*
© 1990 *Kluwer Academic Publishers. Printed in the Netherlands.*

## 2. OBSERVED LINE PROFILES

Presently, four B-type stars are known (Porri and Stalio 1988, Baade 1989, see also Baade 1983) to display curious quasi-emission bumps in several but *not all* of their absorption line profiles. As an example, Fig. 1 shows several profiles of HeI $\lambda$ 667.8, HeI $\lambda$ 447.1, and MgII $\lambda$ 448.1 in $\epsilon$ Cap (B2.5Vpe). It also documents the only known significant variation, the latter appears to be related to a shell phase transition of this emission line star.

## 3. POSSIBLE INTERPRETATIONS

Quite a few different processes could conceivably produce a central hump in an absorption line. The following outlines them only coarsely:

Line emission from circumstellar envelopes is observed in the Balmer lines of three of the four stars so far concerned. However, Balmer, FeII, and HeI emission lines of Be stars nearly always show two peaks whose separation is comparable to twice the $v \sin i$ of the absorption lines. This also holds for the three Be stars of this sample.

Chromospheric emission lines are in slowly rotating stars predicted and observed to be double. In fast rotators, the signature of a spherically symmetric chromosphere should be much diluted unless the chromosphere does not participate in the rotation. Furthermore, there is no other direct evidence of chromospheres in early-type stars.

Under non-LTE conditions, occupation number and/or temperature inversions can occur which show up as central line profile reversals. However, they, too, will be strongly smeared out in the presence of significant rotation.

Binarity requires that the two stars have the same spectrum and luminosity (yet, not all spectral lines are 'double' although their widths do not appear to differ from those that are). RV variations supporting the physical nature of such pairs have not been observed.

Deviations from solid-body rotation can make the center of an absorption line only deeper, not shallower. However, a double-lobe structure may develop still if the surface rotation rate increases from equator to pole in such a way as to weaken the line wings and to deepen the profile at intermediate velocities. The resulting shape is, therefore, in most cases *very* different from 'normal' rotation profiles. Furthermore, in the absence of other effects, it should be roughly the same for all lines of a given star.

Atmospheric velocity fields would need to be symmetric with respect to the rotational axis because there is no indication of rotational modulation. The only field known to have such properties is meridional circulation which, however, has a typical surface amplitude of order 0.1 cm s$^{-1}$ (Tassoul and Tassoul 1982).

Chemically peculiar polar caps cannot be excluded by the present data. But, if some ions are depleted, others may be enhanced and reveal themselves through enhanced central absorption. A systematic observational survey of various ions should unambiguously answer this question. Note that significant variations of composition with stellar longitude (non-polar "spots") are excluded by the lack of rotational modulation.

## 4. SIMULATIONS OF THE EFFECTS OF FAST ROTATION

The primary atmospheric effects of fast rotation were crudely simulated in a schematic model with the following properties:
- o Schuster-Schwarzschild model atmosphere
- o spherical star, visible hemisphere sampled at $\sim$10,000 lines of sight
- o rotation at 70% of break-up velocity

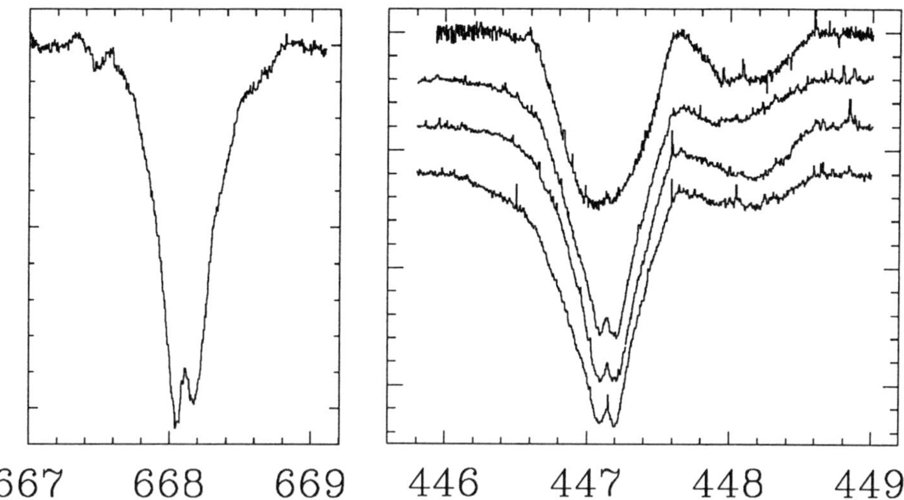

667    668    669      446    447    448    449

Fig. 1a: Profile of HeI $\lambda$ 667.8 observed in $\epsilon$ Cap (= HR8260 = HD205637) on 2 July 1985. The spacing of the small tick marks on the ordinate correspond to 0.02 flux units (continuum = 1.00). Fig. 1b: Dito, except for lines being HeI $\lambda$ 447.1 and MgII $\lambda$ 448.1 and observing dates 9 July 1982, and 7, 8, and 16 June 1983 (from top to bottom). The spectra in Fig. 1b suffer from uncertainties in their rectification. The difference in depth of HeI $\lambda$ 447.1 is real, however. In 1982 July, the H$\alpha$ emission was weak, double (peak separation $\sim$330 km s$^{-1}$), and symmetric with a fairly strong shell absorption superimposed. By 1983 June, the H$\alpha$ emission had weakened, the shell absorption disappeared.

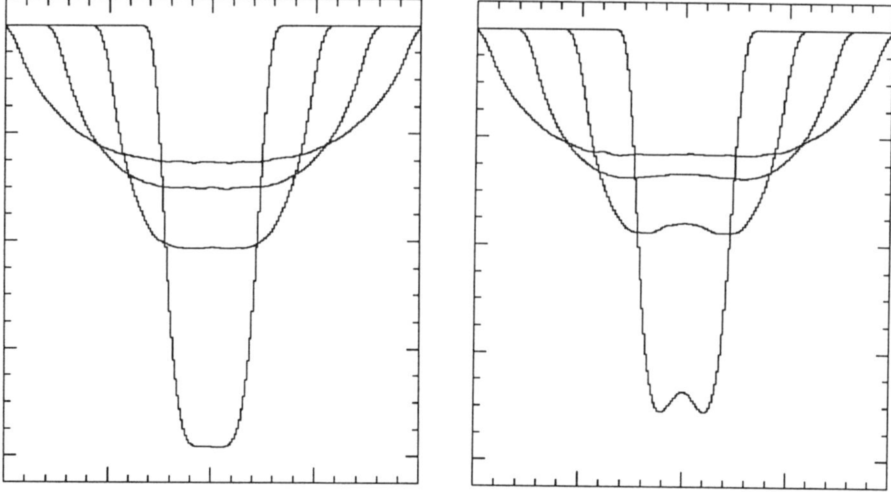

Fig. 2: Computed schematic profiles of a rapidly rotating star. The left panel displays set A, the right one set B as described in Sect. 4. The inclination angles of the rotation axis are 10, 20, 30, and 40 degrees, respectively.

o surface temperature distribution according to von Zeipel's law
o variation of line absorption coefficient arbitrarily parameterized as:

$$\kappa(T) = \kappa_{eq}(1 + \alpha(T - T_{eq})/(T_{pole} - T_{eq}))$$

Two sets of line profiles were computed, each for $i$ being 10, 20, 30, and 40 degrees. In set A, $\alpha = -0.9$; in B, $\alpha = -0.95$. The results (Figs. 2a and b) show that the polar-equatorial contrast in local line strength needs to be large (an order of magnitude) in order to lead to a central hump. At inclinations $i \geq 30°$, the effect vanishes rapidly mainly because of increasing limb darkening and geometrical projection effects.

## 5  Summary

The numerical experiments of this study are limited but clearly suggest that differential rotation will produce central quasi-emission features only at the expense of often extreme deviations of the general line shape from 'normal' rotation profiles. This needs to be ascertained through more systematic investigations. Schematic simulations identify rotational gravity brightening and associated variations of the continuous-to-line absorption ratio as a much more probable explanation; chemically peculiar polar caps obviously have the same basic effect. Unlike differential rotation, but in agreement with the observations, neither of the two models predicts the profiles of all ions to be the like. An unambiguous discrimination will be possible from observations of a wider range of ions.

The computations, in fact, show that (smooth!) equator-to-pole variations of *any* scalar atmospheric quantity will have the effects observed; however, because of limb darkening and geometrical forshortening, only if the star is seen under small to intermediate inclination angles, say $i < 40°$. In this framework, these features therefore place an approximate upper limit on $i$, regardless of the details of their explanation. This permits the selection of individual objects which warrant a more careful analysis of the effects of rapid rotation at high stellar latitudes.

A serious challenge to the proposed model may arise from the detection of central humps also in the shell star $\epsilon$ Cap (Fig. 1) whose line width corresponds to $v_{eq} \sin i \approx 250$ km s$^{-1}$. This star deserves attention also because the only known case of a variation of the features in question involves two different phases of the star's circumstellar shell. Questions to be addressed in more detail concern the geometry of the shell and the amount of masking of stellar spectral features by the shell.

Acknowledgement: I thank Dr. J. Zorec for having drawn my attention to the effects which deviations from solid-body rotation will have on observed line profiles.

## REFERENCES

Baade, D.: 1983, *Astron. Astrophys.* **124**, 283
Carpenter, K.G., Slettebak, A., Sonneborn, G.: 1984, *Astrophys. J.* **286**, 741 Baade, D.:
    1989, *Astron. Astrophys. Suppl. Ser.* **79**, 423
Hutchings, J.B., Stoeckley, T.R.: 1977 *Publ. Astron. Soc. Pacific* **89**, 19
Porri, A., Stalio, R.: 1988, *Astron. Astrophys. Suppl. Ser.* **75**, 371.
Tassoul, J.L., Tassoul, M.: 1982, *Astrophys. J. Suppl. Ser.* **49**, 317

INTENSIVE PHOTOMETRIC CAMPAIGN ON Be STARS:
BEHAVIOUR OF SHORT-TERM PERIODIC VARIATIONS
AND ITS RELATIONSHIP TO PULSATION AND MASS LOSS[*]


L.A. BALONA
South African Astronomical Observatory
J. CUYPERS
Koninklijke Sterrenwacht van België



ABSTRACT:   Results of an intensive photometric campaign on Be stars are
briefly summarized.  There is a highly significant correlation between
the photometric period and the projected rotational velocity which can
be understood in terms of a non-uniform surface brightness distribution
(starspots).  It can also be explained in terms of g-mode non-radial
pulsation with $m = -1$ or $m = -2$ if the frequency in the rotating frame
of reference is practically zero.  In this case nearly all the light
variation must be due to temperature variations and the model is
observationally indistinguishable from the starspot model.  We suggest
that the localized areas responsible for the light variations are
regions of enhanced mass loss.


1.   INTRODUCTION

We have been monitoring a large number of bright Be stars over the last
four years using the Strömgren $b$ filter.  All observations were made at
the SAAO except for two seasons of simultaneous photometry with ESO
(Cuypers, Balona & Marang 1989; Balona Cuypers & Marang 1990).  Of the
54 stars observed (most of them for more than two seasons), periods
could be found for 32 of them.  A total of 37 Be stars have know
periods.
   The fact that the amplitudes are highly variable suggests that
probably all Be stars will show measurable periodic light variations if
observed for a sufficiently long time.  A detailed discussion of these
results will be presented elsewhere (Balona, 1990).

2.   BEHAVIOUR OF THE PERIODIC LIGHT VARIATIONS

Of the 37 periodic Be stars, nearly half show two unequal maxima or
minima per cycle; that it they show *double-wave* light variations.  A few
stars are known where the light curve has changed from a single-wave to

---

* Based on observations made at SAAO and ESO.




*L. A. Willson and R. Stalio (eds.), Angular Momentum and Mass Loss for Hot Stars, 181–184.*
© 1990 *Kluwer Academic Publishers. Printed in the Netherlands.*

a double-wave from one season to the next.  Examples of the converse are also known.  Two stars seem to possess triple-wave light curves.

The peak-to-peak light amplitudes are mostly in the range 0.01 - 0.07 mag. but is as high as 0.12 mag. in a few stars.  The amplitudes are highly variable; cases are known where a significant change has occurred on a time scale of a few weeks.

The light curve often shows considerable scatter which is usually much larger than the expected observational errors.  This scatter is incoherent and not due to multiperiodicity.  We have termed this phenomenon *flickering*.  The time scale of the flickering is between 0.2 d and 5 d and could be explained in terms of orbiting circumstellar blobs.

In spite of the complex and rapidly changing light curves, the period is always constant from season to season.  Because of the changing shape of the light curve, it is difficult to determine whether the maxima and minima retain their phase from year to year, though this seems to be the case in general.  We have looked for, but have never found, evidence for coherent multiperiodicity.  The shortest known period is 0.40 d for *48 Lib*; most periods lie in the range 0.8 - 1.5 d. There is sometimes an ambiguity in deciding whether a light curve should be classified as single- or double-wave.  This can introduce a spurious factor of two in the period.

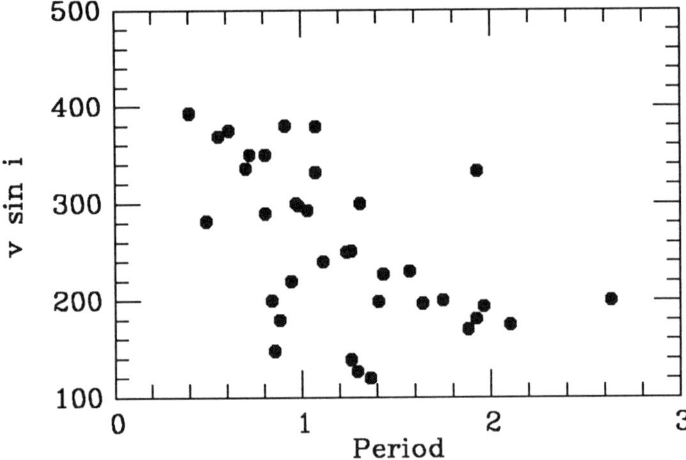

**Fig. 1:** The correlation between projected rotational velocity, $v \sin i$ (in km/s), and photometric period (in days) for periodic Be stars.

An important result is that there is a highly significant correlation between the photometric period and the projected rotational velocity, $v \sin i$ (Fig. 1).  This correlation is not greatly altered if stars with double-wave light curves are omitted.  The period ambiguity in some of these stars is therefore not a serious problem.  We have estimated the stellar radii from the spectral type and have equated the

photometric period with the period of stellar rotation. It is found
that within the expected error in the radius, the distribution of sin $i$
is consistent with all stars having sin $i < 1$ with a maximum at $i = 90^0$.
This implies that the observed light variation is caused by a
non-uniform distribution of surface brightness fixed to the photosphere
(i.e. "starspots").

## 3. RELATIONSHIP TO PULSATION AND MASS LOSS

An equivalent interpretation can be made on the basis of non-radial
pulsation (NRP). In this case the frequency of oscillation in a frame
of reference rotating with the star must be practically zero if the
correlation of Fig. 1 is to be understood. It follows that the
azimuthal spherical harmonic index $m = -1$ for single-wave Be stars and
$m = -2$ for double-wave Be stars. Changes of mode must be proposed to
explain the transition between the two types of light curve.

The long periods and the large light amplitudes imply g-modes of
very high radial order (greater than 50) and rule out purely toroidal
modes which have sometimes been advocated (Osaki 1986). From what is
known about low-order line profile variations in Be stars, it is easy to
show that in the NRP model most of the light variation must be caused by
temperature variations rather than geometrical effects. In any case, no
light variations are to be expected for the $\ell = 1$, $m = -1$ mode due to
geometrical effects alone. This model therefore becomes observationally
indistinguishable from the starspot model.

The presence of low-order line profile variations is known to be a
characteristic of Be stars and is not found in non-Be stars apart from
the pulsating $\beta$ Cep and 53 Per stars. The low order profile variations
in Be stars must be caused by the same phenomenon which is responsible
for the periodic photometric variations. We have monitored quite
intensively several dozens of non-Be stars and confirm that short-period
photometric variability is indeed generally confined to the Be stars.
Such variations were also detected in very few non-emission line stars,
but it is possible that these are quiescent Be stars or close binaries.
We failed to discover any 53 Per stars, with one possible exception,
though 53 Per itself has a rather large light amplitude.

Many B stars (Be and non-Be) show high-order line profile
variations or "moving bumps". These could be NRP modes, but other
interpretations are possible. It has become clear, however, that this
phenomenon is not connected with the presence of emission lines and
probably has no direct role in mass loss in Be stars. There is
therefore no reason to suppose that the high-order and low-order
variations are caused by the same mechanism.

It has been suggested that NRP in conjunction with rotation may be
responsible for mass loss in Be stars, though the mechanism giving rise
to NRP is itself a mystery. Our observations do not necessarily
disprove this idea, but offer the possibility of alternative
explanations. Detailed studies of line profile variations have shown
many effects which cannot be explained in terms of NRP alone (Smith *et
al.* 1987). If we suppose that the "starspots" are localized areas of
enhanced mass loss, the existence of double-wave light curves suggests

the presence of large scale (but presumably weak) dipole magnetic fields. This scenario is not unlike that of the magnetic loop model of Underhill & Fahey (1984) and lends some support to the rotating, magnetic, radiation-driven wind model of Poe & Friend (1986).

An important aspect is the relationship between the shape and amplitude of the light curve and the emission-line profiles. If a correlation can be established, it would lend support to the idea of localized regions of enhanced mass loss. We have not been able to monitor these stars spectroscopically, but it is possible that such observations exist and that such a correlation could be established. A few spectra of Be stars in NGC 3766 suggest that there is a correlation between a *change* in light curve amplitude and a *change* in emission-line strength.

## REFERENCES


Balona, L.A., 1990. *Mon. Not. Roy. astr. Soc.*, submitted.

Balona, L.A., Cuypers, J. & Marang, F., 1989. *Astron. Astrophys. Suppl.*, submitted.

Cuypers, J., Balona, L.A. & Marang, F., 1990. *Astron. Astrophys. Suppl.*, **81**, 151.

Poe, C.H. & Friend, D.B., 1986. *Astrophys. J.*, **311**, 317.

Smith, M.A., Gies, D.R. & Penrod, G.D., 1987. *Physics of Be Stars* (IAU Colloq. 92, eds. A. Slettebak & T.P. Snow), Cambridge Univ. Press, Cambridge, p.464.

Osaki, Y., 1986. *Seismology of the Sun and distant Stars* (ed. D.O. Gough), Reidel, Dordrecht, p.453.

Underhill, A.B. & Fahey, R.P., 1984. *Astrophys. J.*, **280**, 712.

# CONSTRAINTS ON THE THICKNESS OF Be STAR DISKS DERIVED FROM COMBINED IR EXCESS AND OPTICAL POLARIMETRY DATA


J.E. BJORKMAN and J.P. CASSINELLI
*Department of Astronomy*
*University of Wisconsin*
*Madison, Wisconsin 53706*



ABSTRACT. Analytic expressions are developed for the 12 $\mu$m infrared excess and for the optical intrinsic polarization, accounting for occultation of part of the envelope by the stellar disk or depolarization effects due to the finite angular size of the stellar disk. It is assumed that the wind is isothermal and that the electron scattering optical depth is less than unity for all lines of sight. The model is applied to the data for four Be stars which have inclination angles near 90° (equator-on). Fits to the observed IR excess predict a polarization larger than observed unless the model has either of *two* quite different electron density distributions: a) the electrons are in a very narrow equatorial disk (HWHM< 15°) *or* b) the electron envelope is very broad (HWHM∼ 50°). This is because polarization can be reduced relative to the infrared excess by either greatly restricting the volume of the electron envelope, or by having partial cancellation of the equatorial polarization from contributions near the pole. The two solutions have different base densities, $\rho_0$, and may be differentiated by yet other observational data.


## 1. Introduction

Be stars exhibit infrared emission in excess of that expected from photospheric emission (*e.g.* Gehrz, Hackwell and Jones, 1974; Coté and Waters, 1987). This excess is due to free-free emission originating in a dense circumstellar envelope surrounding the star. Be stars also exhibit optical polarization which indicates that this envelope is not spherically symmetric (Poeckert, Bastien and Landstreet, 1979). One model that has been suggested for explaining these observations is that the circumstellar material is confined to a thin equatorial disk, perhaps due to the rotation of the star (Poeckert and Marlborough, 1978); however, little is known about the geometry of this disk.

In this study, we choose a parameterization of the density which permits us to study the effects of a "disk-like" axisymmetric circumstellar shell on the infrared excess and optical polarization as a function of the thickness of the disk and inclination angle of the stellar rotation axis. Using observations of the IR excess and optical polarization we are then able to place constraints on the thickness of the disk.

## 2. Model

We parameterize the density in the equatorial wind by

$$\rho = \rho_0 r^{-n} \sin^m \theta,$$

(1)

185


L. A. Willson and R. Stalio (eds.), Angular Momentum and Mass Loss for Hot Stars, 185–190.
© 1990 Kluwer Academic Publishers. Printed in the Netherlands.

where $r$ is the radial distance (normalized to the stellar radius) and $\theta$ is the angle away from the stellar rotation axis. We define the opening angle of the disk, $\alpha$, by the HWHM of the density distribution.

$$\alpha \equiv \cos^{-1}\left(\frac{1}{2}\right)^{1/m}; \tag{2}$$

therefore, varying, $m$, permits us to examine the effects of changing the thickness of the disk.

The envelope is fully ionized out to some large radius so that the IR excess is produced by thermal free-free and bound-free emission. Following Lamers and Waters (1984) the flux excess is defined by

$$Z_\nu \equiv \frac{F_\nu^w + F_\nu^*}{F_\nu^*}, \tag{3}$$

*i.e.* $Z_\nu$ is the ratio of the total flux to the stellar flux. For an isothermal wind, the IR excess is

$$\begin{aligned}
Z_\nu - 1 = \frac{1}{\pi}\left(\frac{B_\nu^w}{B_\nu^*} - 1\right)\int_0^{2\pi} d\psi \int_0^1 q\,dq\left[1 - e^{-\tau}\right] \\
+ \frac{1}{\pi}\left(\frac{B_\nu^w}{B_\nu^*}\right)\int_0^{2\pi} d\psi \int_1^\infty q\,dq\left[1 - e^{-\tau}\right],
\end{aligned} \tag{4}$$

where $q$ is the impact parameter of the line of sight (LOS), $\psi$ is the position angle of the LOS, and $\tau$ is the optical depth along the LOS.

The optical polarization is produced by electron scattering; therefore the region which produces the IR excess is also responsible for the optical polarization. If we assume that the wind is optically thin at visual wavelengths, then the fractional polarization is given by

$$P = \left(\frac{3\sigma_{\rm T}}{8}\right)\int n_e\left(\frac{3K_\nu - J_\nu}{\pi B_\nu^*}\right)\left(\frac{q^2}{z^2 + q^2}\right)\cos(2\psi)dV, \tag{5}$$

where $z$ is the distance along the LOS and $V$ is the scattering volume. In many cases we have developed analytic solutions to eqs. (4) and (5) including either occultation of the circumstellar material by the stellar disk or corrections due to the finite angular size of the stellar disk. For other cases we have developed a computer code which numerically evaluates the IR excess and optical polarization (Bjorkman and Cassinelli, 1990).

## 3. Model Results

To remove the dependence on the stellar parameters in the optically thick regime, we define a normalized IR excess by

$$\mathrm{Znu} = \frac{Z_\nu}{\Gamma\left(\frac{2n-3}{2n-1}\right)E_\nu^{\left(\frac{2}{2n-1}\right)}}, \tag{6}$$

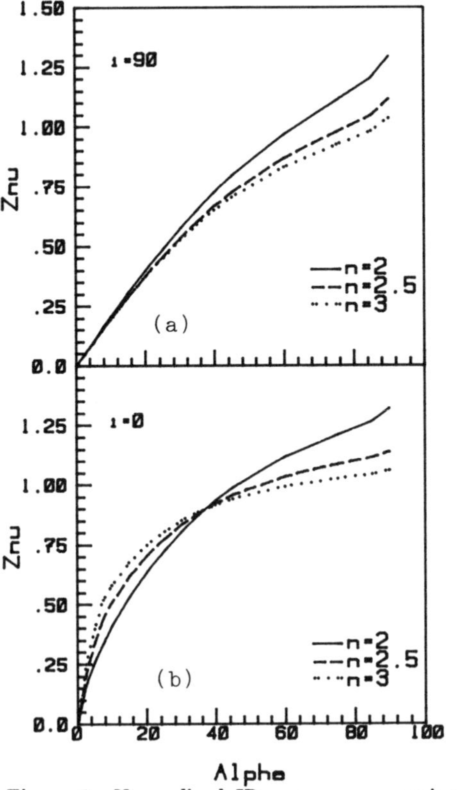

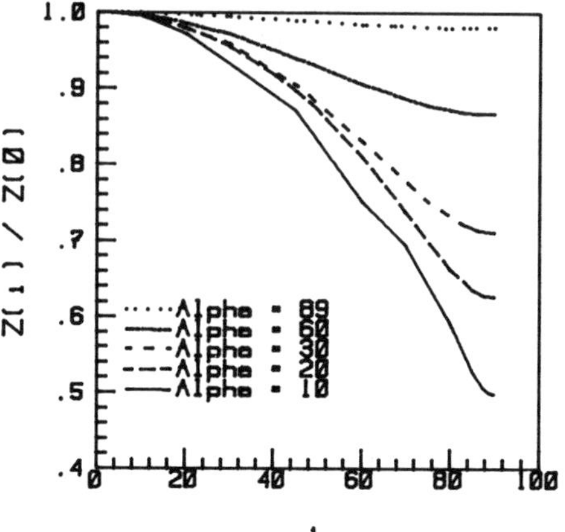

Figure 1. Normalized IR excess *vs.* opening angle for edge-on (top) and pole-on (bottom) cases for various radial exponents, *n*, of the density distribution.

Figure 2. IR excess (relative to pole-on excess) *vs.* inclination angle for various disk opening angles, $\alpha$.

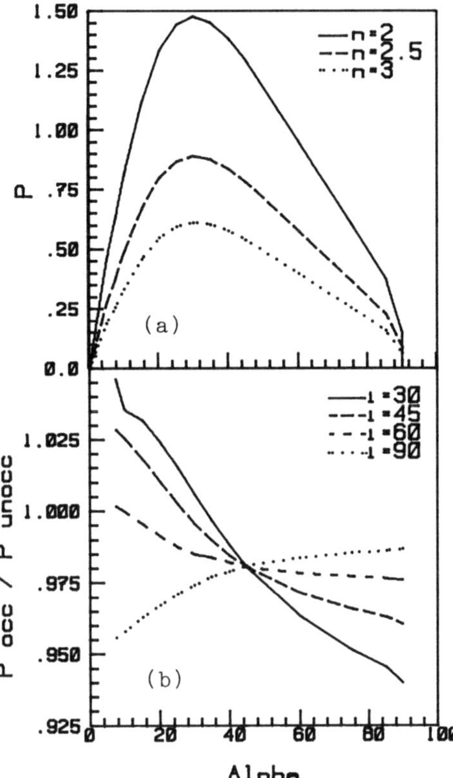

Figure 3. Normalized polarization *vs.* opening angle (top) and ratio of polarization with occultation to that without occultation *vs.* opening angle (bottom) for various inclination angles, *i*.

188

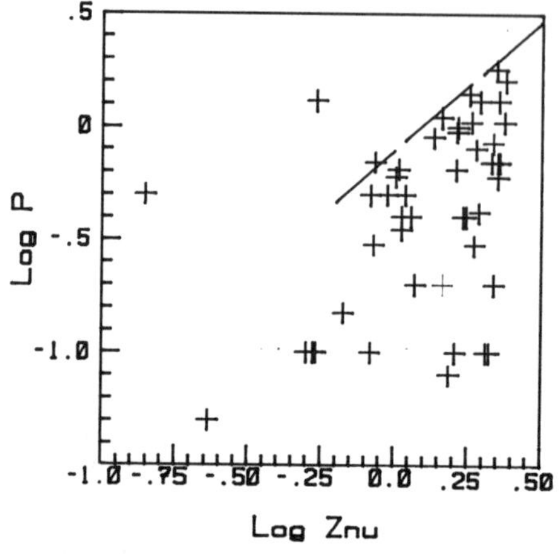

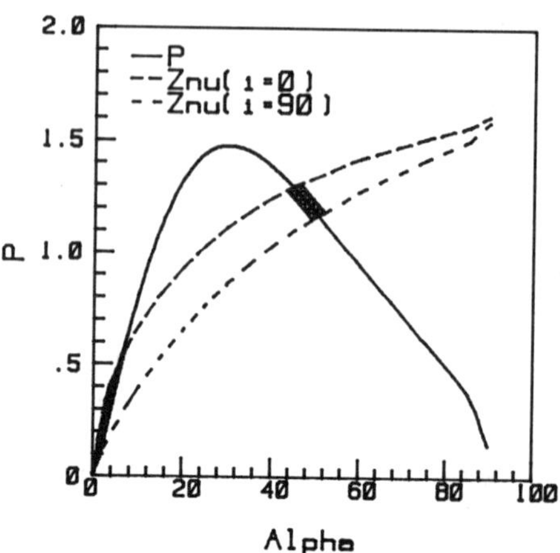

Figure 4. Polarization *vs.* IR excess for 50 Be stars (Figure from Coté and Waters, 1987; and Bjorkman, 1989).

Figure 5. Graphical solution of the opening angle using polarization and IR data for a typical star. The darkened areas indicate solution regions.

where $E_\nu = R_* \kappa_\nu (r = R_*)$ and $\kappa_\nu$ is the bound-free plus free-free opacity (see Lamers and Waters, 1984). We also define a normalized polarization by

$$P = \frac{P}{\left(\frac{3n_e \sigma_T R_*}{16\pi}\right)}. \tag{7}$$

An examination of figures (1) through (3) leads to the following conclusions:

(1a)  The edge-on IR increases approximately as the cross sectional area of the disk.

(1b)  For the pole-on IR excess, the radius at which the envelope becomes optically thin increases with the opening angle of the disk; therefore, the effective cross sectional area of the disk increases with opening angle.

(2)  The IR excess is largest pole-on and decreases approximately as $\cos i$ to the edge-on value. The largest difference between edge-on and pole-on IR excesses occurs for the thinnest disks (*e.g.* for an opening angle of 10° the edge-on excess is a factor of two smaller than the pole-on value).

(3a)  The polarization increases to a maximum at an opening angle of 30°, then decreases to zero as the envelope becomes spherically symmetric. Thus there are two opening angles which produce a given polarization.

(3b)  The contribution of the material occulted by the stellar disk is at most 5 percent of the total polarization; therefore, the analytical expression without occultation is quite accurate, and it is given by

$$P = \left(\frac{3\pi n_e \sigma_T R_*}{64}\right) \frac{\Gamma\left(\frac{n-1}{2}\right)\Gamma\left(\frac{m+2}{2}\right)}{\Gamma\left(\frac{n+2}{2}\right)\Gamma\left(\frac{m+3}{2}\right)} \left(\frac{m}{m+3}\right) \sin^2 i \tag{8}$$

This result is for a finite disk source and the inclination angle dependence is exactly the same found by Brown and McLean (1979) for point source models.

TABLE 1. Disk Density and Opening Angle For 4 Edge-On Be Stars

| HD# | Name | Spectral Class | | $v \sin i$ (km/s) | Thin Disk Solution | | Thick Disk Solution | |
|---|---|---|---|---|---|---|---|---|
| | | | | | $\rho_0$ (g/cm$^3$) | $\alpha$ (degrees) | $\rho_0$ (g/cm$^3$) | $\alpha$ (degrees) |
| 37202 | $\zeta$ Tau | B1 | IV | 220 | 4.6 $(-12)$ | < 3.7 | 1.1 $(-12)$ | $\sim 55$ |
| 137387 | $\kappa^1$ Aps | B3 | IV | 300 | 2.4 $(-12)$ | < 7.4 | 8.4 $(-13)$ | $\sim 50$ |
| 193911 | 25 Vul | B6 | IV | 200 | 1.1 $(-12)$ | < 14.0 | 5.2 $(-13)$ | $\sim 45$ |
| 217050 | EW Lac | B3 | IV | 300 | 4.2 $(-12)$ | < 6.8 | 1.4 $(-12)$ | $\sim 50$ |

We now use our model to determine the thickness of Be star disks from the optical polarization and IR excess data for four nearly edge-on Be stars. Note in figure (4) that there is an upper limit to the polarization vs. IR excess as was first pointed out by Coté and Waters (1987). The maximum polarization is attained edge-on; therefore, stars near this upper limit are likely to have inclination angles near 90°. If we assume that the disk opening angle and temperatures of these stars are similar, then along this upper limit the only difference is the base density of the wind, $\rho_0$. Eliminating $\rho_0$, from equations (4) and (5) gives (in the limit of large IR excess)

$$P(\alpha) = K Z_\nu^{\frac{2n-1}{4}}(\alpha), \tag{9}$$

where $K$ is a constant which depends on the stellar parameters. Therefore the slope of this upper limit depends only on the radial exponent, $n$, and figure (1) yields the result $n \sim 2$, *i.e.* constant velocity. The radial density exponent, $n$, also may be determined from the slope of the IR excess *vs.* wavelength. Using this method Waters, Coté and Lamers (1987) found $n$ in the range of 2 to 3.5.

We may determine the opening angle of the disk in individual cases by solving eq. (9) for $\alpha$. For simplicity we assume that the temperature of the disk equals the stellar temperature. Figure (5) shows the graphical solution to eq. (9) for a typical star. Since there is some uncertainty regarding the inclination angle, we have plotted the solution using both the pole-on and edge-on values of the IR excess in order to bracket the possible solutions. We have estimated the opening angles and base density of the disk for the four stars nearest the upper limit on the plot of $P$ *vs.* $Z_\nu$ plot and the solutions for these stars are listed in table (1). There are two ranges of possible solutions: 1) Thin Disks ($\alpha < 15°$) and 2) Thick disks ($\alpha \sim 50°$). Other observations may resolve this ambiguity.

Bound-free absorption by neutral hydrogen in the envelope as well as extinction by electron scattering will reduce the polarization by

$$P \sim e^{-\tau} P_0 \tag{10}$$

where $P_0$ is the optically thin polarization and $\tau$ is the average total optical depth from the star through the scattering volume to the observer. Longward of the Balmer limit $\tau_{b-f} \sim 0.02$; thus the reduction of the optical polarization will be small. (To estimate this we have used the non-LTE calculations of the neutral hydrogen level populations of Cassinelli, Nordsieck, and Murison, 1987.) Shortward of the Balmer limit $\tau_{b-f} \sim 0.2$; thus the reduction of the polarization may be quite large which produces a "Balmer jump" in the polarization, given by

$$\frac{P_+}{P_-} \sim e^{-\Delta \tau_{b-f}} \sim e^{-K\rho_0^2}. \tag{11}$$

This is quite sensitive to the density of the wind; therefore, the Balmer jump in the polarization may be one way to distinguish which of the two solutions (thin disk *vs.* thick disk) is the preferred geometry.

## 4. Conclusions

We have derived analytic expressions for the IR excess and optical polarization for many cases of interest. The maximum polarization occurs for disks with an opening angle of about 30°. Occultation of the circumstellar envelope by the stellar disk is relatively unimportant in its effect on the polarization, since the occulted material produces at most 5 percent of the total polarization. Simultaneous fits to the infrared excess and optical polarization give a base density of the disk of $\rho_0 \sim 10^{-12}$ g/cm$^3$ and two solutions for the allowed geometry: 1) The disk can be quite thin with opening angles $< 15°$, or 2) The disk can be relatively thick with opening angles of about 50°. The Balmer jump in polarization may distinguish these two possible solutions.

## References


Bjorkman, J.E., and Cassinelli, J.P. (1990), (in preparation).

Bjorkman, K.S. (1989), University of Colorado Ph.D. Thesis.

Brown, J.C., and McLean, I.S. (1979), Astron. Astrophys. **57**, 149.

Cassinelli, J.P., Nordsieck, K.H., and Murison, M.A. (1987), Astrophys. J. **317**, 290.

Coté, J. and Waters, L.B.F.M. (1987), Astron. Astrophys. **176**, 93.

Gehrz, R.D., Hackwell, J.A., and Jones, T.W. (1974), Astrophys. J. **191**, 675.

Lamers, H.J.G.L.M., and Waters, L.B.F.M. (1984), Astron. Astrophys. **136**, 37.

Poeckert, R., Bastien, P., and Landstreet, J.D. (1979), Astron. J. **84**, 812.

Poeckert, R. and Marlborough, J.M. (1978), Astrophys. J. Supp. **38**, 229.

Waters, L.B.F.M., Coté, J., and Lamers, H.J.G.L.M. (1987), Astron. Astrophys. **185**, 206.

# ON THE CORRELATION BETWEEN PULSATION AMPLITUDE AND SHELL ACTIVITY IN THE Be STAR λ ERIDANI


C. T. BOLTON
*David Dunlap Observatory,*
*P.O. Box 360,*
*Richmond Hill, Ontario L4C 4Y6,*
*Canada*

S. ŠTEFL
*Astronomical Institue,*
*Czechoslovak Academy of Sciences*
*251 65 Ondrejov,*
*Czechoslovakia*



ABSTRACT. During the period 1974-1988 the radial velocity of the Be star λ Eridani varied with a period of $0.701\,715 \pm 0.000\,005$ days. The amplitude and mean velocity of the velocity curve change on a time scale of years, but there is no evidence that these changes are correlated with the level of emission activity.


## 1. Introduction

Although λ Eridani ($=$HD 33328, B2IVne) was noted as a possible Be star almost 50 years ago (Mohler 1940) and its emission line variations have been monitored since 1955 (Hubert-Delplace and Hubert 1979), it was not widely known as a Be star until the emission was 'discovered' by Irvine (1975). Subsequently, Bolton (1982) discovered that it is a periodic radial velocity variable with a period of 0.701538 days. He also reported that the light variations, which had been discovered earlier by Percy and Lane (1977), had the same period. The short period and the variations in the amplitude of both the light and radial velocity variations led him to suggest that the variations are due to nonradial pulsation. He also suggested that the variation in the amplitude of the radial velocity curve might be correlated with the emission activity.

λ Eri was included in Penrod's (1987) study of the line profile variations of 25 Bn and rapidly rotating Be stars. He found that the star is pulsating in both low order, $\ell = 2$, and intermediate order, $\ell = 8$, nonradial modes. The low order mode has the same period as the radial velocity variations reported by Bolton. Penrod also suggested that the amplitude of this mode is closely correlated with the emission activity. These results have been confirmed by Smith *et al.* (1987) and Smith (1989), but the latter found no evidence for a correlation between the amplitude of the $\ell = 2$ mode and the level of emission activity.

We have taken 145 photographic blue spectra during the period September 1975 to December 1988 to monitor the long term variations in the He I and H I radial velocities, the Balmer emission lines and the He I equivalent widths and line depths in order to look for connections between variations of the photospheric lines and the shell activity. Some preliminary results are presented below.

191


*L. A. Willson and R. Stalio (eds.), Angular Momentum and Mass Loss for Hot Stars, 191–194.*
© 1990 *Kluwer Academic Publishers. Printed in the Netherlands.*

## 2. Observations and Analysis

The spectra were recorded on a mix of vacuum sensitized IIa-O and IIIa-J plates using the cassegrain spectrograph on the 1.88 m telescope of the David Dunlap Observatory. The dispersion of the spectra is 12 $\text{Åmm}^{-1}$. The spectra were scanned with a PDS microdensitometer and reduced to relative flux normalized to unit continuum *vs.* heliocentric wavelength using standard techniques implemented in a software package developed at DDO.

Relative radial velocities were measured for 144 of the spectra by cross-correlating them with the remaining spectrum of $\lambda$ Eri. The radial velocity of this spectrum was measured by estimating the positions of the lines from a large scale tracing. The He I and H I velocities were analysed separately. The final radial velocities for both groups of lines are accurate to about $\pm 10$ km $\text{s}^{-1}$. Period analyses were done using the phase dispersion minimization (PDM) algorithm (Stellingwerf 1978). The amplitude and mean velocity of the velocity curves for individual seasons were determined formally by fitting a circular orbit to the velocity curve using the program SPEL developed by J. Horn at Ondrejov.

## 3. Short-Term Variability

Both the PDM period analysis and the formal orbital solutions for the different sets of mean He I and H I radial velocities yield a period of $0.701\,715 \pm 0.000\,005$ days with a high significance. We found no other significant period in the radial velocities in the interval between 0.5 and 5 days. The scatter of the equivalent widths ($\approx \pm 10\%$) and depths ($\approx \pm 2\%$) of the He I $\lambda\lambda 4026, 4472$ Å lines about their mean values are consistent with our estimates of the measuring errors. This suggests that any variations in these parameters must be smaller than the measuring errors. We found no significant periods in these data in the range 0.2 to 5 days, and plots of these data *vs.* the 0.7 day period confirm the lack of variability for this period.

## 4. Long-Term Variability.

There are eight "seasons" in which we have obtained an adequate number of spectra with a reasonably uniform coverage of the 0.7 day period to allow us to carry out formal orbital solutions for the velocity curve. The range of Julian dates included in each "season", the number of radial velocities used in the solutions and the amplitudes and mean velocities obtained from these solutions are summarized in Table 1. The numbers in parentheses are the standard errors of the parameters expressed as the number of units uncertainty in the last digit quoted.

We have included a qualitative estimate of the strength of the emission activity in the last column of Table 1. This is based primarily on inspection of the H$\beta$ profiles from our spectra. The classification criteria are as follows: none–deep symmetric absorption with no evidence of variations in line depth or shape; weak–core of line appears slightly filled in and narrower, small variations in line profile; moderate–marked filling and narrowing of line and variations in the line profile, no emission above the continuum, strong–like moderate, only more so, with emission peaks extending above the continuum.

The errors obtained from the formal fits to the velocity curves must be interpreted with caution. The distribution of the observational errors among several parameters may lead to an unrealistically small error estimate for any one of these. Moreover, the results can be biased in a way that is not reflected in the formal error estimates if the observations are not well distributed in phase. When these factors are taken into account, it appears that, with the exception of the 1988 season, any variations in the mean He I and H I velocities are small. The small systematic difference between the velocities for the two groups of lines is probably due largely to the adopted laboratory wavelengths for the He I triplet lines. The large mean velocity for the He I lines in 1988 is clearly real. Ten of the 11 radial velocities are larger than the largest mean in any other season, and there is no evidence that the He I lines are distorted in a way that could account for the discrepancy.

TABLE 1. Amplitude and mean velocity from formal orbital fits to the $\lambda$ Eridani velocity curves

| Year | JD (244 0000+) | no. | He I Amplitude (km s$^{-1}$) | H I Amplitude (km s$^{-1}$) | He I Mean (km s$^{-1}$) | H I Mean (km s$^{-1}$) | Emission Activity |
|---|---|---|---|---|---|---|---|
| 1974 | 2317-2382 | 6 | 8.5(41) | 3.5(21) | 15.8(23) | 7.6(12) | None |
| 1976-77 | 3090-3229 | 33 | 23.8(11) | 19.5(09) | 13.6(08) | 9.4(07) | Strong |
| 1977 | 3470-3509 | 14 | 11.0(19) | 10.6(24) | 15.1(11) | 10.1(13) | Weak |
| 1979-81 | 4136-4663 | 9 | 9.6(12) | 10.3(37) | 18.1(12) | 15.9(29) | Moderate (1979) to Weak (1980-81) |
| 1981-82 | 4889-5006 | 18 | 9.0(08) | 6.5(06) | 14.9(06) | 11.0(05) | None |
| 1982-83 | 5236-5400 | 25 | 13.8(14) | 6.8(10) | 16.2(11) | 11.4(08) | Weak |
| 1983-84 | 5628-5732 | 26 | 11.0(17) | 9.6(12) | 13.5(13) | 7.4(09) | Weak |
| 1988 | 7480-7504 | 11 | 11.0(22) | 11.7(48) | 25.7(19) | 7.7(41) | Weak-Strong |

The amplitudes of the velocity curves derived from both the He I and H I lines in 1976-77 are more than twice the mean of the amplitudes for the other seasons, and the He I amplitude in 1982-83 is also significantly larger than the mean. Otherwise there is no evidence that the amplitudes are variable. There is a tendency for the He I velocity curves to have slightly larger amplitudes than the H I curves, but that effect is dominated by the 1976-77 and 1982-83 seasons when the He I curves are most atypical.

There is no evidence in Table 1 for a correlation between changes in the velocity curve and the emission activity. The two seasons when the amplitude of the He I velocity curve is significantly larger than the mean have very different levels of emission activity. The largest difference in the mean velocities of the He I and H I lines occurred just before, or at the onset of, an episode of strong emission activity (Baade 1989). This might be significant, though it is curious that the change seems to have affected the He I velocities, since one might naïvely suppose that the onset of an emission episode would be related to distortions of the H I lines.

## 5. References


Baade, D. (1989) 'Another chapter of the $\lambda$ Eridani story', *Be Star Newsletter*, **20**, 14-16.

Bolton, C. T. (1982) 'A Preliminary Report on Simultaneous Ultraviolet and Optical Observations of $\lambda$ Eridani', in M. Jaschek and H.-G. Groth (eds.), *Be Stars*, D. Reidel Publishing Co., Dordrecht, pp. 181-183.

Hubert-Delplace, A.-M. and Hubert, H. (1979) *An Atlas of Be Stars*, Paris-Meudon Observatory, Paris.

Irvine, N. J. (1975) 'New Bright Hydrogen Emission Stars', *Astrophysical Journal*, **196**, 773-775.

Mohler, O. (1940) 'Six New Be Stars', *Astrophysical Journal*, **92**, 315.

Penrod, G. D. (1987) 'Nonradial Pulsations and the Be Phenomenon', in A. Slettebak and T. P. Snow (eds.), *The Physics of Be Stars*, Cambridge University Press, Cambridge, pp. 463

Percy, J. R. and Lane, M. C. (1977) 'Search for $\beta$ Cephei Stars. I. Photometric and Spectroscopic Studies of Northern B-Type Stars', *Astronomical Journal*, **82**, 353-359.

Smith, M. A. (1989) 'Transients in the He I $\lambda6678$ Line of $\lambda$ Eri (B2e): Magnetic Quasi-Cycles?, *Astrophysical Journal Supplement*, **71**, 357-386.

Smith, M. A., Gies, D. R. and Penrod, G. D. (1987) 'Spectral Transients in the Line Profiles of $\lambda$ Eridani', in A. Slettebak and T. P. Snow (eds.), *The Physics of Be Stars*, Cambridge University Press, Cambridge, pp. 464-465.

Stellingwerf, R. F. (1978) 'Period Determination Using Phase Dispersion Minimization', *Astrophysical Journal*, **224**, 953-960.

B[e] SUPERGIANTS: CONTINUUM POLARIZATION BY ELECTRON SCATTERING
IN ROTATIONALLY DISTORTED, RADIATION DRIVEN STELLAR WINDS


Cameron J. Boyd and J. M. Marlborough
Department of Astronomy
The University of Western Ontario
London, Ontario, N6A 3K7, Canada


ABSTRACT. A rotationally distorted stellar wind of the CAK type for a
star rotating at 80% of the critical speed yields a polarization of only
0.05% to the red of the Balmer limit, two orders of magnitude smaller
than the observed values. Some directions for possible improvement of
the model are suggested.


## 1. INTRODUCTION

B[e] stars are luminous stars of spectral type B, which have prominent
continuous spectra. Their optical spectra show strong Balmer emission
lines together with many emission lines, both permitted and forbidden,
from metals in low stages of ionization. As a group they have large
infrared excesses, but are not generally associated with regions of star
formation. The low excitation lines are narrow with widths of 5-50
km/s. In the UV the spectra show broad absorption lines of N V, C IV
and Si IV with edge velocities in the range $10^3$ - 1.5 x $10^3$ km/s. A
brief summary of properties of B[e] supergiants is given by Zickgraf
(1989).
   Zickgraf et al. (1986) have suggested that the hybrid spectra of
B[e] supergiants can be understood in terms of a model in which the
circumstellar envelope is a two component wind. In the polar regions a
high speed, low density wind of the CAK type (Castor et al. 1975) exists
with a kinetic temperature consistent with that expected from the polar
radiation field of a B star. The equatorial wind is of low speed and
high density and is cool enough to allow the existence of neutral and
singly ionized metals and dust.
   Recently Zickgraf and Schulte-Ladbeck (1989) have presented linear
polarization data for several B[e] supergiants. They concluded for at
least two of these stars, MWC 645 and MWC 939, that the polarization is
intrinsic and its wavelength dependence is consistent with electron
scattering modified by bound-free absorption by hydrogen in the wind.
For MWC 645 the polarization just longward of the Balmer jump is 6-7%,
while for MWC 939 it is 1-2% at the same wavelength.
   In this paper we describe the results of our attempt to model the

195

*L. A. Willson and R. Stalio (eds.), Angular Momentum and Mass Loss for Hot Stars*, 195–198.

wind of a B[e] supergiant in terms of a radiation driven wind of the CAK type.


## 2.  THE WIND MODEL

For ease of description imagine a spherical polar coordinate system with origin at the centre of the star and rotation axis the z axis.  We choose the star to have the following properties:  $M = 30\ M_\odot$, $R = 75\ R_\odot$, $L = 8.11 \times 10^5\ L_\odot$, $T = 20,000$ K, and $\Gamma = 0.5$, and we neglect the distortion of the star by rotation.  These parameters are similar to those deduced by Lamers (1986) for P Cyg.  The electron temperature of the wind is assumed to vary as $r^{-1/2}$, independent of the polar angle $\theta$. We assume the wind is of the CAK type, with the finite angular size of the star included (Friend and Abbott 1986).  For the radiation force we assume constant k and $\alpha$ parameters with $k = 0.28$ and $\alpha = 0.56$ as suggested by Abbott (1982).

We assume the structure of the wind along the rotation axis is given by the solution of the equation for a spherically symmetric wind, with the requirement that the electron scattering optical depth through the wind to the stellar surface be 2/3.  We include rotation, assuming the rotational speed $\propto 1/r$, and solve for the structure of the wind in the equatorial plane using the same requirement for the electron scattering optical depth.  Our procedure is similar to that of Friend and Abbott (1986).  The density structure of the complete wind model is obtained from the simple assumption that

$$\rho(r,\theta) = \rho_{\rm p}\ (r) + [\rho_{\rm e}\ (r) - \rho_{\rm p}\ (r)]\ \sin^2 \theta\ ,$$

where the subscripts p and e refer respectively to the polar and equatorial solutions, and in each ease $\rho_{\rm p}$ and $\rho_{\rm e}$ are obtained from the velocity v(r) and the mass loss rate given by the wind solution.

Complete solutions between the surface and 300 stellar radii were obtained for equatorial rotation speeds up to 0.7 of the critical speed. At larger rotation speeds solutions could only be otained to some maximum distance which decreased with increasing rotation speed.  For a rotation speed of 0.8 of the critical speed the numerical solution was obtained only to 3.5 R.  The solution was extrapolated to larger radii using a simple mathematical form wich was tested on the complete solutions at smaller rotation speeds.  Friend and Abbott (1986) report analogous mathematical difficulties.

We find that the ratio of equatorial to polar mass loss rate as a function of the ratio of rotation speed to the critical speed increases from 1 at zero rotation to 2.2 at 0.8 of the critical speed.  Our results agree well with those of Friend and Abbott (1986, Fig.4).  The ratio of equatorial to polar terminal speed decreases from 1 at zero rotation to ~ 0.45 at 0.8 of the critical speed; at zero rotation the terminal speed = 637 km/s.  For r > 1.5 R the ratio of equatorial to polar density exceeds a factor of 4.

## 3. IONIZATION EXCITATION EQUILIBRIUM

The wind was assumed to consist of hydrogen and helium, with hydrogen
having 8 bound levels, including 2S and 2P separately.  For helium only
the ground states were included.  Both radiative and collisional
ionization and both radiative and three-body recombination were included
for all levels.  In the radiative ionization terms both the stellar and
the diffuse radiation from the entire wind were considered.  Collisional
transitions between all bound levels were included also.  Radiative
transitions to lower energy levels were treated under the nebular case A
assumption, i.e. all such line radiation escapes.  The nebular case B
gave essentially the same electron number density in the region of the
wind where scattering of photospheric radiation occurred.  Both free-
free emission and absorption were included.  For the continuum linear
polarization single scattering was assumed.  The procedure for solving
the transfer equation is described by Poeckert and Marlborough (1978).

## 4. CONTINUUM LINEAR POLARIZATION

The predicted continuum linear polarization as a function of wavelength
is shown in Fig. 1 for 3 rotation speeds for the case that the line of
sight is perpendicular to the rotation axis.  This arrangement maximizes
the polarization for a given rotation speed.  Even for a rotation speed
of 0.8 of critical the predicted polarization on the longward side of
the Balmer limit is only ~ 0.05%, about two orders of magnitude smaller
than what is observed.

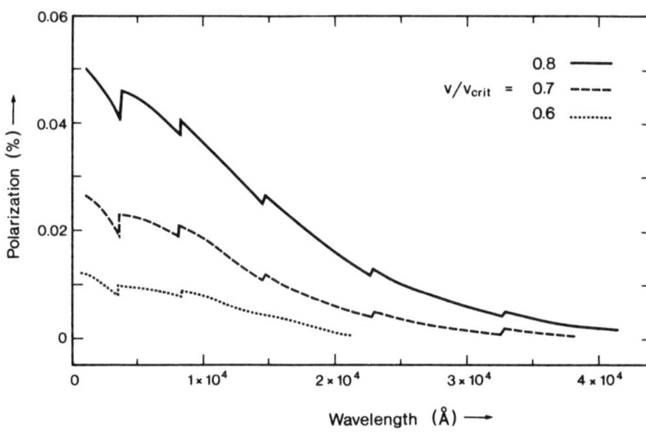


Figure 1.  Continuum
polarization (%) as a
function of wavelength
for models rotating at
0.6, 0.7 and 0.8 of
the critical speed.
For all cases the line
of sight is perpendi-
cular to the rotation
axis.

There are two reasons for the very low value of the predicted
polarization.  First, the contribution to the Stokes parameter Q, the
polarized scattered flux, from the region of the wind $45 < \theta < 90$ is
partially cancelled by scattering in the region $0 < \theta < 45$,  However,

inspection of the detailed radiative transfer solution indicates that removal of the portion of the wind with $\theta < 45$ will yield an increase in polarization by no more than a factor of 2.  Second, not enough stellar radiation is scattered in the wind so that the fraction of this which is polarized at any wavelength is much too small relative to the total radiation emitted by the system at that wavelength.  The small amount of scattered radiation is a direct consequnce of the rapid acceleration of a CAK type wind, i.e. a wind driven by strong lines, and thus of the rapid decrease in density which accompanies this, even in the equatorial regions of a star rotating at 80% of the critical speed.  This appears to be the major effect responsible for the small polarization.

To increase the polarization produced in the type of wind model described here one must increase significantly the density in regions near the equatorial plane.  A slower wind will yield a higher density even for the same equatorial mass loss rate as the CAK wind.  Two ideas for producing slow winds, which deserve to be investigated further for B[e] supergiants, are:  i) a wind driven radiatively by a large number of weak lines, as has been proposed by Lamers (1986) for P Cyg, and ii) a wind driven by acoustic wave dissipation (Pijpers and Hearn 1989). Whether either approach can yield polarizations of order 5% remains to be seen.

**ACKNOWLEDGEMENT**

We thank L.B.F.M. Waters for his comments.  This research was supported by NSERC, the Natural Sciences and Engineering Research Council of Canada.

**REFERENCES**

Abbott, D.C.: 1982, Ap.J. 259, 282.
Castor, J.I., Abbott, D.C., and Klein, R.I.: 1975, Ap. J. 195, 157.
Friend, D.B., and Abbott, D.C.: 1986, Ap. J. 311, 701.
Lamers, H.J.G.L.M.: 1986, Astr. Ap. 159, 90.
Pijpers, F.J., and Hearn, A.G.: 1989, Astr. Ap. 209, 198.
Poeckert, R., and Marlborough, J.M.: 1978, Ap. J. 220, 940.
Zickgraf, F.-J.,: 1989, in *Physics of Luminous Blue Variables*, eds.
  K. Davidson, A.F.J. Moffat, and H.J.G.L.M. Lamers (Kluwer), p. 117.
Zickgraf, F.-J., and Schulte-Ladbeck, R.E.: 1989, Astr. Ap. 214, 274.
Zickgraf, F.-J., Wolf, B., Stahl, O., Leitherer, C., and Appenzeller,
  I.: 1986, Astr. Ap. 163, 119.

# ROTATIONAL EVOLUTION OF HOT STARS DUE TO MASS LOSS AND MAGNETIC FIELDS


D. B. FRIEND
Department of Physics
Weber State College
Ogden, Utah 84408
USA



ABSTRACT. To obtain estimates of magnetic field strengths in hot stars, we have computed the evolution of rigidly rotating 15 and 30 solar mass stars, including the effects of magnetically-coupled, line-driven mass loss. Using mean rotational velocities derived from observations of main sequence stars of these masses to specify the initial state, we followed the variations in time of the surface rotation rate due to internal changes and wind-related braking. The initial magnetic field strength was varied until the calculated rotational velocity at the onset of the blue supergiant phase was in reasonable agreement with the mean value derived from observations. We find that, for both models, the computed rotational velocity decrease in the presence of a magnetic field of even modest strength (100 gauss) exceeds the limits set by observations.


## 1. Introduction

Magnetic fields can play an important role in the rotational evolution of stars, because a magnetic wind can carry away a significant amount of angular momentum. Hot, luminous, stars have massive stellar winds, but it is not known whether they have magnetic fields large enough to cause significant magnetic braking. Direct observation of magnetic fields (through Zeeman broadening) on hot stars is very difficult, since the spectral lines in these stars are broadened so much by rapid rotation, turbulence, and the stellar wind (Barker *et al.* 1981). Also, if the geometry of the magnetic field is complex enough, even large fields will not be observable.

One way to get an estimate of magnetic field strengths in hot stars would be to calculate the amount of spin-down due to magnetic braking in a stellar wind, and compare with observations of the rotational velocities of hot stars at different stages of their evolution. In order to do this, we would need a model for the wind from a rotating, magnetic, hot star. Such a model was made by Friend and MacGregor (1984). This model assumes that line radiation pressure is the dominant force driving the wind (as in Castor, Abbott, and Klein 1975), but also includes rotational and magnetic forces as in the solar wind model of Weber and Davis (1967). The model treats only the equatorial plane, and assumes axial symmetry, so that the wind equation of motion is one-dimensional. The magnetic field is assumed to be in the form of open field lines in the equatorial plane. The rotational velocity and surface magnetic field strength can be specified as inputs, and the mass loss rate and velocity of the wind are determined by solving the wind equation of motion. As described by Friend and MacGregor (1984), the mass loss rate in the wind can be enhanced by rapid rotation, and the terminal velocity can be




*L. A. Willson and R. Stalio (eds.), Angular Momentum and Mass Loss for Hot Stars, 199–203.*
© 1990 *Kluwer Academic Publishers. Printed in the Netherlands.*

enhanced by a large magnetic field.

## 2. Model and Results

The angular momentum loss rate in the wind depends on the rotation rate of the star, the mass loss rate in the wind, and the magnetic field strength. For a main sequence star of a given spectral type, we can assume an initial rotational velocity and magnetic field strength and follow the evolution of the rotational velocity as the star evolves. In order to do this we need an evolutionary model to specify the stellar parameters that affect the wind as the star evolves. We have used the code of Eggleton (1971, 1972) for this purpose. To calculate the rotational velocity as a function of time, we have assumed that the star rotates like a rigid body, though the changing moment of inertia is used from the interior model. We also assume that magnetic flux is conserved, so that the surface field strength decreases as the star grows larger. We made models for stars with initial masses of 15 and 30 solar masses, and followed the evolution of rotational velocity for two values of magnetic field strength: zero and 100 gauss. The evolutionary tracks for these two model stars are shown in figure 1.

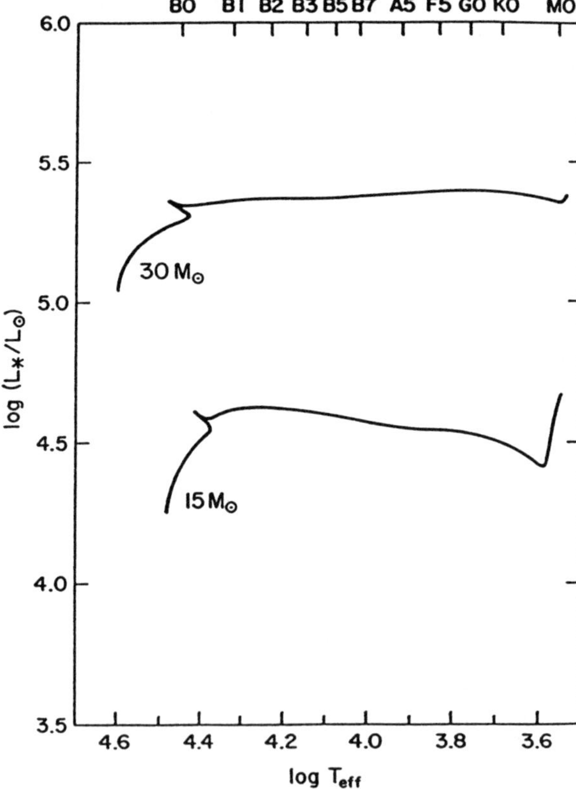


Figure 1. Evolutionary tracks for 15 and 30 solar mass stars undergoing mass loss. The spectral type calibration at the top of the figure is that of Humphreys (1978) for luminosity classes I and II.

In order to compare our computed rotational velocities with observations, we need a large sample of rotational velocities of stars at various stages of evolution. Fukuda (1982) has tabulated the rotational velocities of a very large number of stars of all spectral types and luminosity classes, and has binned them according to spectral type and luminosity class. We have chosen our initial rotational velocities from Fukuda's averages for main sequence stars of the appropriate spectral types. In order to compare our rotational velocity as the star evolves with observations, we have followed the evolution of the model stars up to the point where they would be classified as supergiants. We then can compare our computed rotational velocities to the average observed values from Fukuda for supergiants of the appropriate spectral types. Figures 2 and 3 are plots of the rotational velocity as a function of time for our two model stars, both for zero magnetic field and for a magnetic field of 100 gauss. At the onset of the blue supergiant phase, we have also plotted the average observed value from Fukuda. We see that in both cases the observed value lies slightly above the curve, even for the zero magnetic field case. Even a 100 gauss field seems to make the star spin down too rapidly. Since, for the 30 solar mass star, the calculated values of rotational velocity are well below the observed value, we also ran a zero magnetic field model for a 30 solar mass star with a smaller mass loss rate. This smaller mass loss rate is taken from the empirical relation of Garmany and Conti (1984), and is more consistent with observations of O star winds. The calculated rotational velocity curve is plotted in figure 4, and we see that this model (with no magnetic field) is very consistent with Fukuda's observed value of rotational velocity.

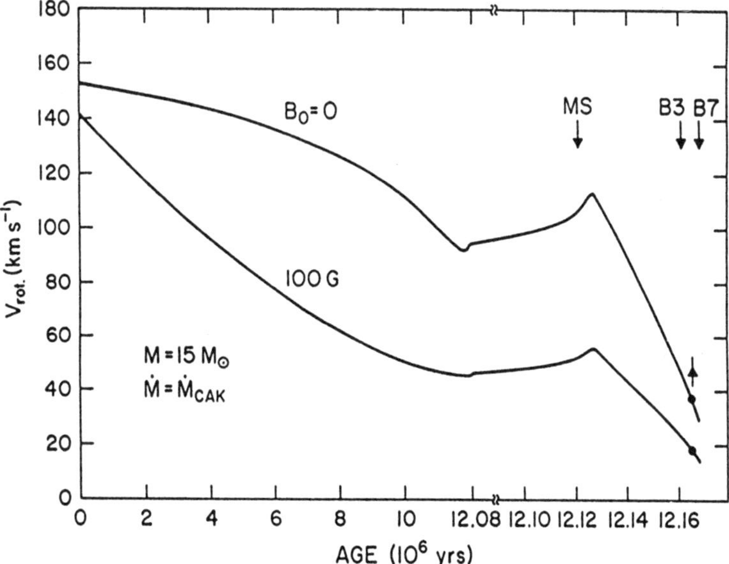


Figure 2. The rotational evolution of a 15 solar mass star with the mass loss rate taken from the wind model, for initial surface magnetic field strengths of 0 and 100 gauss. Arrows are drawn at times marking the end of main sequence evolution and the period during which the star belongs to luminosity class I and has a spectral type in the indicated range. The heavy dot on each curve denotes the average value of the rotational velocity during the supergiant phase, while the triangle represents the observed value of v sin i from Fukuda (1982).

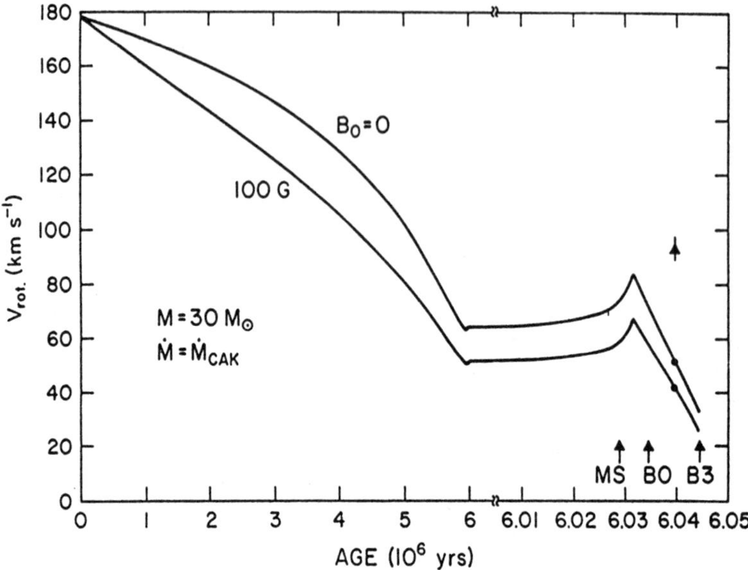



Figure 3.  The rotational evolution of a 30 solar mass star.  All symbols are as in figure 1.

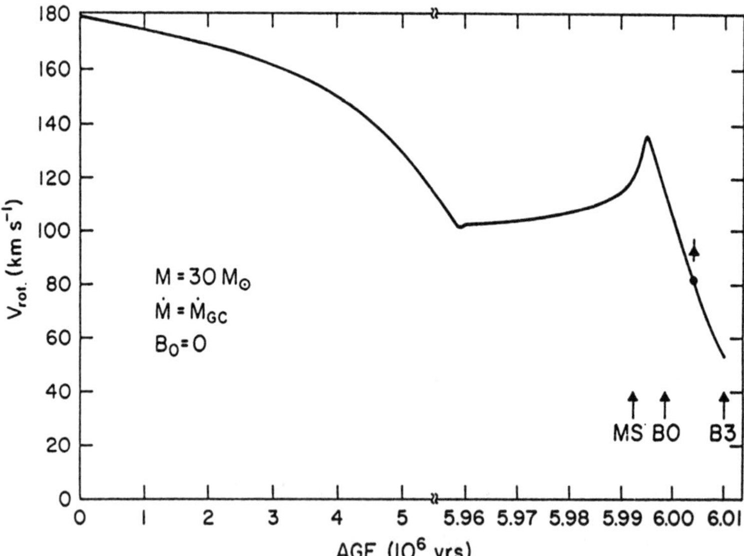



Figure 4.  The rotational evolution of a 30 solar mass star, but with the mass loss rate taken from the empirical relation of Garmany and Conti (1984).

## 3. Conclusions

We see that a magnetic field strength of even 100 gauss spins down 15 and 30 solar mass stars more rapidly than the observations indicate. We conclude that the average open magnetic field on the equator of OB stars must be less than 100 gauss, and the calculations are most consistent with no magnetic fields at all. We must remember that this analysis says nothing about *closed* magnetic fields, or fields at the *poles* of hot stars, since these fields do not affect angular momentum loss through a stellar wind.

There are a few provisos which should also be noted about these models. Fukuda's bins include not only the stars we are interested in, but also stars of different initial masses which happen to pass through the same point on the H-R diagram. Since these stars will be at different stages of evolution, the averages do not truly represent stars of a unique age or spin-down state. Another problem with our model is the assumption of rigid body rotation. Stars probably do not rotate as rigid bodies, and the envelope could well spin down more rapidly than the core. This effect would lower the computed rotational velocites even more, and make our conclusion even stronger. However, the presence of substantial closed field regions on the equator could offset this tendency. We should also note that we are only dealing with the averages for a large number of stars, and this analysis does not preclude individual stars from having large open magnetic fields.

## 4. Acknowledgements

This work is a condensation of a paper by MacGregor, Friend, and Gilliland (1990), and I thank Keith MacGregor and Ron Gilliland for permission to use our results in this conference prior to publication. I also thank the Physics Department at Weber State College for graciously allowing me to miss a week of classes to attend this meeting.

## 5. References

Barker, P. K., Landstreet, J. D., Marlborough, J. M., Thompson, I., and Maza, J. (1981) 'The Magnetic Field of Zeta Puppis', *Ap. J.*, **250**, 300.

Castor, J. I., Abbott, D. C., and Klein, R. I. (1975) 'Radiation-Driven Winds in Of Stars', *Ap. J.*, **195**, 157.

Eggleton, P. P. (1971) *M. N. R. A. S.*, **151**, 351.

Eggleton, P. P. (1972) *M. N. R. A. S.*, **156**, 361.

Friend, D. B., and MacGregor, K. B. (1984) 'Winds from Rotating, Magnetic, Hot Stars. I. General Model Results', *Ap. J.*, **282**, 591.

Fukuda, I. (1982) 'A Statistical Study of Rotational Velocities of the Stars', *Pub. A. S. P.*, **94**, 271.

Garmany, C. D., and Conti, P. S. (1984) 'Mass Loss in O-Type Stars: Parameters Which Affect It', *Ap. J.*, **284**, 705.

Humphreys, R. M. (1978) 'Studies of the Luminous Stars in Nearby Galaxies. I. Supergiants and O Stars in the Milky Way', *Ap. J. Suppl.*, **38**, 309.

MacGregor, K. B., Friend, D. B., and Gilliland, R. L. (1990) 'Winds from Rotating, Magnetic, Hot Stars: Consequences for the Rotational Evolution of O and B Stars', *Astr. Ap.*, submitted.

Weber, E. J., and Davis, L., Jr. (1967) 'The Angular Momentum of the Solar Wind', *Ap. J.*, **148**, 217.

# NEW FACTS ABOUT THE VARIABILITY OF 45 $\epsilon$ Persei


P. HARMANEC[1] and A.E. TARASOV[2]
1) Astronomical Institute, 251 65 Ondřejov, Czechoslovakia
2) Crimean Astrophysical Observatory, P.O. 334413,
Nauchnyj, Crimea, U.S.S.R.


45 Per ($\epsilon$ Per, HR 1220, HD 24760, ADS 2888A), a bright B0.5V star, is an archetype of the OB line-profile variables (for a definition of the group, see, e.g., Bolton 1987). Its large and rapid line-profile variations in the form of narrow sub-features travelling from blue to red accross the line profiles were first recognized by Bolton (1983). Before that, some observers were misled by the sub-features and classified the star as a double-line spectroscopic binary (c.f. Petrie 1958).

First detailed studies based on electronic spectrograms were published by Smith (1985, 1986) and Smith et al. (1987), who concluded that the object is a non-radial pulsator with $l=4$ and $l=6$ sectorial modes excited, the respective observed periods being $3.85\pm0.02$ and $2.25\pm0.03$ hours. Gies and Kullavanijaya (1988) carried out a power-spectrum period analysis of the Si III line intensities in their new series of Octicon data and recovered four periods, 4.47, 3.84, 3.04 and 2.26 hours, which they identified with the NRP modes $m=-3$, $-4$, $-5$, and $-6$, respectively. Harmanec (1987) pointed out that all four periods may in fact be sub-multiples of one period of 1.12 days. Later on, Harmanec (1989) re-analyzed a selection of the line profiles published by Smith (1985), Smith et al. (1987) and Gies and Kullavanijaya (1988) and concluded that the line-profile variations are due to six different features which re-appeared in the line profiles with a period of 0.567 days - about a half of the originally suspected value of 1.12 days. Analysing available RV observations of the star, he noted that 45 Per could be a 14-day spectroscopic binary.

We obtained 117 new 3 Å/mm high S/N spectrograms with a red-sensitive CCD camera in the coudé focus of the 2.6-m telescope of the Crimean Astrophysical Observatory on 11 nights in 1988 (JD 2447375-410). Given the instrumentation available, we had to monitor the He I 6678 line. Additionally, we obtained some Hα spectrograms.

Our principal findings are the following:

## 1. STELLAR RV VARIATIONS

Centroid radial velocity was measured for all He I 6678 line profiles. To avoid spurious RV variations due to line-profile changes, only RV's values corresponding to a few apparently symmetric profiles were selected from each series. It was found that the velocity of the star *does* vary by about 30 km/s. Combining our data with 5 velocities published by Gies and Kullavanijaya (1988), we arrived at three possible periods, 14.1, 15.2 or 15.9 days (and their one-day aliases of 0.931, 0.936 and 0.939 days). Since neither Smith et al. (1987) nor Gies and Kullavanijaya (1988) reported velocity variations during their 0.4-day long series, we tentatively assume that the RV of the star varies with about a two-week period. An

205


*L. A. Willson and R. Stalio (eds.), Angular Momentum and Mass Loss for Hot Stars, 205–212.*
© 1990 *Kluwer Academic Publishers. Printed in the Netherlands.*

orbital solution for the value of $14.1^d$ (which is close to that found by Harmanec 1989 from the historical data) gives the following elements (the corresponding RV curve is shown in Fig. 1):

$$P = 14.140{\pm}0.007 \text{ days}, \quad T_{periastr.} = \text{HJD } 2447389.8{\pm}0.4, \quad T_{max.RV} = \text{HJD } 2447386.8$$
$$e = 0.41{\pm}0.10, \quad \Omega = 124°{\pm}11°, \quad K = 15.6{\pm}1.6 \text{ km/s}, \quad \gamma = 0.7{\pm}1.3 \text{ km/s}.$$

## 2. RAPID LINE-PROFILE VARIATIONS

To study the line-profile variations, we formed a mean He I 6678 line profile (corrected for stellar RV variations) of all 117 profiles and subtracted it from the individual profiles. One series of the original and residual spectra is shown in Fig. 2a,b. The variations of several measured quantities along the series are also displayed (Fig. 2c). Strong travelling sub-features are clearly visible. The amplitude of the bumps varies slowly from night to night. It was largest shortly after the binary passed through periastron, but only one periastron passage was observed. The transition of each sub-feature across the line profile is clearly reflected in the variation of the cetral intensity and to some extent also the equivalent width and centroid RV of the line (see Fig. 2c).

Using the measured accelerations of the individual sub-features, and following the procedures recommended by Harmanec (1989), we tried to establish whether the particular sub-features re-appear periodically in the line profiles. We were able to obtain some reasonable fits for periods $P_1=1.1184^d$ and $P_2=2.2368^d$ rather than for a period close to 0.56 days, but also for periods of $P_3=1.2390^d$ and $P_4=1.9188^d$. The accelerations of the sub-features are often nearly linear, even near the edge of the line, and apparently do not follow a sine curve with an amplitude close to v sin i.

One of the possible periods we found, $P_1$, is close to 1.12 days derived by Harmanec (1987) from Gies' and Kullavanijaya's (1988) results, and about twice as long as the period of 0.567 days, derived later by Harmanec (1989) from the published selection of Smith et al. (1987) data. Without re-analysing the original data, we are unable to say whether this indicates a secular decrease of the period or some unresolved aliasing problems in the period search. A new finding is that the following relations hold:

$$P_1^{-1} - P_3^{-1} \doteq P^{-1}, \qquad P_4^{-1} - P_2^{-1} \doteq P^{-1};$$

P being the orbital period. In other words, if $P_1$ ($P_4$) is identified with the *rotational* period of the star, then $P_3$ ($P_2$) may correspond to the *synodic* period with which a given meridian of the primary "sees" the orbiting secondary. (Note that for the stellar radius of R $\approx$ 7 $R_\odot$ estimated from cluster membership and the observed v sin i of about 150 km/s, $P_{rot.} \leq 2.4$ days.) Again, more accurate knowledge of all the periods is needed to say whether the above relations hold exactly.

## 3. THE EQUIVALENT WIDTH OF He I 6678 AND H$\alpha$

The disadvantage of the standard procedure of subtrating the *mean observed* profile from the individual profiles is that one is unable to say whether the sub-features represent additional absorption, emission of combination of both. To get some insight into the problem, we overlaid all He I profiles and measured the equivalent width(EW) of the upper and lower envelope of this superposition. We obtained 0.679, and 1.154 Å, respectively. The measured EW's of the individual profiles range from 0.85 to 0.96 Å. The EWs of the theoretical NLTE He I 6678 profiles of non-supergiant O and B stars published by Auer and

Mihalas (1972, 1973) attain the maximum value of 0.679 Å for $T_{eff} = 35000$ K, $\log g = 4.5$ (GCS), which by chance coincides with our value for the upper envelope of the line. The EWs of all individual profiles are higher than the theoretical maximum value. Our measured EW of the H$\alpha$ line, 2.50 Å, seems to differ significantly from the EW of 4.62 Å obtained by McDonald (1953).

Our tentative conclusions are:

i. Although our result still needs further verification, there are now good reasons to believe that 45 Per is the primary component of a spectroscopic binary and there are some indications that the line-profile variations observed could be somehow *causally* related to the motion of the star in an eccentric orbit.

ii. The detection of the period of 1.12 days (and its first overtone) in the new data indicates that the same periodic physical variation has dominated the line-profile changes of 45 Per all the time since their first monitoring some six years ago. Whether the 1.12-day period undergoes slight secular variations (similarly as it is the case for some late-type emission-line stars) or whether it is constant remains to be investigated. Additionally, the detection of the periods related to the 14-day RV period in the line-profile variations seems to support the reality of the (supposedly orbital) RV changes.

iii. One can conjencture, then, that the observed line-profile and other variations result from an interplay of two basic physical causes: rotation of the primary and the periodically changing tidal force of the orbiting secondary.

iv. The large difference between the EW of H$\alpha$ from McDonald's (1953) and our spectra warrants further investigation to see whether 45 Per is not a mild Be star in fact, similarly as 13 Oph, another archetype line-profile variable. More generally, our findings concerning the EW of both studied lines show that a detailed modelling of many different spectral lines of 45 Per would be very desirable.

(A detailed study will appear in Bull. Astron. Inst. Czechosl.)

### *Acknowledgements*

We feel very obliged to Drs. C.T. Bolton, I. Hubený and G.A.H. Walker for their valuable critical comments on the subject.

### References

Auer L.H., Mihalas D. 1972 Astrophys. J. Suppl. **24**, 193
Auer L.H., Mihalas D. 1973 Astrophys. J. Suppl. **25**, 433
Bolton C.T. 1983 Hvar Obs. Bull. **7**, 141
Bolton C.T. 1987 Publ. Astron. Inst. Czechosl. Acad. Sci. No. 70, 176
Gies D.R., Kullavanijaya A. 1988 Astrophys. J. **326**, 813
Harmanec P. 1987 Inf. Bull. Var. Stars No. 3097
Harmanec P. 1989 Bull. Astron. Inst. Czechosl. **40**, 201
McDonald J.K. 1953 Publ. Dom. Astrophys. Obs. Victoria **9**, 269
Petrie R.M. 1958 Mon. Not. Roy. Astron. Soc. **118**, 80
Smith M.A. 1985 Astrophys. J. **288**, 266
Smith M.A. 1986 Astrophys. J. **307**, 213
Smith M.A., Fullerton A.W., Percy J.R. 1987 Astrophys. J. **320**, 768

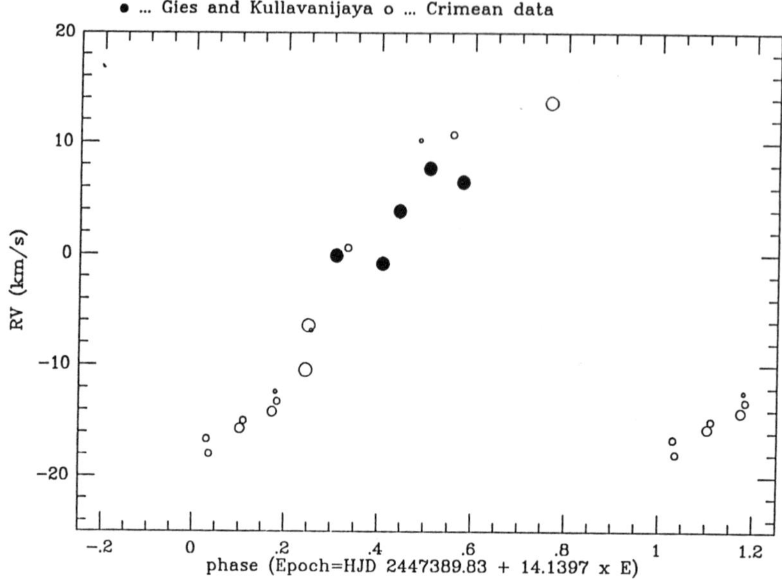



**Fig. 1  Orbital RV curve of 45 Per.**

**Fig. 2a.  Fig. 2b.  Fig. 2c.  Fig. 2d.  (on following pages)**

Fig. 2  Original (a) and residual (b) He I 6678 CCD spectra of 45 Per obtained on **JD 2447406** at Crimea.  Fig. 2c and 2d show the variations of the EW, central intensity, centroid radial velocity and velocity of the sub-features of apparent absorption and emission measured along the series.

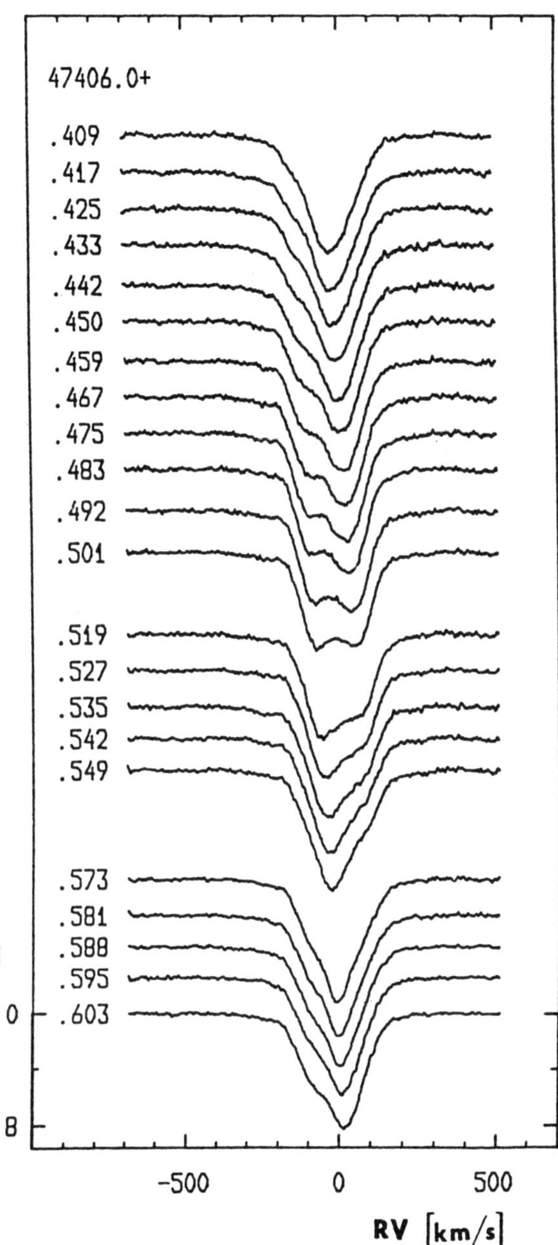

Fig. 2a.

47406.0+
.409
.417
.425
.433
.442
.450
.459
.467
.475
.483
.492
.501
.520
.527
.535
.542
.549
.573
.581
.588
.595
.603
I
0.03
0.00
-0.03
-500
0
500
RV [km/s]

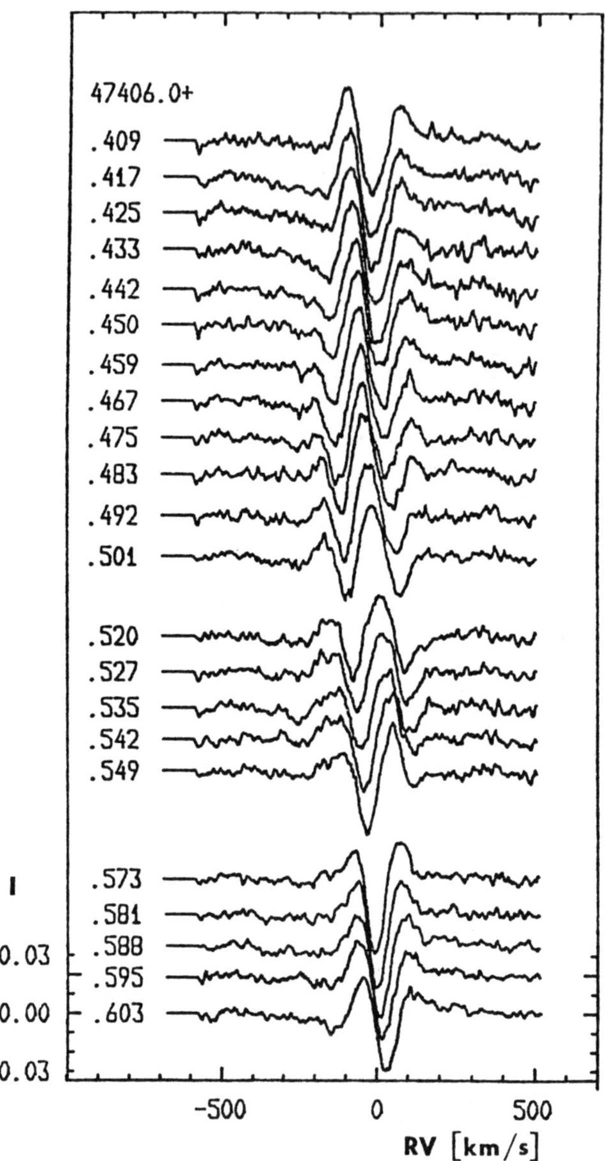

Fig. 2b.

45 Per   equivalent width  Crimean He I 6678 observations
EW[A]
0.86
0.90
0.94
0.98
47406.4
47406.5
47406.6

45 Per   central intensity  Crimean He I 6678 observations
Ic
3.
2.
1.
0.
-1.
-2.
-3.
47406.4
47406.5
47406.6

45 Per   centroid velocity  Crimean He I 6678 observations
RV(km/s)
10.0
0.0
-10.0
-20.0
47406.4
47406.5
47406.6

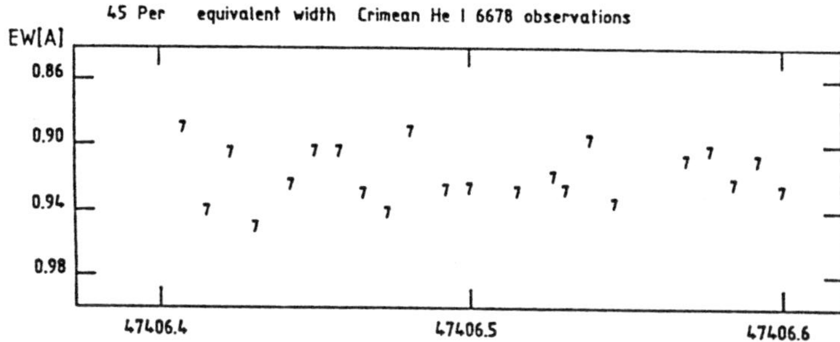

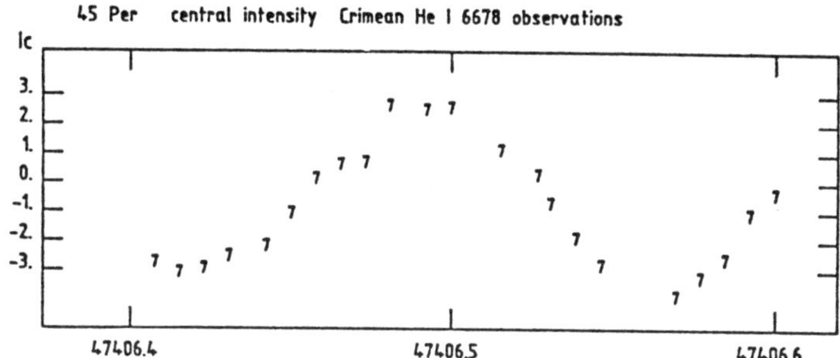

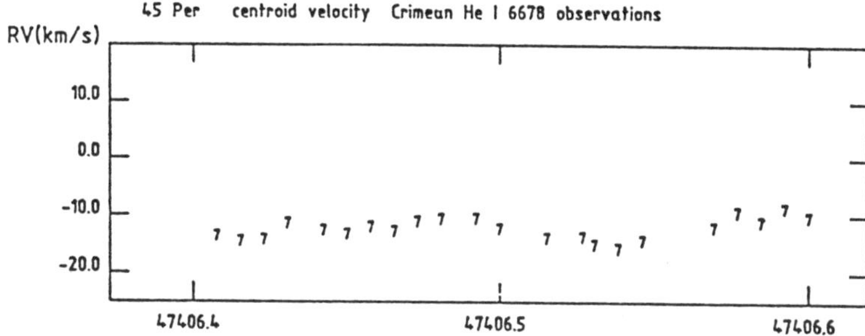

**Fig. 2c.**

45 Per    absorption blips   Crimean He I 6678 observations
RV(km/s)
150.
100.
50.
0.
-50.
-100.
-150.
47406.4
47406.5
47406.6

45 Per    emission blips   Crimean He I 6678 observations
RV(km/s)
150.
100.
50.
0.
-50.
-100.
-150.
47406.4
47406.5
47406.6
HJD - 2400000.0

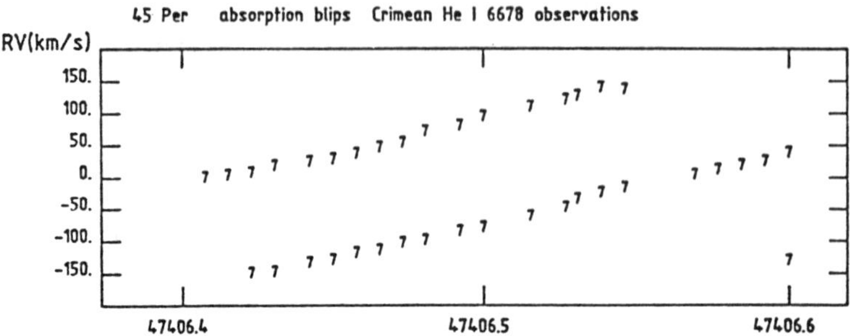

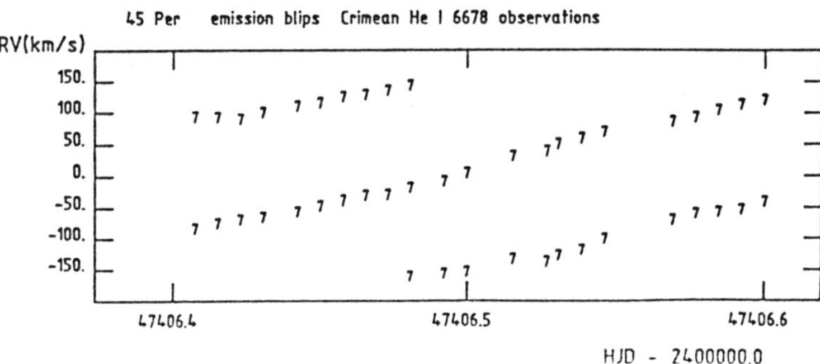

**Fig. 2d.**

# Long-term study of stellar-wind variability of O stars

L. Kaper[1], H. F. Henrichs[1,2], G. A. A. Zwarthoed[1] and J. Nichols-Bohlin[3]

[1]Astronomical Institute, University of Amsterdam,
   Roetersstraat 15, 1018 WB Amsterdam, The Netherlands
[2]Universitäts Sternwarte, München, West-Germany
[3]Astronomy Programs, Computer Science Corporation, Greenbelt, MD

ABSTRACT. As part of our study of rapid variability in UV P Cygni profiles of early-type stars, we present results from 3 successful observing campaigns with the *International Ultraviolet Explorer* in 1986, 1987 and 1988. About 215 high-resolution spectra of 4 O-stars are presented in the form of an atlas of gray-scale pictures, facilitating a rapid overview of subtle and systematic changes in the spectra as a function of time. The morphological behavior of this variability is described for each star.
The main conclusions are: (1) the behavior of the "discrete absorption components" is different for each star, and (2) for a given star the behavior is rather similar over a timescale of years. This is consistent with a rather constant mechanism that controls the rapid structural changes in the wind. The ultimate goal of this study is to understand the origin of the widely-observed variable nature of stellar winds in early-type stars.

## 1. Introduction

From the many observations of early-type stars in the ultraviolet region it is evident that stellar winds are not at all stationary. The UV P Cygni profiles, which give information about the velocity distribution of the scattering ions in the stellar wind, show variability on a wide variety of timescales. In particular, large variations are observed in the absorption parts of unsaturated lines (Si IV) and at the steep edge of the saturated lines (C IV, N V). These two types of variations are correlated (Henrichs *et al.* 1988). See figure 2.

The variability in the unsaturated P Cygni profiles is mainly in the form of the presence (or absence) of the so-called *Discrete Absorption Components* (DACs). These features are now detected in more than 90% of all galactic O stars and represent therefore a fundamental property of the stellar wind of early-type stars (for reviews see: Henrichs, 1988 and Howarth and Prinja, 1989).

Typical DAC behavior is characterized by the sudden appearance of a broad absorption enhancement, starting at low, but supersonic, velocity, which subsequently moves towards higher velocities. During the acceleration phase the absorption component becomes narrower and eventually disappears. The acceleration of the feature is largest at the beginning and vanishes when the DAC reaches its asymptotical velocity. This asymptotical velocity has been identified with the terminal velocity of the stellar wind (Henrichs *et al.* 1988).

The many spectra taken during a period of 2.2 years enabled us to study the 'stability' of the DAC behavior over long time intervals, which is the subject of this paper.



*L. A. Willson and R. Stalio (eds.), Angular Momentum and Mass Loss for Hot Stars, 213–218.*

## 2. Reduction and presentation of results

In figures 1 – 4 we present the Si IV profiles of 68 Cyg, $\xi$ Per, $\lambda$ Cep and 19 Cep, and C IV for $\xi$ Per. All spectra were reduced in an as homogeneous as possible way, using the *Starlink* IUEDR software package (Giddings, 1981). The spectra are grouped per year and displayed by means of gray-scale pictures. Time is running upwards along the y-axis, with the same scale factor for all figures to facilitate a comparison of timescales. The added velocity scale on the x-axis is given with respect to the rest wavelength of the strongest doublet component. The intensity is represented in levels of gray with 32 steps from black (corresponding to the deepest absorption) to white (just below the continuum). Individual remarks about the profile changes for each of the program stars are given in the captions to the figures.

## 3. Conclusions

(1) Typical timescales of variability in the absorption parts of P Cygni profiles are in the order of hours to days.
(2) The short-time variability can be resolved in separated events, with duration and evolution different (and therefore characteristic) for each star.
(3) The recurrence timescales of new absorption features range from 0.7 day to more than 5.8 days, depending on the star. This recurrence timescale is possibly correlated with the stellar rotation period.
(4) The acceleration of the discrete absorption components varies from star to star, as well as during a given event for a given star.
(5) The observations are consistent with a quasi-regular pattern over more than two years without substantial variations for a given star.

## 4. Discussion

Many of the points above have been addressed in earlier discussions (see *e.g.* Henrichs 1988, Henrichs *et al.* 1988, Prinja 1988, Howarth and Prinja 1989). The most relevant new conclusion is that each star has got its own very characteristic pattern of DAC behavior, and that this pattern does not vary substantially over a number of years. This implies a rather constant controlling mechanism responsible for the rapid structural changes in the winds of early-type stars. The nature of this mechanism, however, remains unknown.

**Acknowledgements.** LK received a travel grant from ASTRON. HFH gratefully acknowledges support from the NWO Constantijn en Christiaan Huygens Program and the warm hospitality at the Sternwarte in München under directorship of R.-P. Kudritzki.

## References

Giddings, J. R. 1981, *ESA IUE Newsletter*, No.12, p.22

Henrichs, H. F. 1988, *in O, Of and Wolf-Rayet stars, Eds. P.S. Conti and A.B. Underhill NASA/CNRS monograph series*, p. 199

Henrichs, H. F., Kaper, L. and Zwarthoed, G. A. A. 1988 *in Proc. Celebratory Symp.: A decade of UV astronomy with IUE*, ESA SP-281, Volume 2, 145

Howarth, I. D. and Prinja, R. K. 1989 *Astrophys. J. Suppl.* **69**, 527

Prinja, R. K., Howarth, I. D. and Henrichs, H. F. 1987, *Astrophys. J.* **317**, 389

Prinja, R. K. and Howarth, I. D. 1988 *Monthly Not. Roy. Soc.* **233**, 123

Prinja, R. K. 1988, *Monthly Not. Roy. Soc.* **231**, 21P

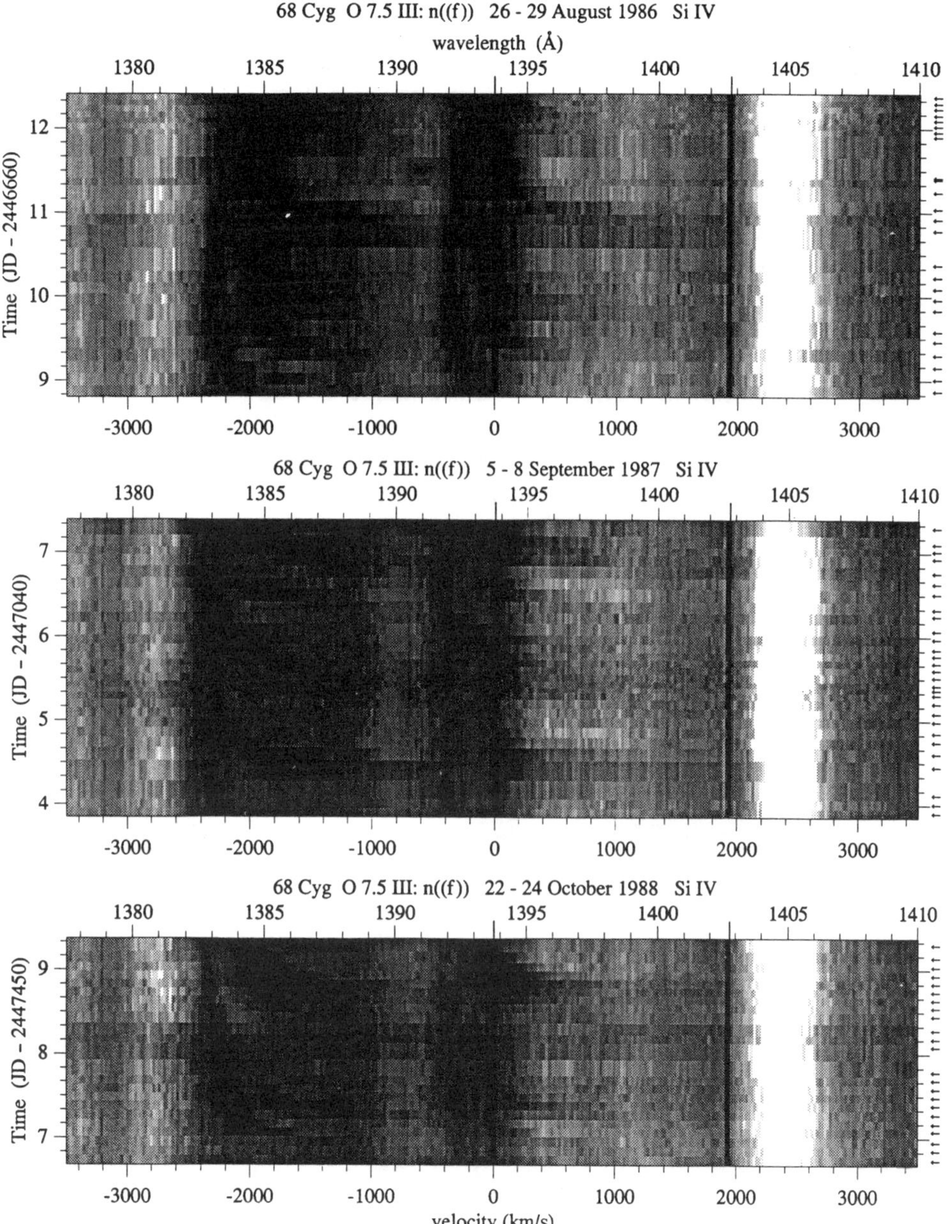

FIGURE 1. **68 Cyg O7.5 III:n((f))**
This star shows the most regular DAC behavior among the sample stars. The recurrence timescale is approximately one day. The asymptotical velocity of $-2350 \pm 50$ km/s is well defined. It is evident that this star shows far more DAC episodes than the slow rotator 19 Cep (see figure 4).

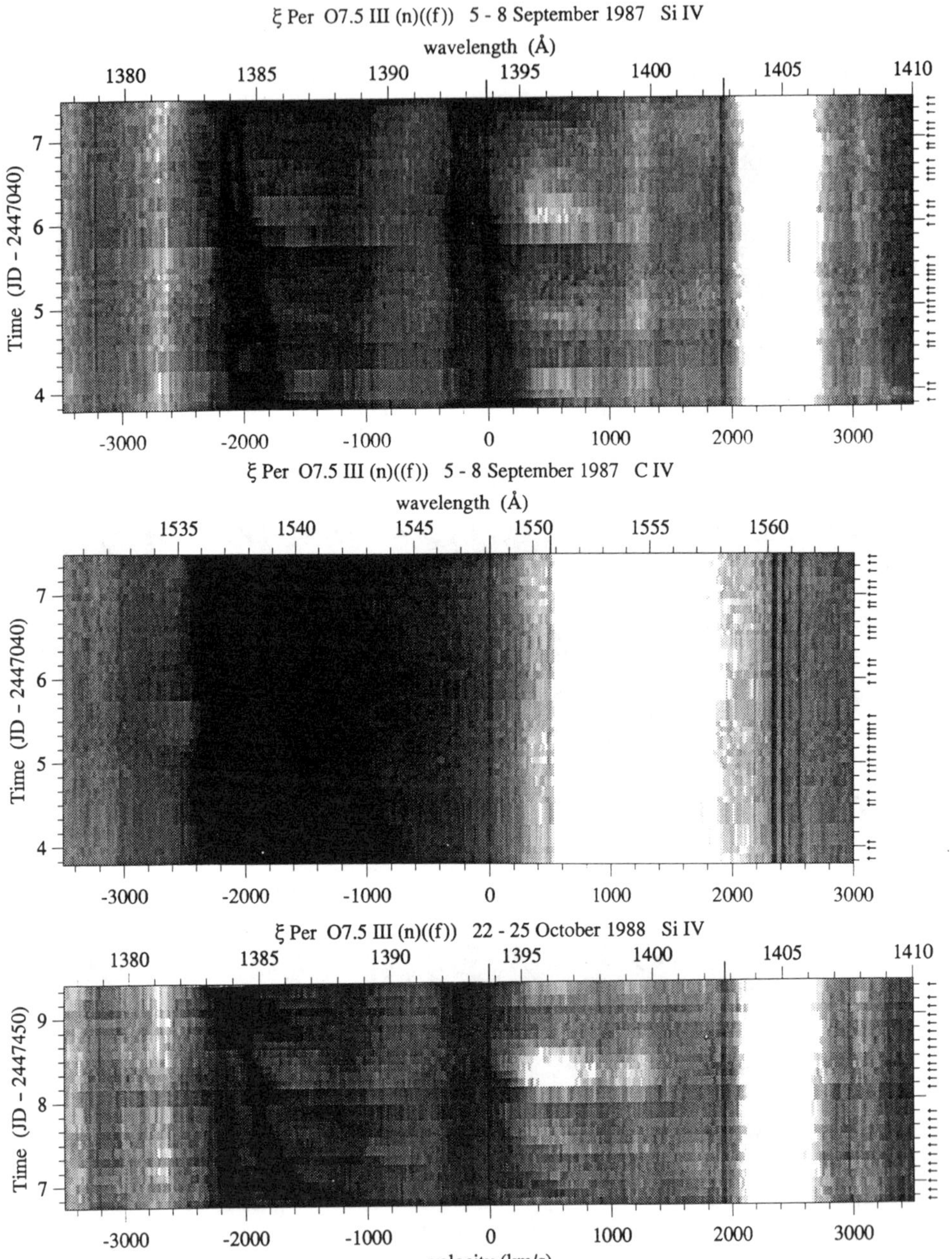



FIGURE 2. ξ **Per O7.5 III(n)((f))**
The behavior of the DACs in this star is more irregular than in 68 Cyg, in spite of similar spectral types. The recurrence timescale is about 1 day. The asymptotic velocity is −2100 km/s, much less than the steep C IV edge velocity. The figure in the middle shows the variability of the C IV steep edge in 1987. The minimum edge velocity corresponds to the beginning of a new DAC episode.

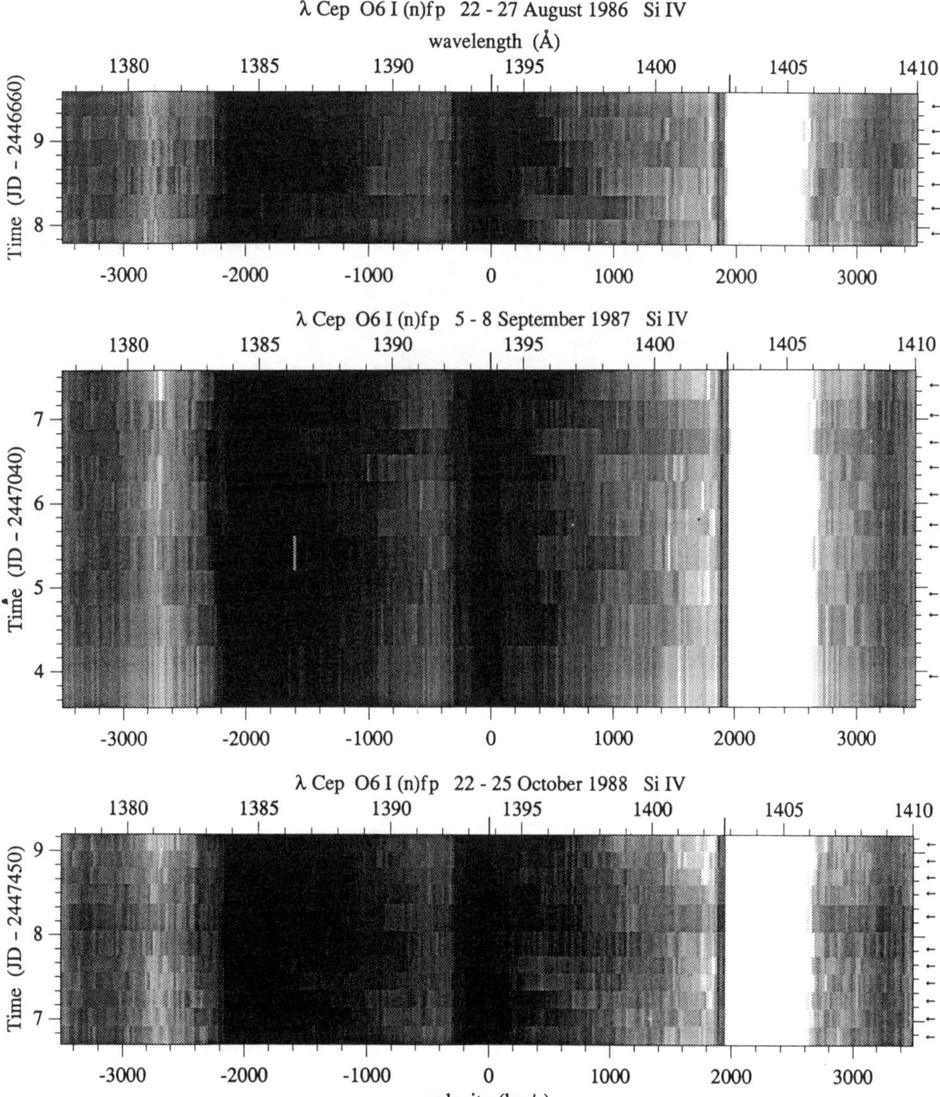



FIGURE 3. $\lambda$ **Cep O6 I(n)fp**
Due to the relatively high level of saturation in the Si IV profiles of this star the subtle changes
are more difficult to detect. The best example can be found in 1986. The recurrence timescale is
probably of the order of one day. The asymptotical velocity is around $-2000$ km/s. Lack of time
resolution in 1987 and 1988 prevents the precise determination of the DAC behavior. The overall
behavior seems nevertheless to be very similar.

FIGURE 4. *(See next page.)* **19 Cep O9.5 Ib**
This is the best example of 'slow' behavior of DAC's, consistent with the slow rotation rate of the
star. From the 1986 data we derive an asymptotic velocity of $-1800$ km/s. The set of spectra during
the relatively short study periods in 1987 and 1988 do not contradict the pattern observed in 1986.

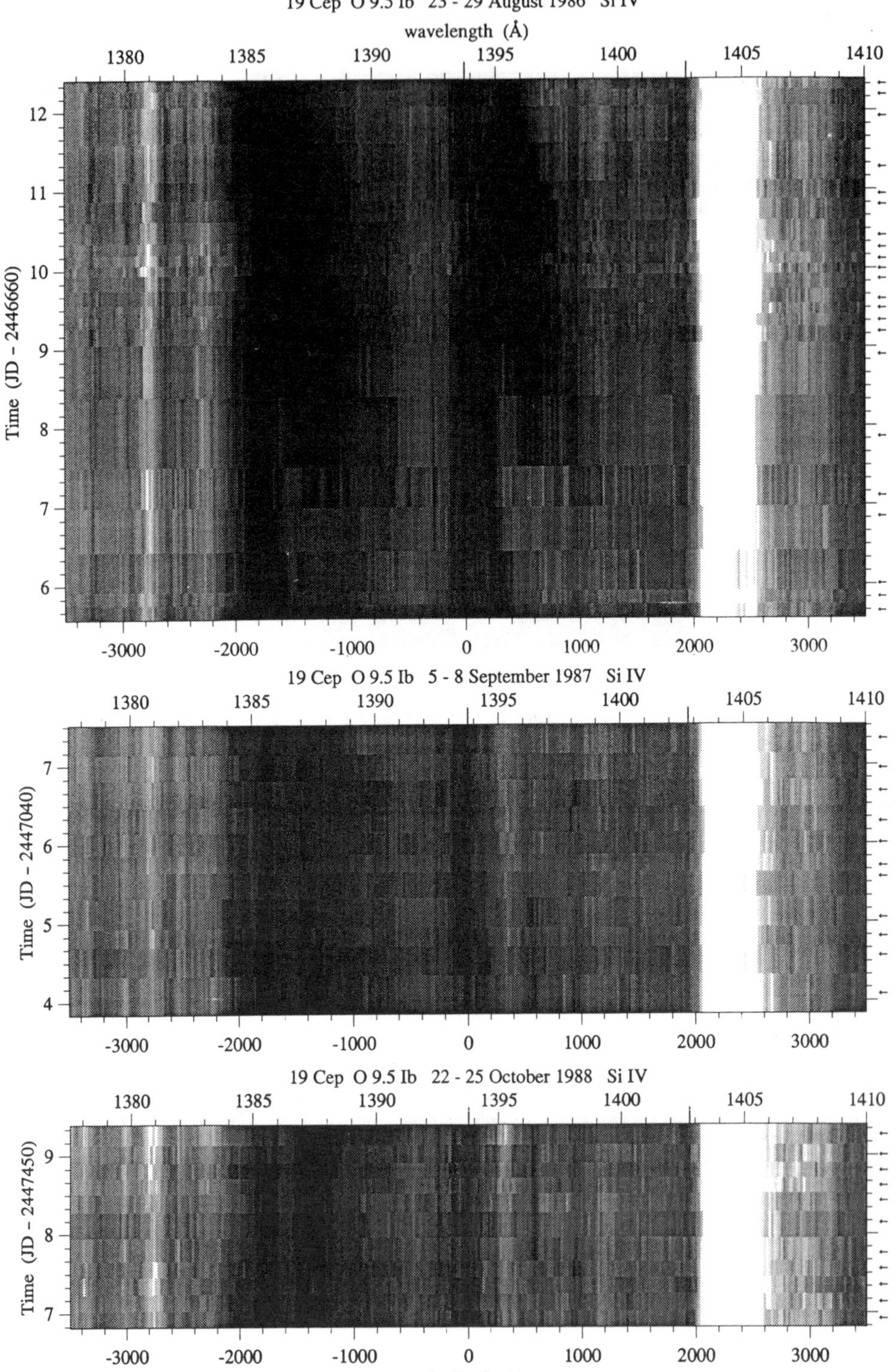
19 Cep  O 9.5 Ib   23 - 29 August 1986   Si IV
wavelength  (Å)
1380   1385   1390   1395   1400   1405   1410
12
11
10
9
8
7
6
Time (JD - 2446660)
-3000   -2000   -1000   0   1000   2000   3000
19 Cep  O 9.5 Ib   5 - 8 September 1987   Si IV
1380   1385   1390   1395   1400   1405   1410
7
6
5
4
Time (JD - 2447040)
-3000   -2000   -1000   0   1000   2000   3000
19 Cep  O 9.5 Ib   22 - 25 October 1988   Si IV
1380   1385   1390   1395   1400   1405   1410
9
8
7
Time (JD - 2447450)
-3000   -2000   -1000   0   1000   2000   3000
velocity (km/s)

He II $\lambda$1640 AS A DIAGNOSTIC FOR ASSESSING THE EXTENT OF RAPID ROTATION
IN Be STARS

GERALDINE J. PETERS
Space Sciences Center
University of Southern California
Los Angeles, CA   90089-1341
U.S.A.

ABSTRACT.   To look for evidence of a substantial enhancement in the
polar temperature of Be stars, which one might expect if these stars
were rotating near their critical velocities, the strength of the He II
$\lambda$1640 absorption line has been investigated in a restricted group of
objects (B1.5-B2.5 IV-V).   The fact that no correlation between the
equivalent width of He II $\lambda$1640 and $v \sin i$ was found argues against
the presence of significant polar brightening.   He II $\lambda$1640 is variable
in an individual star, but does not correlate well with the strength of
the wind.   It may be a signature of enhanced photospheric activity.

## 1. INTRODUCTION

When line profiles are computed for determining the projected
rotational velocity or the inclination in Be stars, it is usually
assumed that Von Zeipel's theorem is correct or that the flux at any
point on the stellar surface is proportional to the local gravity
($F \propto g$ or $T_{eff} \propto g^{0.25}$).   This is the familiar concept of a "red"
equator and a "blue" pole.   Although polar brightening (or equatorial
darkening) is assumed to exist in Be stars (as a consequence of their
rapid rotation), its presence has never been confirmed independently
from observation.

This project was undertaken to look for evidence of an elevated
temperature at the poles of Be stars, which one would expect if these
objects are rotating at velocities in excess of 0.90 $V_{cr}$.   A search was
made for a diagnostic line whose strength is very sensitive to the
ambient temperature and thus would be more localized in latitude than
the usual classification diagnostics (the Balmer lines, several lines
of He I, and Mg II $\lambda$4481).   The line found to be ideally suited for
such an investigation was He II $\lambda$1640 (the counterpart of H$\alpha$).

The stars selected for this study are main-sequence band objects
with spectral classes in the range B1.5 - B2.5.   In this group of stars
(18,000 < $T_{eff}$ < 23,000 K) the He I lines, which are the principal
classification diagnostics, display a broad maximum in strength.   Spectrum synthesis calculations employing the Kurucz (1979) model atmospheres show that maximum He I absorption occurs near $T_{eff}$ = 22,500 K,



*L. A. Willson and R. Stalio (eds.), Angular Momentum and Mass Loss for Hot Stars, 219–222.*
© 1990 *Kluwer Academic Publishers. Printed in the Netherlands.*

log $g$ = 4. He I lines are 10% weaker at 18,000 and 24,000 K and show a 30% decline at 16,000 and 29,000 K. Alternatively, He II $\lambda1640$ is barely detectable in a 20,000 K star (B2.0 V), but dominates the spectrum with an equivalent width of nearly 0.4 Å in a 24,000 K object (B1 V). Therefore, if any significant polar brightening exists due to rapid rotation, the strengths of the He I lines should be *relatively* independent of latitude (except perhaps within ±10° of the equator), but the contribution to He II $\lambda1640$ should be primarily from the pole. If the star is "pole-on" (seen at an inclination <30°), the line should appear stronger.

## II. OBSERVATIONS AND MEASUREMENTS

Data on the He II $\lambda1640$ line were obtained from the *IUE* archives (cf. Boggess, *et al.* 1978 for a description of the *IUE* spacecraft). Examination of the spectra of sharp-lined standard stars reveals that the region around the He II line is moderately blended. In particular, there is a strong line to the violet (blend of Fe II, Cr III, Zn III, and others) that displays a complex variation in strength with $T_{eff}$ in the B1.5-B2.5 range. Since in most of the Be stars, rapid rotation causes this feature to be blended with the He II line, all measurements intentionally included both components. The equivalent widths given in this paper represent absorption from $\lambda\lambda1638.5$–$1641.5$.

## III. RESULTS

The He II $\lambda1640$ line was measured in 27 Be stars with ground-based spectral types in the range B1.5 - B2.5 and $50 < v \sin i < 350$ km s$^{-1}$. In Figure 1 the equivalent width of this line is plotted versus spectral class. The *dashed* line is a least squares fit to similar measurements for non-emission line B stars. Note the wide range in the strength of He II for the Be stars.

A plot of the equivalent width of the He II line versus the $v \sin i$ of the star is shown in Figure 2. It is very apparent that there is *no* trend with the inclination of the star to our line of sight. It should be mentioned that in the region from 50-150 km s$^{-1}$ the sample includes nearly equal contribution from all sub-types. The lack of an enhancement in stars with low values of $v \sin i$ argues against the presence of significant polar brightening in Be stars and suggests that *at best* the temperatures of the equators of these stars are <2000 K cooler than their poles. The results imply that the rotational velocities in Be stars (which display exceptionally variable mass loss) are less than about 0.85 $V_{cr}$.

From a cursory examination of several stars for which there are a large number of *IUE* images, it became apparent that the He II $\lambda1640$ feature is variable. For example, in $\lambda$ Eri (B2 III-IVe) the equivalent width ranges from 0.35-0.65 Å. Since $\lambda$ Eri has also displayed a variable wind during the interval of time that it has been observed with *IUE*, an attempt was made to learn whether the strength of He II $\lambda1640$ correlates with that of the C IV wind line. From the display and fit illustrated in Figure 3, it would appear that the correlation is at

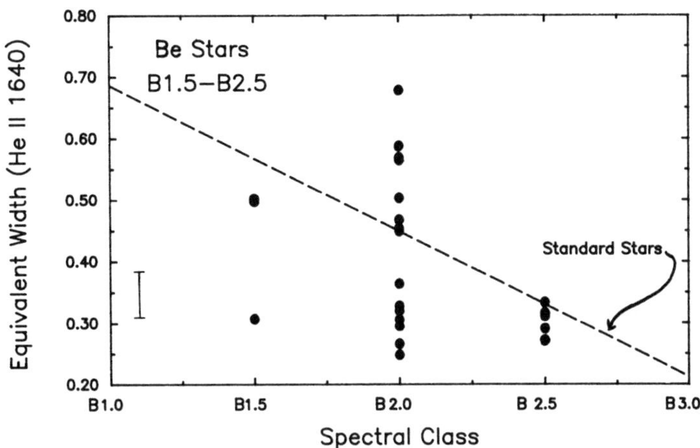



Figure 1. The equivalent width of He II λ1640 in early Be stars versus spectral class. The *dashed* curve is a linear fit to the data for the standard stars.

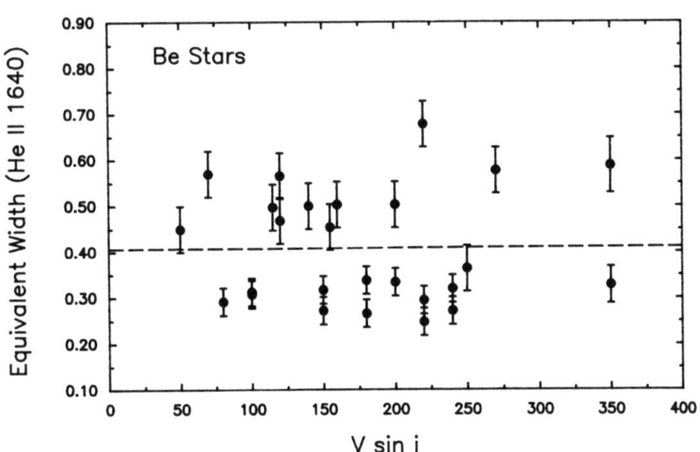



Figure 2. The equivalent width of He II λ1640 versus *v* sin *i*. The *dashed* line is a linear fit to the data. No correlation is observed.

best weak and the line is probably not simply formed at the base of the wind. The data suggest a possible connection between the strength of the He II line and the onset of activity, but further study is needed to confirm this and assess its significance. The behavior of He II and C IV in selected observations of 66 Oph was also studied.

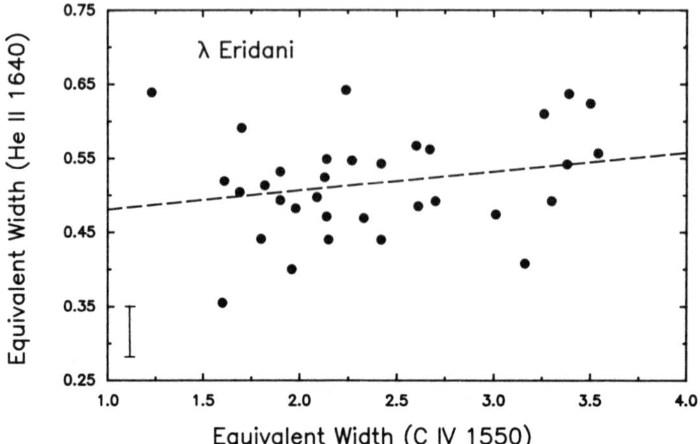



Figure 3. The equivalent width of He II λ1640 versus that of the C IV λ1550 wind line in λ Eri from 1982-1988. The *dashed* line is a linear fit to the data (correlation coefficient is 0.23).

---

The wind activity in this star is impressive (Grady *et al.* 1987, Peters 1988) spanning the gamut from being strong to off! Two episodes during which the wind disappeared then reestablished itself, all within a few months, were examined. The He II line appears to be variable during both episodes, but it does not correlate with C IV. From limited data it seems that the He II line is strongest during the period of early reestablishment of the wind when narrow components to the wind lines with velocities of -1000 km s$^{-1}$ are seen. As in the case of λ Eri, the data must undergo more scrutiny.

This study was supported in part by NASA grant NSG 5422 and involved the use of the *IUE* Regional Data Analysis Facilities (RDAFs) at the Goddard Space Flight Center and the University of Colorado.

REFERENCES


Boggess, A. *et al.* (1978) 'The IUE spacecraft and instrumentation' *Nature*, 275, 2-7.

Grady, C. A., Sonneborn, G., Wu, C.-C., and Henrichs, H. F. (1987) "Recurrent episodic mass loss in a B2e star: 66 Ophiuchi, 1982-1985', *Ap.J.Suppl.*, 65, 673-694.

Kurucz, R. L. (1979) 'Model atmospheres for G, F, A, B, and O stars', *Ap.J.Suppl.*, 40, 1-340.

Peters, G. J. (1988), "The loss and reestablishment of the wind in 66 Ophiuchi in 1985', *Publ.A.S.P.*, 100, 207-216.

# SOME EXAMPLES OF THE ROLE OF STELLAR ROTATION IN HOT STAR WINDS

RAMAN K. PRINJA
*Dept. of Physics & Astronomy,*
*University College London,*
*Gower St., London WC1E 6BT, England, U.K.*

ABSTRACT.   Direct observational evidence of the role of stellar rotation in hot star winds is highlighted, based on our recent ultraviolet studies of mass-loss in the upper left-hand part of the H-R diagram.

The characteristics of 'discrete absorption components' in OB stars, stellar winds velocities in Be stars, and the mass-loss rates of O stars are discussed as a function of stellar rotation.

## 1.   Discrete absorption components in OB stars

It is now widely recognised that the stellar winds of OB stars (probed via the UV resonance lines) are variable on timescales down to $\leq 1$ hour, and that a significant portion of this variability may be attributed to the progressive displacement of discrete absorption (opacity) enhancements superimposed on a relatively fixed 'underlying' P Cygni profile (see *e.g.* Henrichs, 1984; Prinja, Howarth & Henrichs, 1987; Prinja & Howarth, 1988). The variability may be interpreted in terms of recurrent episodes of low-velocity broad absorption enhancements evolving into high-velocity narrow absorption components.

Recent intensive campaigns of short-period *IUE* observations of selected OB stars have enabled the nature of the wind variability to be addressed as a function of stellar rotation. These data have provided direct and reliable evidence for rotationally modulated variability in OB star winds (see *e.g.* Prinja, 1988; Henrichs, Kaper & Zwarthoed, 1988).

The modelled central velocities of 7 (high and low $v_e \sin(i)$) stars are plotted in Figure 1 as a function of time. The data illustrate the recurrence and dissipation of separate opacity enhancement episodes. These data suggest
(i) The timescale of progression for low-velocity opacity enhancements evolving into high-velocity discrete absorption components is correlated with the projected rotation velocity, $v_e \sin(i)$.



*L. A. Willson and R. Stalio (eds.), Angular Momentum and Mass Loss for Hot Stars, 223–226.*
© 1990 *Kluwer Academic Publishers. Printed in the Netherlands.*

224

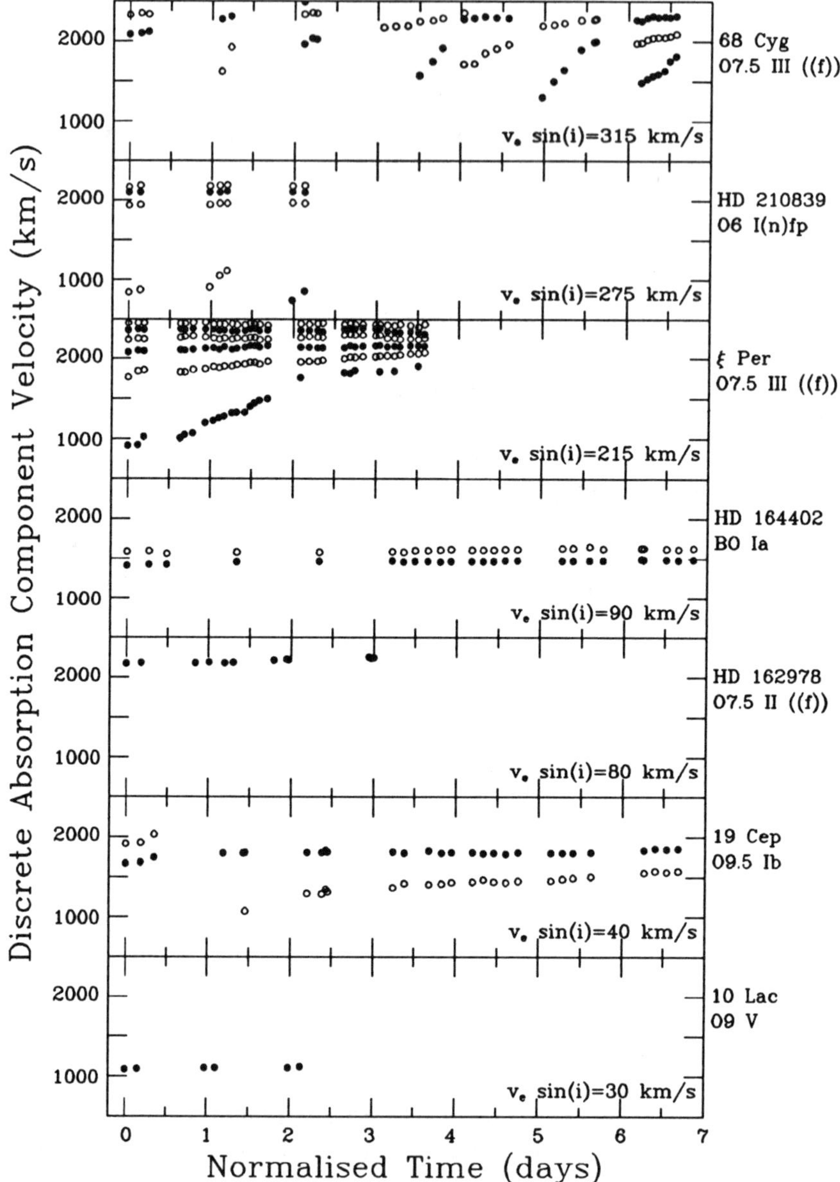


**Figure 1** − The central velocities of consecutive sequences of discrete absorption components as a function of time. The recurrence and dissipation timescales of the discrete features are correlated with the projected rotation velocity.

(ii) The recurrence timescale for the development of the opacity enhancements is considerably longer ($\sim$ many days) for the low $v_e \sin(i)$ cases. There are at least six separate features evident over $\sim$ 4 days of *IUE* observations of 68 Cyg and $\xi$ Per in Figure 1, whereas, for example, only two consecutive features are apparent for HD164402 and 19 Cep over the same period.

These data emphasise the role of stellar rotation as a key element in physical models of wind variability in OB stars. The data are consistent (for example) with the morphological model described by Prinja & Howarth (1988) where the variability is viewed not in terms of 'shells' or 'puffs' of radially propagating matter, but instead as material passing *through* perturbations in the flow. These perturbations may be related to the (evolving) radiative instabilities described by Owocki *et al.* (1988).

Future studies concerning rotationally modulated variability in OB stars must address at least three important points;
(i) The physical connection between radiative instabilities and stellar rotation to relate, for example, to the observations presented in Figure 1.
(ii) Co-ordinated, multi-wavelength observations in order to explore possible connections between the discrete component variability and changes in the near-star wind structure (*e.g.* H-$\alpha$).
(iii) The connection (if any) to these data of exceptional, high $v_e \sin(i)$, stars like $\zeta$ Oph, HD93521 and HD60848 where the discrete absorption components are remarkably stable in velocity (see *e.g.* Howarth, Prinja & Willis, 1984).

## 2. Stellar wind velocities in Be stars

The mostly large projected rotation velocities in Be stars have led to interest in the possibility that the stellar winds of these stars may be distorted and enhanced by rapid stellar rotation (see *e.g.* the review by Marlborough, 1987).
The ratio of $v_{\rm edge}/v_{\rm esc}$ versus $v_e \sin(i)/v_{\rm crit}$ is plotted in Figure 2 from the UV sample of 40 Be stars studied by Prinja (1989). ($v_{\rm edge}$ is the maximum wind velocity observed in absorption in a given line).
The data for Be stars show a significant trend of increasing $v_{\rm edge}/v_{\rm esc}$ as a function of $v_e \sin(i)/v_{\rm crit}$. (The linear correlation coefficient = 0.7, with a percentage probability $\sim$ 0.02 of the parent sample being uncorrelated).
These results may be pertinent, for example, to the calculations of Poe & Friend (1986) who incorporate magnetic fields into models of the structure of a radiation pressure-driven stellar wind. They note, for example, that a magnetic field of $\sim$ 100G forces the terminal velocity to *rise* with increasing rotation velocity.

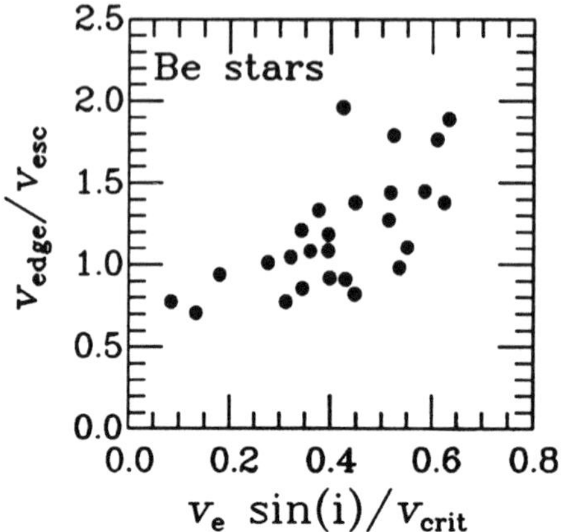


**Figure 2** – The effect of (projected) stellar rotation on the maximum observed wind velocities ($v_{edge}$) in Be stars.

### 3. Mass-loss rates in O-type stars

In their recent quantitative survey of the stellar winds of 203 galactic O stars, Howarth & Prinja (1989) present some tentative evidence of an increase in the O star mass-loss rates ($\dot{M}$) with increasing stellar rotation (an effect predicted, for example, by the models of Friend & Abbott, 1986). Howarth & Prinja (1989) derive a parameter ($\Lambda$) which has the principal trends in $\dot{M}$ due to luminosity ($\dot{M} \propto L_*^{1.69}$) and luminosity class removed. Comparisons of $\Lambda$ values for the slowest and fastest (projected) rotators show a clear difference between them, with the faster rotators having larger values of $\Lambda$ compared to the mean for the sample.

### References

Friend, D. B. & Abbott, D. C. 1986, *Ap.J.*, **311**, 701.

Henrichs, H. F. 1984, 4th European *IUE* Conference, ESA SP-218, p.43.

Henrichs, H. F., Kaper, L. & Zwarthoed, G. A. 1988, *Decade of UV Astronomy with IUE*, ESA SP-281, vol. 2, p.145.

Howarth, I. D., Prinja, R. K. & Willis, A. J. 1984, *M. N. R. A. S.*, **208**, 525.

Howarth, I. D. & Prinja, R. K. 1989, *Ap. J. Suppl.*, **69**, 527.

Marlborough, J. M. 1987, *Physics of Be stars*, IAU Colloq. 92, p. 316.

Owocki, S., Castor, J. I. & Rybicki, G. B. 1988, *Ap. J.*, **335**, 914.

Poe, C. H. & Friend, D. B. 1986, *Ap. J.*, **311**, 317.

Prinja, R. K. 1988, *M. N. R. A. S.*, **231**, 21P.

Prinja, R. K. 1989, *M. N. R. A. S.*, in press.

Prinja, R. K., Howarth, I. D. & Henrichs, H. F. 1987, *Ap. J.*, **156**, 609.

Prinja, R. K. & Howarth, I. D. 1988, *M. N. R. A. S.*, **233**, 123.

# HD193077—A FAST ROTATING WOLF-RAYET STAR

W. SCHMUTZ
*Joint Institute for Laboratory Astrophysics*
*University of Colorado and National Institute of Standards and Technology*
*Boulder, CO 80309-0440, USA*

ABSTRACT. Evidence is presented that the unshifted absorption lines in the spectrum of HD193077 are intrinsic to the Wolf-Rayet star. This makes HD193077 an interesting peculiar object, possibly rotating close to breakup velocity and evolving homogeneously.

## 1.    Introduction

Unshifted absorption lines that are intrinsic to a Wolf-Rayet star are not uncommon for WN7 subtypes but this phenomenon appears to be much rarer for other WR subtypes. In the newest list of Galactic WR stars (van der Hucht et al. 1988) only 2 out of 62 non-WN7 WN stars might have absorption lines that are intrinsic to the WR. Interestingly, the emission line spectra of both WNE+abs candidates show clearly that hydrogen is present in their atmospheres (Massey 1980), which is exceptional for WNE's also. Here we discuss one of them, HD193077, classified as WN5+abs.[1]

## 2.    Observations

The spectra shown in Figure 1 were obtained by W.-R. Hamann between August 18 and 21, 1986 with the 2.2m telescope at Calar Alto. The resolution is 0.5 Å and the range shown is a composite of three overlapping settings of the Coudé spectrograph. The relative system response was determined from the observed continuum counts of the narrow lined O9V star 10 Lac and the spectra were rectified by adjusting a straight line to each of the three recorded wavelength ranges. The slopes of these lines were allowed to differ for each individual wavelength setting in order to compensate for differences in atmospheric and interstellar extinction between the target and the comparison star, 10 Lac.

## 3.    Are the Absorption Lines Intrinsic to the WR?

As noted by Massey (1980), the He I absorption lines are so strong that there are only two possibilities: "Either the OB star's continuum dominates ..., or there is only one star present ... ." In order to determine to which star the absorption lines belong (the WR or

---

[1]Lamontagne et al. (1982) favor a WN6 classification. We retain the older WN5 designation since in our spectrum (Fig. 1) NIII $\lambda4634$–41 $\approx$ NV $\lambda4603/19$.



*L. A. Willson and R. Stalio (eds.), Angular Momentum and Mass Loss for Hot Stars, 227–230.*

an OB companion), we have considered the following points:

*Spectral appearance.* It is remarkable to what degree in line strength and width (expansion velocity) the spectrum of HD193077 resembles that of the binary V444 Cygni for which there is no question that an O star is present (Fig. 1).[2] This might indicate that there is an O star contributing to the spectrum of HD193077 also. In line with Massey's remark cited above, we have found that 3/4 of the (broadened to 450 km s$^{-1}$) absorption line strengths of 10 Lac is needed in order to fill in the unshifted absorption lines of HD193077. Subtracting 3/4 of 10 Lac's absorption lines from the spectrum of HD193077, we obtain the spectrum indicated by the dotted line overplotted on the original spectrum of HD193077 in Fig. 1. If we compare this modified spectrum to a strong-line WR, e.g. HD192163, we notice a subtle but striking difference: The line profiles of HD193077 are more pointed than those of HD192163 which are clearly rounded. This is most pronounced for the weak NV 4945 line, indicating that the expansion velocity just above the photosphere is smaller in HD193077 than in WR-types with strong lines. Interestingly, the same remarks apply to V444 Cyg also, implying that the WR star in this system is also not identical to a strong-line WR. Taken together, there is no convincing evidence from the spectral appearance to decide the question of the origin of the absorptions lines.

*Model atmospheres.* Spectral analysis does also not allow to answer whether the absorption lines are intrinsic to the WR or not. The proof that it is possible to model the spectrum assuming that only one star is contributing to the spectrum has been given by Schmutz et al. (1988; see also Fig. 2 in Schmutz 1988). They presented model calculations that predict correctly emission and intrinsic unshifted absorption lines. No detailed model has been constructed for the other alternative. But a coarse analysis shows that it would be possible as well. If we assume that the OB star contributes 3/4 of the continuum flux, the equivalent widths of the WR are 4 times larger than measured, e.g. $W_\lambda$(He II$\lambda$4686) = 265 Å, $W_\lambda$(He II$\lambda$5411) = 56 Å, $W_\lambda$(He I$\lambda$5876) = 16 Å, and the absolute visual magnitude would be $M_v = -3.8$ mag. Analyzing these numbers with the diagrams given in Schmutz et al. (1989) we find the following stellar parameters for the (strong-line) WR star: $T_* = 55$ kK, $L = 10^{4.9}$ L$_\odot$, $R_* = 3.1$ R$_\odot$, $\dot{M} = 10^{-4.5}$ M$_\odot$ yr$^{-1}$. A model atmosphere with these parameters would look similar to the one of HD50896 (Hillier 1987; Hamann et al. 1988) and no intrinsic unshifted absorption line is predicted.

*Mass estimate.* We may obtain an estimate of the upper limit to the WR mass from the mass-luminosity relation predicted from stellar evolution models (Maeder and Meynet 1987, Fig. 16). A star with the spectral appearance of a WR is certainly not within the main sequence band anymore. Therefore the evolutionary mass of a star at the end of the main sequence band with the luminosity of the WR is an upper limit to the WR mass. We have found that if the absorption lines are due to an OB companion the WR luminosity is $L \approx 10^{4.9}$ L$_\odot$. Thus, an upper limit for the WR mass is $M_{max} = 17$ M$_\odot$. From the orbital solution Lamontagne et al. (1982) derived $M_{WR} \sin^3 i = 41^{+37}_{-24}$ M$_\odot$. Considering the errors of both mass determinations these two results do not really disagree but the agreement has to be considered marginal, at best. We prefer to interpret this mass-comparison as an inconsistency between evolutionary mass and orbital mass.

*Radial velocities.* In Table 1 we list the radial velocities measured in our spectra of August

---

[2]Even the interstellar sodium absorption lines and the diffuse interstellar bands are virtually identical in V444 Cyg and HD193077, indicating that the two stars probably have the same distance. Membership in Cyg OB 1 is likely for both.

19, 1986. If we compare the velocity of He II $\lambda4686$ with the orbital solution for this line given by Lamontagne et al. (1982, Fig. 6b) we find reasonable agreement. However, we cannot confirm their suggestion that the absorption lines vary in anti-phase with the emission lines. The RV's of our absorption lines are accurate to about 20 km s$^{-1}$, clearly outside the solution proposed by Lamontagne et al. (1982). From our RV values we rather conclude that the absorption lines move in phase with the emission lines.

The first two considerations do not allow us to decide whether or not there is an OB companion, but the latter two points favor the absorption lines being intrinsic to the Wolf-Rayet star. However, we have to admit that with only one good set of radial velocities at hand, we cannot exclude the presence of an OB companion. The period proposed by Lamontagne et al. (1982) is demonstrably not unique, e.g. 4.0 yr fit as well as the proposed 4.8 yr, and we should also consider smaller velocity amplitudes and an eccentric orbit. Obviously more high S/N observations of this system are needed.

## 4. Discussion

Our evidence that the absorption lines are intrinsic to the WR star makes HD193077 a peculiar object. The absorption lines are rotationally broadened to about 450 km s$^{-1}$ (Massey 1980) which in turn would have two interesting implications: First, this high rotation velocity might be fast enough to mix the star completely and cause it to evolve homogeneously (Maeder 1987) and second, it is possible that the star rotates close to breakup velocity. The second possibility comes from the following comparison: The breakup mass for a star with a radius of 16 R$_\odot$ (Schmutz et al. 1988) and a rotation velocity of 450 km s$^{-1}$ is of the order of $M_{break} = 17$ M$_\odot$. According to Schmutz et al. (1988) the luminosity of HD193077 is about $L = 10^{5.4}$ L$_\odot$ and a homogeneous star with this luminosity, a 1:1 helium to hydrogen number-ratio, and solar metal abundance would have a mass of about 20 M$_\odot$ (C. J. Hansen, personal communication).

ACKNOWLEDGEMENTS. W.S. acknowledges support through the Swiss National Foundation and contributions from NSF Grant AST-8802937. We thank Carl Hansen for computing models of homogeneous stars and Lorraine Volsky for editorial assistance. Special thanks go to Wolf-Rainer Hamann for valuable comments and suggestions and for obtaining the observations this paper is based on.

REFERENCES

Hamann, W.-R., Schmutz, W., and Wessolowski, U. 1988, *Astr. Ap.*, **194**, 190.
Hillier, J. 1987, *Ap. J. Suppl.*, **63**, 965.
van der Hucht, K. A., Hidayat, B., Admiranto, A. G., Supelli, K. R., and Doom, C. 1988, *Astr. Ap.*, **199**, 217.
Lamontagne, R., Moffat, A. F. J., Koenigsberger, G., and Seggewiss, W. 1982, *Ap. J.*, **253**, 230.
Maeder, A. 1987, *Astr. Ap.*, **178**, 159.
Maeder, A. and Meynet, G. 1987, *Astr. Ap.*, **182**, 243.
Massey, P. 1980, *Ap. J.*, **236**, 526.
Schmutz, W. 1988, *IAU Colloq.* **108**, 133.
Schmutz, W., Hamann, W.-R., and Wessolowski, U. 1988, *IAU Colloq.* **108**, 143.
Schmutz, W., Hamann, W.-R., and Wessolowski, U. 1989, *Astr. Ap.*, **210**, 236.

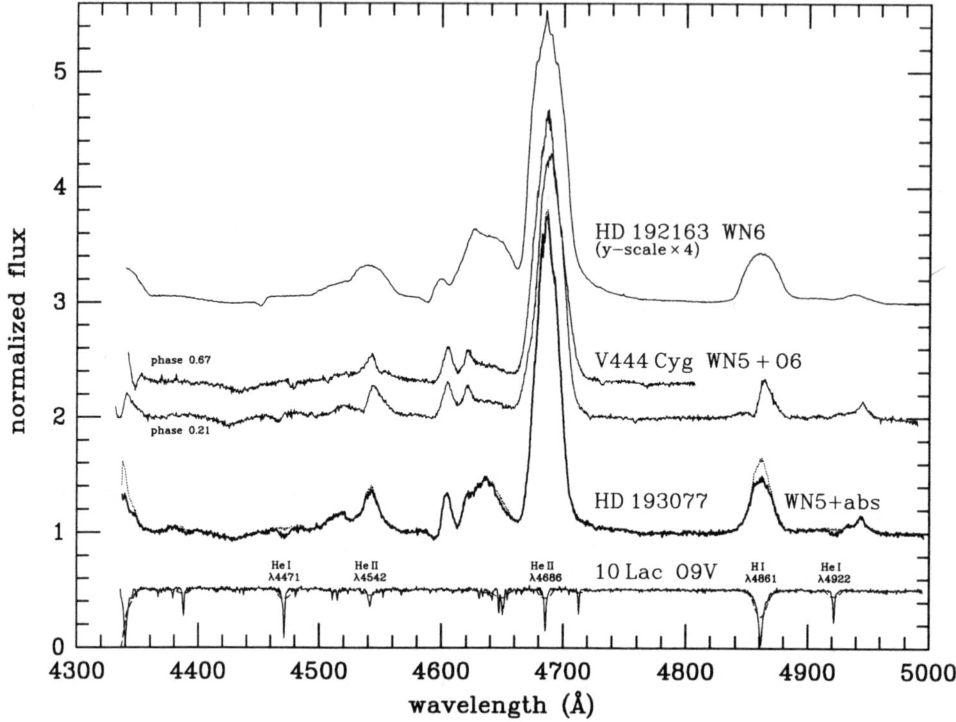



Figure 1.    Continuum normalized spectra. All spectra are drawn on the same y-scale except for HD192163 which is 4 times compressed. Two observations of the binary V444 Cygni are shown, taken at about opposite elongations. The two spectra are shifted in wavelength so that the N V λ4604 emission lines are aligned with each other and with the N V emission line of HD193077. The dots overplotted on the spectrum of HD193077 represent the resulting spectrum if 3/4 of the absorption lines (rotationally broadend to 450 km s$^{-1}$) of 10 Lac are subtracted. For a discussion of these spectra see text, Sect. 3.

TABLE 1.    Heliocentric radial velocities measured in the spectrum of HD193077 obtained August 18/19, 1986 UT23:40–UT00:20 (JD2446661.5).

| Spectral line | He I 4471.5 abs | He II 4541.6 em | N V 4603.7 em | He II 4685.7 em | H I/HeII 4860.0 em | He I 4921.9 abs | N V[a] 4944.5 em | He II 5411.5 em |
|---|---|---|---|---|---|---|---|---|
| RV[b] | −25 | +15 | +50 | +50 | +70 | +5 | −70 | +70 |
| predicted[c] | +53 | | | +46 | | +53 | | |

[a] only the peak at 4544 Å is measured, not the whole line complex.

[b] in km s$^{-1}$ relative to the wavelength given in the table. The accuracy is estimated to be of the order of 20 km s$^{-1}$.

[c] from the orbital elements given in Table 6 of Lamontagne et al. (1982). The phase of the observing time is 0.63 for the absorption lines and 0.61 for the He II λ4686 emission.

# HOW EFFECTIVE IS ROTATION IN ENHANCING THE RATE OF MASS LOSS IN EARLY TYPE STARS?


M. S. VARDYA
Tata Institute of Fundamental Research
Bombay 400 005, India



Abstract: Conflicting results on the effect of rotation on mass loss of OB stars have been briefly discussed. It is concluded that rotation definitely increases mass flux but a better value of the enhancement factor will have to await larger sample and improved data than considerd so far.


Rotation, one expects, should enhance the rate of stellar mass loss, by reducing the effective gravity and hence, escape velocity. This is a direct effect. Indirectly, rotation produces other subtle effects, which may also increase mass loss.

Early theoretical attempts, say, by de Greve et al (1972) and by Marlborough and Zamir (1984) showed that

$$\dot{M}(\text{rotation}) \gtrsim \dot{M}(\text{no} - \text{rotation}).$$

However, observational results are somewhat nebulous. Furenlid and Young (1980) found that $H_\alpha$ line asymmetry, which is a measure of mass loss, in 60 normal main sequence B0-B3 stars (excluding Be and peculiar stars) is always large when projected linear rotational velocity, $v \sin i \gtrsim 200$ km/s; however, they did not consider $\dot{M}$ itself. Snow (1981) analysed 22 B stars from B 0.5 to B6, including 19 Be-like stars, Doazan et al (1982) 21 Be, B shell and normal stars, and Slettebak and Carpenter (1983) 12 Be and standard stars, but failed to find correlation of $\dot{M}$ with $v \sin i$. Gaithier et al (1981) did find a qualitative dependence of $\dot{M}$ with $v \sin i$ in 25 high luminosity OB stars, but not in early B supergiants.

This lack of definitiveness may be due to $\sin i$ factor as theory demands only $v$. Can one randomize $\sin i$ to circumvent it?

A large mixed sample of stars is perhaps the answer. The above studies have been limited covering a small spectral class range. Therefore, we considered 81 stars (Vardya 1985) covering temperature spectral type from O3 to B9, luminosity class $Ia^+$ to $V, v \sin i$ from 15 to 505 km s$^{-1}$, log $\dot{M}$ from $-10.5$ to $-4.7$; this sample has 21 O stars with f, (f) and ((f)) spectral characteristics, 10 Be or 18 Be-like stars, and 6 peculiar stars and a range of log $L/L\odot$ from 2.5 to 6.4, $M/M\odot$ from 7 to 136, and $R/R\odot$ from 5 to 86.

231


L. A. Willson and R. Stalio (eds.), Angular Momentum and Mass Loss for Hot Stars, 231–234.

No relation was found between $\dot{M}$ and $v \sin i$, but mass flux, $\dot{M}/R^2$ was found to correlate with $v \sin i$, but resulted in two relations — one for 63 normal OB stars, and another for 18 Be-like stars.

This was encouraging, but surprising. Rotation, in a way, is an extrinsic property. Then why should there be two relations? In addition, note that

$$< \dot{M} >_{\text{Be-like}} \simeq 10^{-4} < \dot{M} >_{\text{normal OB}},$$

and

$$< M/R^2 >_{\text{Be-like}} \simeq 10^{-3} < \dot{M}/R^2 >_{\text{normal OB}}$$

but

$$< v \sin i >_{\text{Be-like}} \simeq 3 < v \sin i >_{\text{normal OB}} .$$

To achieve a single relation for the two groups, the dominant cause of mass loss, i.e. radiation pressure or luminosity effect was differenced out by using the semi-empirical relation (Vardya 1984):

$$\dot{M} = A L^2 (R/M)^{3/2}$$

where $A$ is a scaling factor. The effect of other parameters, like rotation, magnetic field, chemical composition etc. is contained in $A$. Therefore, we considered a relation between $A/R^2$ and $v \sin i$. This resulted in a single linear relation for all the 81 stars, with a high correlation coefficient. This improved further when $v$ was replaced by the angular rotational velocity, $\Omega$. It can be further improved by taking a quadratic term. Thus it was shown conclusively for the first time using observational data that rotation enhances mass flux in OB stars.

Is this increase in log $A/R^2$ due to rotation commensurate with theoretical expectation? Log $|A/R^2|$ increases by 2 for an increase of 1.5 in log $(v \sin i)$, or by 2.8 for an increase of 2.6 in log $(\Omega \sin i)$. Theoretically, Pauldrach et al (1986) found 26% increase in $\dot{M}$ as $v$ goes from 0 to 350 km s$^{-1}$ in a O5 V star; Poe and Friend (1986) found 62% enhancement for $v$ from 125 to 400 km s$^{-1}$ (with B=200 G) for a O6ef star, and 370% for a B1.5 Ve star as $v$ goes from 125 to 540 km s$^{-1}$ (B=50 G). Friend and Abbott (1986) found 100% increase in $\dot{M}$ as $\Gamma_{\text{rot}} \equiv v$ (rotation)/$v$(break-up) goes from 0 to 0.8; however, their final conclusion, using observational data for OB stars, but excluding Be stars, is "that there is currently no evidence for a dependence of the mass loss rates on rotational velocity..." ; they have also made the cryptic statement that "A correlation between mass-loss rate and rotational velocity has been sought by Vardya (1985), but the evidence is weak at best".

Nieuwenhuijzen and de Jager (1988) have discussed the discrepancy between our (Vardya 1985) results and the theoretical conclusions by considering 142 non-emission early type stars, excluding Be and shell stars. For their sample, they find correlation between $\dot{M}$ and $L$ as well as between $L$ and $v \sin i$, leading to an artificial correlation between $\dot{M}$ and $v \sin i$, which they imply, what we have found. To avoid this, they have fitted the data of $\dot{M}$ in terms of product of three Chebychev polynomials, which are function of L, $T_{eff}$, and $v \sin i$; this has 39 coefficients, 20 independent of $v \sin i$, and have concluded that $\dot{M}$ depends only slightly on $v \sin i$.

A critical examination of these criticisms and discrepancies leads to:

(a) In our sample (Vardya 1985), $(v \sin i, L)$ are not correlated except in a limited region (see Fig. 1). Fig. 2 gives a check where $\log \dot{M}$ vs $\log v \sin i$ is plotted for $\log L = 5.0 \pm .2$ for 23 stars; it shows no correlation. Further, we have considered not only $\dot{M}$ but also $A$, from which luminosity effect has been taken out. Note that the range of $\log A/R^2$ is larger than that of $\log A$ by a factor of $\sim 1.5$.

(b) Nieuwenhuizen and de Jager (1988) have excluded Be and shell stars. This prevents proper randomization of $\sin i$.

(c) Chebychev polynomial with 22 positive and 17 negative coefficients, with half of the same order of magnitude, provide a good numerical fit, but using it for physical interpretation is dangerous. Further, we are interested in $v$ and not $v \sin i$; such a good fit will incorporate even $\sin i$.

(d) Note that in a restricted sample, e.g., limited to a small range in spectral class, the scaling or constant factor will absorb similar dependence, thus preventing an explicit manifestation of real dependence . Analogously, when a large varied sample is fitted with a very large number of coefficients, the real dependence gets absorbed in these coefficients and one sees only residual dependence.

(e) A comparision between theory and observational results is beset with the problem that theory demands $v$ but observations provide $v \sin i$. Further, theoretical models incorporate rotation as reduction in effective gravity, whereas observationally there is no such inhibition, and indirect effects due to rotation will also be manifested. Besides, $A$ contains dependence of other parameters, not incorporated explicitly.

Recently, Howarth and Prinja (1989) considered 163 O stars with $v \sin i$ from 5 to 435 km s$^{-1}$, $\log \dot{M}$ from $-4.6$ to $-7.8$, $\log L/L_\odot$ 4.5 to 6.4, $M/M_\odot$ 18 to 150 and $R/R_\odot$ 5 to 36, and found a maximum change $\Delta \log \dot{M} \simeq 0.4$; they also considered a quantity similar to our $A$. Note that though they have taken a large sample, it is a restricted sample compared to ours. Incidentally, the authors have claimed that "the present result is the first reliable indication that such an effect actually exists in nature."

In conclusion, rotation definitely enhances mass flux in early type stars though it is a secondary effect, the primary effect being luminosity. Due to $\sin i$ factor, and the errors in the various quantities including $\dot{M}$, it is difficult to fully assess the enhancement factor. The answer must await a larger varied sample of OB stars with better randomizations of $\sin i$, and improved values of the basic data of $\dot{M}, L, M$ and $R$.

I am thankful to Professors I. Furenlid and S.P. Tarafdar for useful comments. An earlier version of this work was presented at the "International Workshop on Binary Stars and Stellar Atmospheres" held at Hyderabad.

## References


de Greve, J.P., de Loore, C., and de Jager, C. 1972, *Ap. Sp. Sci.* **18**, 128.

Doazan, V., Franco, M.L., Stalio, R., and Thomas, R.N. 1982, In *IAU Symposium No 98*, Be Stars, eds. M. Jaschek and H.G. Groth (Dordrecht: Reidel), p. **318**

Friend, D.B., and Abbott, D.C. 1986, *Ap. J.* **311**, 701.

Furenlid, I., and Young, A., 1980, *Ap. J.* **240**, L59.

Gathiers, R., Lamers, H.J.G.L.M., and Snow, T.P. 1981, *Ap. J.* **247**, 173.

Howarth, I.D., and Prinja, R.K. 1989, *Ap. J. Suppl.* 69, 527.

Marlborough, J.M.M., and Zamir, M. 1984, *Ap. J.* **276**, 706.

Nieuwenhuijzen, H., and de Jager, C. 1988, *Astr. Ap.* **203**, 355.
Pauldrach, A., Puls, J., and Kurdritzki, R.P. 1986, *Astr. Ap.* **164**, 86.
Poe, C.H., and Friend, D.B. 1986, *Ap. J.* **311**, 317.
Slettebak, A.E., and Carpenter, K.G., 1983, *Ap. J. Suppl.* **53**, 869.
Snow, T.P. 1981, *Ap. J.* **251**, 139.
Vardya, M.S. 1984, *Ap. Sp. Sci.* **107**, 141.
Vardya, M.S. 1985, *Ap. J.* **299**, 255.

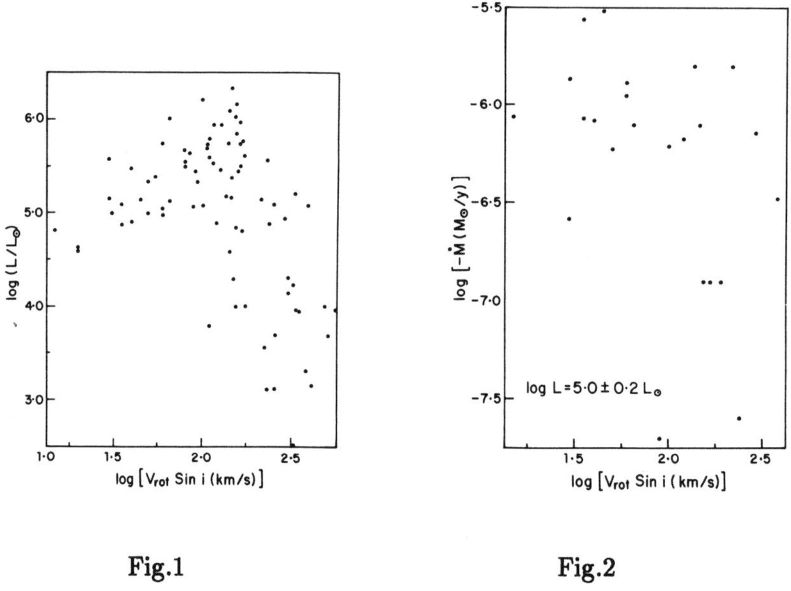

Fig.1          Fig.2

Fig. 1: Luminosity vs. *v sin i* for O and B stars.
Fig. 2: Rate of mass loss vs. *v sin i* for O and B stars with $\log L/L_\odot = 5.0 \pm 2$

# ROTATION AND PULSATION–MODE–SELECTION IN B–TYPE STARS


C. WAELKENS*
*Astronomisch Instituut Katholieke Universiteit Leuven*
*Celestijnenlaan 200B*
*3030 Leuven*
*Belgium*



ABSTRACT. Evidence is presented that the nonradial oscillation modes in B stars tend to to be observed with a spherical harmonic order $l$ that increases with the axial–rotation velocity of the star. This result is here for the first time shown for the $\beta$ Cephei stars. A similar relation holds for the later–B–type pulsators. It cannot yet be ruled out that the non–detection of high–order modes in slow rotators is a selection effect. Our results give a new explanation for the small photometric amplitudes in broad–lined variables.


## 1. Introduction

The possible connection between rotation and pulsation in early–type variables is a debatable and often debated one. A major observational problem is constituted by the selection effects that large rotation induces on spectroscopic studies.

In the present paper, we discuss the line–profile variations of several early–type pulsating stars. Our sample is assembled in a way so that rotation velocity does not enter as a selection criterion. It turns out that a systematic trend is observed between the spherical harmonic order $l$ of the oscillation modes and the projected rotation velocities of the stars.

## 2. The $\beta$ Cephei stars

The $\beta$ Cephei stars form a group of early–B variables that occupy a well defined part of the upper HR diagram and so probably all pulsate because of the same physical mechanism (Lesh, 1982). Initially, it was thought that all $\beta$ Cephei stars are slow rotators, but this turned out to be an artefact of the spectroscopic detection techniques of these stars, velocity variations being harder to detect in broad–lined stars. Nevertheless, it is probably significant that the largest–amplitude $\beta$ Cephei stars are narrow–lined objects (Jakate, 1979).


* Research Associate of the Belgian National Fund for Scientific Research (NFWO)





*L. A. Willson and R. Stalio (eds.), Angular Momentum and Mass Loss for Hot Stars, 235–238.*

Although the $\beta$ Cephei stars form a fairly homogeneous group, the pulsation modes are different from star to star: there is no unique period—luminosity relation, so that the radial wavenumbers can take different values for different stars; also the angular wavenumbers differ, and radial and nonradial oscillations both occur. It is not known what makes a star choose a particular mode.

We have obtained line—profile observations of several $\beta$ Cephei stars with the CES spectrograph at ESO. We show in Figure 1 a superposition of the profiles of the SiIII—line at 455 nm taken during one night for two well—known narrow—lined variables, $\delta$ Ceti (radial mode) and $\beta$ Canis Majoris (nonradial mode with $l=2$). In Figure 2 we show individual lines profiles for two broad—variables, $\kappa$ Scorpii and $\lambda$ Scorpii; clearly, nonradial oscillation modes witlarge $l$—values are present in these stars. Our results are summarized in Table 1; the listed projected rotation velocities were taken from the literature.

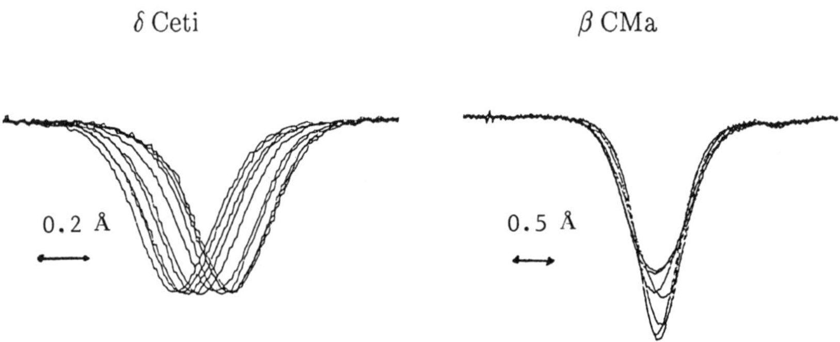


Figure 1: Line profiles for two slowly rotating $\beta$ Cephei stars

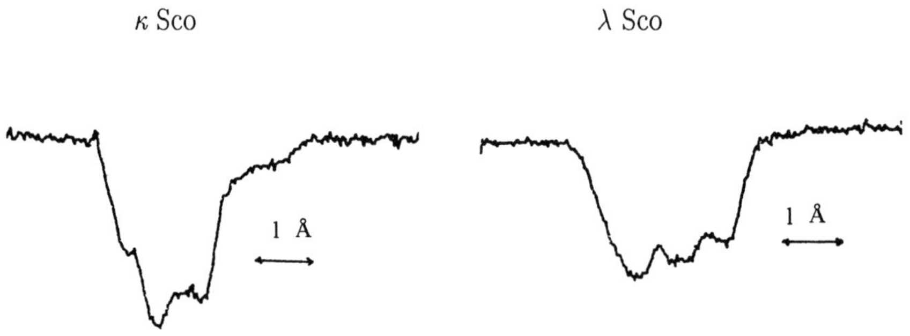


Figure 2: Line profiles for two rapidly rotating $\beta$ Cephei stars

Table 1: Projected rotation velocities and wavenumber $l$
for the observed $\beta$ Cephei stars.

| Star | V sin i | $l$ |
|------|---------|-----|
| $\delta$ Cet | 25 | 0 |
| $\nu$ Eri | 31 | 0 and 2 |
| $\beta$ CMa | 38 | 2 |
| $\beta$ Cru | 32 | 0 or 2 |
| $\beta$ Cen | 128 | $\geq 4$ |
| $\alpha$ Vir | 172 | 8 |
| $\lambda$ Sco | 300 | 8 |
| $\kappa$ Sco | 99 | 8 |

Inspection of Table 1 reveals a trend that the larger the v sin i, the higher the wavenumber $l$ of the observed mode. This could mean that stellar rotation influences the selection of a particular pulsation mode in a $\beta$ Cephei star. There may also be an observational selection effect, since the Doppler broadening tends to favor the observability of high–order modes in rapid rotators. An argument for a positive correlation may be the mentioned trend that slower rotators are also larger–photometric–amplitude variables (Jakate, 1979): modes with large $l$–values have low photometric amplitudes.

### 3. 53 Persei stars and broad–lined line–profile variable stars

53 Persei stars were first detected from line–profile analysis (Smith and Karp, 1976) and later recovered in a purely photometric search (Waelkens and Rufener, 1985). That these stars are pulsating stars is shown by the detection of multiple periods in some of them (Waelkens, 1987; Manfroid and Renson, 1989). All seem to oscillate in modes of low $l$, and all are slow rotators, also those that were detected without preconception about rotation velocities (Waelkens, 1987).

Line–profile variations are also observed for broad–lined B stars and Be stars (for a review, see Baade, 1987). If these variations are to be interpreted as pulsations, then most often non–radial modes of high order are implied; however, Baade contends that in Be stars a low–$l$ mode is always present. Such a mode should show up in photometry too. It would be interesting to verify whether it corresponds to the periodic variations reported for Be stars by Balona and Cuypers (1989).

### 4. Discussion

Various pieces of information suggest the general trend that rotation influences the pulsation of early–type stars, in the sense that higher–order modes are selected in more rapid rotators. So far the results are somewhat preliminary, since an important observational caveat has to be checked still further: low−$l$ modes are harder to detect in the line profiles of broad−lined stars, and high−$l$ modes are not so easily seen for sharp−lined objects. For this and other similar projects, we strongly recommend a coordinated spectroscopic and photometric approach.

So far our working hypothesis is consistent with the data in the cases of the $\beta$ Cephei stars, the 53 Persei stars, and the broad−lined B stars. If the effect is real, it would once more set the problem of the classification of early−type variable stars. Indeed, from a line−profile analysis one could have the impression that rapidly and slowly rotating $\beta$ Cephei stars are different kinds of objects, but other types of observations show that they are not. Similarly, it may be asked whether the distinction we make between 53 Persei stars and other B−type variables is not merely a superficial one which masks the real issue of the mechanism which makes these stars pulsate. Along similar lines, the fact that a group of rapidly rotating stars such as the Be stars has a charasteristic pulsation pattern does not necessarily imply that rotation has anything to do with the mechanism that makes these stars pulsate.

## Acknowledgement

This research was supported by the Belgian Fund for Collective Fundamental Research, under grant Nr. 2.0098.87.

## References

Baade, D., 1987, in Proc. IAU Coll. No. 98 Physics of Be stars, A. Slettebak and T.P. Snow (eds.), Cambridge Univ. Press, Cambridge, p. 361.
Balona, L.A., Cuypers, J., 1989, these proceedings.
Jakate, S.M., 1979, Astron. J. 84, 552.
Manfroid, J., Renson, P., 1989, Astron. Astrophys. 223, 187.
Smith, M.A., Karp, 1976, *Los Alamos Conf. Solar and Stellar Pulsation*, p. 289.
Waelkens, C., 1987, in Stellar Pulsation, Lecture Notes in Physics 274, 75.
Waelkens, C., Rufener, F., 1985, Astron. Astrophys. 152, 6.

# THE ANGULAR MOMENTUM-LOSS AND THE DIFFERENTIAL ROTATION IN B AND Be STARS


J. Zorec, R. Mochkovitch, A. Garcia

*Institut d'Astrophysique de Paris, CNRS*
*98$^{bis}$, Bd. Arago - 75014 Paris, France*



**Abstract**: In this paper we compare the angular momentum of B and Be stars calculated using their mean true rotational velocities $<V>$ at luminosity classes V and III and models of rigidly rotating stars. The inferred mass-loss rates associated with the loss of angular momentum needed to explain the change of $<V>$ from the main sequence to the mean luminosty class III are larger than the observed ones. This might be the result of some angular momentum transport in the stars which introduces a supplementary deceleration to the stellar surface angular velocity and enhances a radial differential rotation.


## 1. The method

We study the distribution of true rotational velocities of normal B and Be stars separately (normal-B stars are considered those not classified Be and/or Bp), as function of their masses and luminosity class. Then, we calculate barotropic models of rigidly rotating stars, to obtain an estimate of their total angular momentum corresponding to the observed mean true equatorial velocities. Comparing the angular momentum calculated for stars having the same mass and being of different luminosity classe, we deduce the mass-loss rates associated with the loss of angular momentum which would be responsible for the observed change of the stellar surface rotational velocities from the mean main sequence to the mean sequence of luminosity class III. The comparison of these mass-loss rates with the observed ones, will finally allow us to discuss if the studied stars evolve as rigid or as differential rotators.

### 1.1 Distribution of true rotational velocities

We call "true rotational velocities" (TRV), the V sin i parameters which were statistically corrected for the geometrical factor sin i supposing that the stellar rotational axis are oriented at random in space. The V sin i parameters used are from the Uesugi and Fukuda's (1982) compilation, which were transformed to the new Slettebak's et al. (1975) system. We completed the Uesugi and Fukuda's list of Be stars with the Slettebak's (1982) one. We used as much as possible the BCD (Barbier-Chalonge-Divan) spectrophotometric spectral types. Otherwise, the spectral types given in the Bright Star Catalogue were used, or the catalogues of selected spectral types of CDS in Strasbourg. The masses of B and Be stars are from the calibration of the BCD $\lambda_1$ and D parameters of the Balmer discontinuity in fundamental stellar parameters (Divan and Zorec, 1982; Zorec, 1990).

239


*L. A. Willson and R. Stalio (eds.), Angular Momentum and Mass Loss for Hot Stars, 239–243.*
© 1990 *Kluwer Academic Publishers. Printed in the Netherlands.*

240

## 1.2 Models of rotating stars

We have supposed that the stars are rigid rotators. From the Poincaré-Wavre theorem (Tassoul, 1978) it is known that in this case we can separate in a first approximation, the study of the stellar energy problem from that of its geometrical structure. The Poisson's equation and the equation of hydrostatic equilibrium were solved using the cells algorithm described by Clement (1974). To represent the stellar inertial momentum at each luminosity class, we used barotropic relations $P = P(\rho)$ (P: pressure; $\rho$: density), derived for non-rotating stars at different evolutionary stages. This approximation may be justified, because for rigid rotators the energy ratio $\tau = K/|W|$ is very low: $\tau < \tau_C = 0.008$ (K: rotational kinetic energy; W: gravitational potential energy; the limit $\tau_C$ corresponds to critical rotators just following the ZAMS). It was shown by Maeder (1974) that in this case the change of the slopes $\partial P/\partial\rho$ for stars at evolutionary stages before the core He-ignition is less than 0.5% from that of non-rotating stars, which do not change sensitively the estimate of the stellar radius and that of its total angular momentum.

The barotropic relation $P = P(\rho)$ for each mass at each mean luminosity classe was determined using the stellar models and the evolutionary tracks calculated by Becker (1981) for the initial chemical composition X=0.28 and Y=0.70, together with the above mentioned calibrations of $\lambda_1 D$ parameters in $T_{eff}$, log g, $M_{bol}$ and $M/M_\odot$.

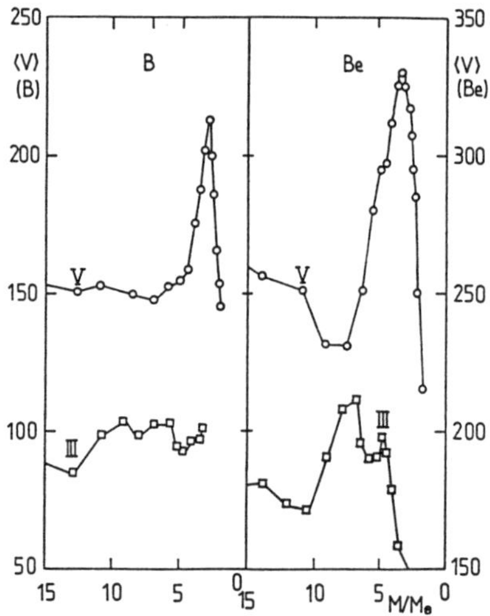

**Fig. 1:** *Distribution of mean true rotational velocities $<V>$ (km $s^{-1}$) of B and Be stars against the mass and for mean luminosity classes V and III.*

## 2. Results

The distributions of mean TRV $<V>$ of B and Be stars against the stellar mass and for mean luminosity classes V and III are shown in Fig. 1. Apart from the known difference that $<V(Be)>$ are 1.5 to 2 times larger than $<V(B)>$, there seem to exist some differences in the shape of the distributions. Writing the standard deviation of true rotational velocities as: $\sigma_V = \mu <V>$; $\mu = (\mu_2/\mu_1^2 - 1)^{1/2}$, where $\mu_1$ and $\mu_2$ are respectively the moments of first and second order of distributions of TRV. The factors $\mu$ averaged over all B spectral types in each luminosity sequence for normal-B stars are: $\mu_{V,IV,III}(B) = 0.50, 0.55, 0.61$ and for Be stars: $\mu_{V,IV,III}(Be) = 0.29, 0.31, 0.35$. We see that $\mu(Be) < \mu(B)$. Using the values of $<V>$ from Fig. 1, we also obtain that: $\sigma_V(Be) < \sigma_V(B)$ for each stellar mass and luminosity class. This may indicate that Be stars have a tendency to be rotating more closely to a given velocity V than the normal-B ones.

From models of rigidly rotating stars corresponding to velocities $<V>$ of Fig. 1 and for different masses, we obtained an estimate of the stellar angular momentum J. The change $\Delta J$ of this angular momentum for a given mass and from one luminosity class to another was represented by: $\Delta J \simeq < (2/3)j_e > \times \Delta M$, where $j_e = <V.R>_{\Delta T}$ is the equatorial specific angular momentum averaged over the studied period of time of evolution $\Delta T$ (R: stellar radius); $\Delta M$ is the mass lost associated with $\Delta J$. The stellar ages $T$ were deduced using the Becker's (1981) models. The mean mass-loss rates which would be associated with the observed changes of $<V>$ are then estimated by: $\dot{M}_{\Delta J} = \Delta M/\Delta T$. These mass-loss rates are plotted in Fig. 2. For comparison the mass-loss rates $\dot{M}_{obs}(UV)$ inferred from the far-UV spectral line observations and those derived using far-IR data: $\dot{M}_{obs}(IR)$ are also shown. The observed mass-loss rates $\dot{M}_{obs}(UV)$ are from the compilation made by Vardya (1985) and the $\dot{M}_{obs}(IR)$ were obtained by Waters (1986).

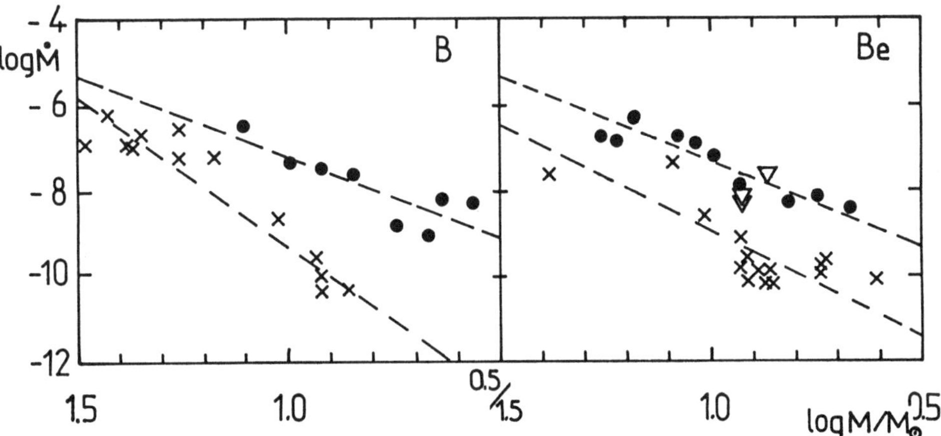

**Fig. 2:** *Mass-loss rates (in $M_\odot yr^{-1}$) of B and Be stars against the stellar mass.* $\bullet$: $\dot{M}_{\Delta J}$ = *mass-loss rates associated with the angular momentum loss $\Delta J$;* $\times$: $\dot{M}_{obs}(UV)$ = *mass-loss rates determined from far-UV lines;* $\bigtriangledown$: $\dot{M}_{obs}(IR)$ = *mass-loss rates determined from far-IR excesses.*

## 3. Discussion

Concerning the property of Be stars being all rotating closely to a given velocity, let notice that the force ratio $Q = (F_\Omega + A_R)/|g|$, where $F_\Omega = (2/3)(V^2/R)$ is the mean centrifugal acceleration (R: stellar radius) and $A_R$ is the maximum radiative acceleration in a stellar atmosphere as given by the Kurucz' (1979) models for a given effective temperature $T_{eff}$ and a mean effective gravity $|g_{eff}| = |g|( 1 - F_\Omega/|g|)$, does not depend on the luminosty class and it depends very slightly on the stellar mass: $Q = (0.22 \pm 0.05) + (0.013 \pm 0.004)(M/M_\odot)$. The maximum value of Q is at the stellar equator: $Q_{MAX} \approx 0.43$. Similar ratios Q were already obtained by Vardya (1985). From the fact that the rotational velocities of Be stars have a rather low dispersion $\sigma_V$ around the mean $<V>$ and the constancy of Q with the luminosity class and rougly also with the mass, we conclude that the Be phenomenon tends to appear at a rather precise under-critical ratio $V/V_C < 1$ ($V_C$: critical velocity for a rigid rotator) which, as seen from the expression of Q, is slightly lower for hot stars where the radiation pressure is higher.

In Fig.2 we see that the inferred mass-loss rates $\dot{M}_{\Delta J}$ of normal-B stars are one to two orders of magnitude larger than $\dot{M}_{obs}(UV)$. For Be stars, the $\dot{M}_{\Delta J}$ rates are one order of magnitude larger than $\dot{M}_{obs}(UV)$, and about the same as $\dot{M}_{obs}(IR)$.

For normal-B stars, we may conclude that the fraction of angular momentum which is not lost by $\dot{M}_{obs}$, has to be redistributed inducing or enhancing a radial differential rotation between the interior and the stellar surface. The interpretation of the results obtained for Be stars depends howerver on which mass-loss rate: $\dot{M}_{obs}(UV)$ or $\dot{M}_{obs}(IR)$ we use to better represent the mass just leaving the stellar surface. The confusion on the origin of $\dot{M}_{obs}$ may be particularly important when Be stars with far-IR excesses have a companion. In such cases, general disc-like structures having densities varying as $\exp(-z^2/h^2)$ (z: coordinate vertical to the orbital plane; h: height scale) can easily be formed (Prendergast 1964). The disc is then not due to the mass lost mainly through the stellar equator, but to the mass emitted from the whole stellar surface which is trapped into a disc-like structure by the force-field of the binary system. For single Be stars, first, strong dis-like circumstellar envelopes are not supported by the polarization observations (Höflich and Zorec, 1989), and second, it is not clear how the mass-flux velocity entering the determination of $\dot{M}_{obs}(IR)$ can became $v_{wind} < 10$ km s$^{-1}$ at distamces $R \simeq 2R_*$ from the initial ejection value $v_0 \sim 10^2 - 10^3$ km s$^{-1}$ if it is not decelerated by a shell of mass amassed around the star and probably due to earlier generations of stellar mass-ejections (Zorec, 1981). Again, the mass stored in this way cannot be used to represent an instantaneous mean stellar mass-ejection.

From this we are tempted to think, that $\dot{M}_{obs}(UV)$ may better represent than $\dot{M}_{obs}(IR)$ the instantaneous mean mass outflow from the surface of Be stars, and that also for Be stars we have: $\dot{M}_{\Delta J} > \dot{M}_{obs}$. However, in Be stars other mechanisms or other types of angular momentum redistributions have to be operating, which make them all rotating very fast in the surface and at the limit given by the ratio Q, value at which the stars seem to become "unstable" against the Be phenomenon.

It has to be finally noted that the calculated models of rotating stars are concerned by definition only by low values of $\tau = K/|W|$ [$\tau(B) \lesssim 0.0016$; $\tau(Be) \lesssim 0.0049$], which introduce a strong limitation on the estimates of J. In this respect it is worth keeping in mind that at least two cases of differential rotation may exist for the same value of the "effective" surface rotational velocity V: a) the "under-critical" differential rotation corresponding to $\tau < \tau_C$, and b) a "neat" differential rotation, characterized by $\tau > \tau_C$ and where $\tau$ can be as high as $\tau \simeq 0.14$ for secularly stable models (Bodenheimer, 1971; Zorec et al. 1988a,b). First rough estimates of $\tau$, deduced considering that the observed stellar parameters are function of 6 unknowns: M(mass); chemical composition; the energy ratio $\tau$; the distribution law of the angular momentum; the evolutionary stage and the inclination i, show that $\tau$ might be higher than $\tau_C$ of rigid rotators, and that they are about the same for B and Be stars (Zorec et al. 1987; Zorec, 1989b). The origin of the Be phenomenon has then to be seen as a problem related to the stellar structure determined by the amount of stored rotational energy and the associated hydrodynamical instabilities (Zahn, 1983; Ando, 1986), which manage the loss and the redistribution of the angular momentum in Be stars, perhaps in a different way than in normal B stars.

**References:**

Ando, H.: 1986, *Astron. Astrophys.* **163**, 97

Becker, S.A.: 1981, *Astrophys. J. Suppl.* **45**, 475

Bodenheimer, P.: 1971, *Astrophys. J.* **167**, 153

Clement, M.: 1974, *Astrophys. J.* **194**, 709

Divan, L., Zorec, J.: 1982, *The Scientific Aspects of the Hipparcos Astronomy Mission*, ESA SP-177, p.291

Höflich, P., Zorec, J.: 1989, *Modeling the Stellar Environment: how and why?*, IAP Astrophysics Meeting N° **4**, eds. Ph. Delache, S. Laloë, C. Magnan and J. Tran Thanh Van, p.257

Maeder, A.: 1974, *Astron. Astrophys.* **34**, 409

Prendergast, K.H.: 1960, *Astrophys. J.* **132**, 162

Slettebak, A., Collins, G., Boyce, P., White, N., Parkinson, T.: 1975, *Ap. J. Suppl.* **29**, 137

Slettebak, A.: 1982, *Astrophys. J. Suppl.* **50**, 55

Tassoul, J.L.: 1978, *Theory of Rotating Stars*, Princeton Univ. Press

Uesugi, A., Fukuda, I.: 1982, *Revised Catalogue of Stellar Rotational Velocities*, Kyoto University

Vardya, M.S.: 1985, *Astrophys. J.* **299**, 255

Waters, L.B.F.M.: 1986, *Astron. Astrophys.* **162**, 121

Zahn, J.P.: 1983, *Astrophysical Processes in Upper Main Sequence Stars*, 13th Advanced Course of Saas Fee, Observatoire de Genève, p.253

Zorec,J.: 1981, *Effects of Mass Loss on Stellar Evolution*, IAU Coll. N⁰ **59**, eds. C. Chiosi and R. Stalio, p.539

Zorec, J., Divan, L., Mochkovitch, R., Garcia, A.: 1987, *Physics of Be Stars*, IAU Coll. N⁰ **92**, eds. A. Slettebak and Th. Snow, p.67

Zorec, J., Mochkovitch, R., Garcia, A.: 1988a, *C.R. Acad. Sci. Paris* **306**, 1225

Zorec, J., Mochkovitch, R., Divan, L.: 1988b, *C.R. Acad. Sci. Paris* **306**, 1265

Zorec, J.: 1990, *Astron. Astrophys.*, submitted

Zorec, J.: 1989, *Problems of Stellar Hydrodynamics*, Third Cycle of Astronomy and Astrophysics, Belgium

THE ROLE OF AXIAL SYMMETRY IN THE UPPER PART OF THE HRD: B[e]
SUPERGIANTS AND LBVs



Franz-Josef Zickgraf
Department of Astronomy
University of Minnesota
116 Church Street S.E.
Minneapolis, MN 55455




ABSTRACT.  In this review the general characteristics of B[e]
supergiants and LBVs and the observational evidence for the presence
of axial symmetry in these very massive and luminous hot stars are
discussed.



## 1. INTRODUCTION


The existence of a temperature dependent upper limit of stellar
luminosity in the Hertzsprung-Russell-diagram (HRD) (Humphreys and
Davidson, 1979) is observationally a well-established fact. Unstable
stars like $\eta$ Car, the Hubble-Sandage and S Doradus variables, and the
P Cygni type stars located near the observed limit were summarized by
Conti (1984) in the class of Luminous Blue Variables (LBVs). However,
not only the highly unstable LBVs are found in this area of the HRD
but also the peculiar B[e] supergiants (Zickgraf et al. 1989) as well
as the normal B supergiants.
   The reason for the existence of the stability limit is not yet
clear though a large variety of possible explanations has been
presented.  A modified Eddington limit has been suggested by several
authors (Humphreys and Davidson 1984, Appenzeller 1986, 1989, Lamers
1986, Davidson 1987, Lamers and Fitzpatrick 1988). Recent model
calculations by Pauldrach et al. (1989) for the stellar wind of P
Cygni revealed that the mass loss of this star is highly unstable with
respect to small changes of the stellar radius and the luminosity. The
physical reason for this mechanism is the existence of a critical
density above which the Lyman continuum becomes optically thick. This
causes a drop in the wind ionization and a strong increase of the
radiative acceleration.  Other mechanisms discussed for the
empirically observed instability involve the interior evolution or
effects of atmospheric turbulence (cf. Maeder, 1983, 1989, de Jager,
1980, 1984).
   There is now increasing observational evidence that another
parameter might be involved in these objects, namely rotation or
angular momentum. Observations of $\eta$ Car, AG Car and the newly defined
class of B[e] supergiants revealed evidence for the existence
of bipolar or disk-like structures. It should be noted that for WR




L. A. Willson and R. Stalio (eds.), Angular Momentum and Mass Loss for Hot Stars, 245–263.
© 1990 Kluwer Academic Publishers. Printed in the Netherlands.

stars rapid rotation is now being taken also into account (e.g., Poe et al. 1989). Based on models for a rotating WR star Cassinelli et al. (1989 and this volume) made a first attempt to model the mass loss of B[e] supergiants by including rotation and magnetic fields.

Sreenivasan and Wilson (1989) studied the role of rotation in LBVs from a theoretical point-of-view. They showed that differential rotation with conservation of specific angular momentum has large effects on stellar models. Shear is generated bringing the stars into a dynamical equilibrium. The core size increases and mixing between core and envelope leads to bluer models.

The intention of this contribution is to draw attention to the observational evidence of axial symmetry in LBVs and B[e] supergiants. In chapter 2 I will give a short summary of the basic properties of the LBVs. I will mainly refer to the most recent reviews on this matter given at the IAU Colloqium No. 113 in Val Morin 1988, the first meeting solely dedicated to the physics of LBVs, particularly by Humphreys (1989) and Wolf (1989a). These and other contributions at that meeting comprehensively reviewed our present knowledge of the observed properties of LBVs. The following chapter 3 will discuss basic characteristics of the B[e] supergiants. In the last section the observational evidence for the existence of axial symmetry in both classes of stars is discussed in detail. It will become clear that there is certainly a need for the inclusion of hydrodynamical effects caused by rotation in an attempt to understand the nature of these stars.


## 2. BASIC PROPERTIES OF LUMINOUS BLUE VARIABLES


A list of confirmed LBVs and the stellar parameters of some well-studied cases can be found in Humphreys (1989). Recently Stahl and Wolf (1989) discovered a further S Doradus variable, R110 (=HDE269662) in the LMC.

It is now widely accepted that LBVs represent an evolved shortlived evolutionary stage of massive stars. Maeder (1989) discussed possible evolutionary sequences for massive O stars in different mass ranges. For initial masses between 50 and 120 $M_\odot$ the sequence

$$O - Of - (BSG) - LBV - WR - SN$$

has become increasingly clear (see also Maeder 1983, Maeder and Meynet 1987). As will be discussed below some LBVs show in fact Of/WN transition characteristics and enhanced CNO abundances which fit well to the suggested scenario. For lower masses around 40 to 50 $M_\odot$ a sequence through the stage of OH/IR sources

$$O - Of - (BSG) - LBV - OH/IR - WR - SN$$

was alternatively suggested as possible scenario for objects like Var A in M33 and the OH/IR source IRC+104020 (see Humphreys 1987).

In the following sections I will briefly discuss some of the basic observed characteristics of LBVs. The designation LBV already gives a very short summary of what these stars are alike. They are intrinsically very luminous hot stars with $M_{bol}$ between -9 to -11.

Their most distinguishing characteristic is the occurance of irregular eruptions.

## 2.1 Light Variability

The photometric variations occur on different time scales and with correspondingly different amplitudes. Enormous outbursts or eruptions with $\Delta V > 3$ mag are observed rarely but make the most outstanding events. $\eta$ Car's outburst during the 1840's is the best example. Before this outburst the star varied between 2nd to 4th magnitude. In 1837 $\eta$ Car brightened to about 1st magnitude and for a short period in 1843 even to -1 mag. After 1860 the brightness decreased to 8th magnitude due to obscuration by a dust shell originating in the outburst. It slowly recovered until today to about 6th mag. $\eta$ Car's historical light curve and a discussion of the event can be found in Davidson (1989). Related events of this kind are probably the outbursts of P Cygni in the 17th century (see de Jager 1980, Lamers 1986) and of V12 in NGC2403 in the 1950s (Tammann and Sandage 1968). Though the time scale of these "plinian" outbursts (Davidson 1989) is not known, their low frequency suggests a value of typically 1000 years (Lamers 1989).

Outburst phenoma with amplitudes of about 1 to 2 magnitudes, the moderate or "normal" variations occur on time scales of 10 to 40 years. These normal outbursts are often referred to as "S Doradus type" outbursts. The variations of $\eta$ Car before the "Great Eruption" could also have been such a normal variability.

Even smaller variations of about 0.5 mag on top of these normal variations have time scales of months to years. All these variations are overlayed by a small scale microvariablity (van Genderen et al. 1985) of about 0.1 mag which is also found in normal hot supergiants.

One of the most important findings was that during the "normal" outbursts the luminosity remains essentially constant, i.e. $M_{bol} =$ const (see e.g. the reviews of Wolf 1986, and Lamers 1986). It thus appears that this type of light variablity is due to variations of $T_*$ and $R_*$. The star is brighter when $T_*$ is lower and $R_*$ larger and vice versa. This is in accordance with the observed redder color during maximum when the star appears to be cooler. The great outburst of $\eta$ Car is different in this respect as it required an additional brightening (Davidson 1989).

## 2.2 Spectroscopic Variations and Mass Loss

The normal brightness variations are due to a flux redistribution in an expanding circumstellar envelope as evidenced by the spectroscopic appearance. Spectroscopic variations occur in correspondence to the light variations. In general LBV spectra are characterized by emission lines of HI, HeI, FeII, and [FeII]. During minimum state the photospheric spectrum of a hot B supergiant or an even hotter Of/WN star (see below) is visible. [FeII] is usually strong during minimum. The spectrum during the maximum phase can be characterized as that of an extended low-gravity optically thick pseudo-photosphere resembling a late A or early F type star (cf. Leitherer et al. 1985).

Two well-studied examples are R71 and R127 in the LMC. In both cases observations during light minimum and maximum are available (cf. Wolf 1989a and reference therein). R71 was studied in detail by Wolf et al. (1981). During the minimum the star exhibited a typical B supergiant spectrum with additional [FeII] emission lines. The maximum spectrum showed a cooler A-type envelope spectrum with numerous P Cygni profiles of singly ionized ions. This is the typical spectrum during the normal S Doradus outbursts. A recent NLTE-analysis by Kudritzki and co-workers (see Kudritzki et al. 1989) of high quality spectra of R71 obtained during the present minimum phase yielded the following stellar parameters: $T_{eff}$ = 14000 K, log g = 1.5, helium abundance y = 0.30. The present mass is about 8 to 10 $M_{\odot}$, the ZAMS mass being around 25 $M_{\odot}$.

R127 was detected by Stahl et al.(1983a) to have undergone an outburst. Before this event R127 has been classified by Walborn (1977, 1982, 1989) as transition type Ofpe/WN9 ($T_{eff}$ = 33000K, $M_{bol}$ = -10.3). Other LBVs with Of/WN characteristics during minimum light are MWC112 and AG Car (Stahl 1987). The spectrum of R127 shows that during the present light maximum a cool, T = 8000 K, pseudo-photosphere has developed. It strikingly resembles the maximum spectrum of S Doradus (Wolf 1989a). The same behaviour during outburst was found for the Hubble-Sandage variable Var C in M33 by Humphreys et al. (1988).

It has been pointed out by Wolf (1989b) that in the H-R-diagram the LBVs during maximum populate a vertical strip at about log $T_{eff}$ = 3.9. This coincides with the opaque wind limit of Davidson (1987) (see also Maeder 1989). At minimum light according to Wolf the LBVs are lined up in an inclined "instability strip" between about (log T, $M_{bol}$) = (6.4,4.5) and (5.5,4.1). He found a brightness-amplitude-relation and suggested to use the LBVs as cosmic distance indicators.

The described variations are closely related to variability of the mass-loss rates. A summary of the mass-loss characteristics of the LBVs has been given by Lamers (1989). The essential points are: a) During maximum light the mass-loss rates are approximately the same for all LBVs: log $\dot{M}$ = -4.43±0.15. They are typically 10 to 100 times larger than expected from the observed $M_{bol}$-$\dot{M}$-relation for normal stars. b) The mass-loss rates during minimum light appear to agree with the mentioned $M_{bol}$-$\dot{M}$-relation. c) As pointed out by Lamers radiation pressure is sufficient to drive even these massive winds, i.e. log $\dot{M}v_{\infty}c/L$ < 0. d) The averaged mass-loss rate during the normal outbursts is found to be log $\dot{M}$ = -4.7±0.3. Additional mass ejection during eruptions of the $\eta$ Car type removes about $10^{-2}$ to 1 $M_{\odot}$ or even 2 to 3 $M_{\odot}$ in the special case of $\eta$ Car. Taking into account the estimated recurrence time for these eruptions of $10^3$ years and an estimated life time of the LBV phase of $10^4$ years (see below) the total averaged mass-loss rate adds up to about log $\dot{M}$ = -4 or more if the contribution of eruptions with $\Delta M$ = 1 $M_{\odot}$ is significant (Lamers, 1989).

2.3 Circumstellar Ejecta

In some cases circumstellar ejecta are directly visible. $\eta$ Car e.g. is embedded in the dusty homunculus (Gaviola 1950). Other examples are AG Car which is surrounded by a ring nebula (Thackeray 1950, Stahl 1987), and R127 (Stahl 1987). The nebula of $\eta$ Car and AG Car will be discussed in more detail in chapter 4.2. In these cases evidence for axial symmetry has been found. From direct imaging of some nebulae Stahl (1987) estimated a kinematical life time of $10^4$ years for the ejecta and a mass in the nebulae of about 1 $M_{\odot}$. This lifetime agrees well with other estimates of about $10^4$ to $10^5$ years (see the review of Lamers 1989, for details). Details about circumstellar ejecta of LBVs can be found in the review of Stahl (1989).

The ejecta at least in the few analysed cases show CNO processed matter (see e.g. Davidson et al. 1982, 1986, for $\eta$ Car and Dufour and Mitra 1987, for AG Car) giving evidence that the LBVs  in fact represent a post-main sequence evolutionary phase of massive stars. Also the high helium abundance of R71 argues for this conclusion.

Interestingly some LBVs are not only surrounded by gaseous envelopes but also by circumstellar dust shells. $\eta$ Car e.g. is obsured by dust produced during the 1840s outburst (see chapter 4). Dust has also been found in the ring nebula around AG Car (McGregor et al. 1988a). Waters and Wesselius (1986) found an excess in the IRAS fluxes of P Cygni at 60$\mu$m and 100$\mu$m.  They interpreted the excess as thermal radiation from a cool dust shell at about $10^4$ $R_*$ distance. The dust shell is probably related to the 17th century outburst of P Cygni.  A similar finding was made by Wolf and Zickgraf (1986) in the case of R71. This star shows an IR excess at 10$\mu$m and at 12, 25 and 60$\mu$m in the IRAS fluxes though it was not detected below 10$\mu$m from the ground. The IR excess was ascribed to a cool dust envelope (T = 120K, R = 8000 $R_*$) with an estimated kinematical age of 400 years (Wolf and Zickgraf 1986).


## 3. BASIC PROPERTIES OF B[e] SUPERGIANTS


The B[e] supergiants are a second class of very peculiar emission-line stars in the upper left part of the HRD. Though in some respects they are similar to the LBVs there are, however, also profound differences, namely the absence of LBV-type variability and the hybrid spectra of the B[e] supergiants.

Apart from being luminous hot stars we regard the following three characteristics as typical for B[e] supergiants:

a) The optical spectra are dominated by extremely strong Balmer emission lines (mostly with P Cygni profiles) showing equivalent widths of the order of 100 to up to more than 1000Å.

b) Hybrid spectra: narrow low-excitation emission lines of singly ionized metals like FeII, [FeII] and [OI]  are contrasted by high velocity absorption components of UV-resonance absorption lines and/or P Cygni absorption components of HI and HeI.

c) A strong infrared excess gives evidence for the presence of hot (T = 1000K) circumstellar dust.

The presence of the strong IR excess in some of the peculiar emission-line stars in the Magellanic Clouds (MCs) was discovered by Allen and Glass (1976) and Glass (1977).  Due to its wavelength dependence the excess was ascribed to hot circumstellar dust. More stars with circumstellar dust were found by Stahl et al. (1983, 1984) and McGregor et al. (1988c) (see also Stahl et al. 1989).  A recent IR survey of bright emission-line stars in the SMC lead to the detection of a further B[e] candidate (Zickgraf and Stahl, 1989, in preparation). Infrared excesses had been found earlier in galactic emission-line objects.  Many display a similar spectroscopic appearance as the MC objects (e.g. Geisel 1970, Allen 1973, Allen and Swings 1976). They were called "B[e] stars" due the presence of a B-type continuum and forbidden emission lines.  However, distances and luminosities are known only for a few of them. For the objects in the MCs, however, there is no doubt about their high luminosity.

Therefore the designation "B[e] supergiant" was chosen for these stars (Zickgraf et al. 1986) and their luminous galactic counterparts.

A list of B[e] supergiants is given in Table 1. It is of course certainly not complete in particular for our Milky Way. Here the difficulties in determining the distances to B[e] supergiant candidates and hence their luminosities are serious. Stellar parameters of the MC stars are given in Table 2. For parameters of galactic stars see e.g. McGregor et al. (1988b) and Winkler and Wolf (1989).

Table 1  List of B[e] supergiants; references for the MC stars are e.g. Henize (1956), Azzopardi and Vigneau (1982), Feast et al. (1960), Zickgraf et al. (1985, 1986, 1989), McGregor et al. (1988c). For the galactic stars cf. e.g. Allen and Swings (1976), Hartmann et al. (1980), Cohen et al. (1985), McGregor et al. (1988b), Winkler and Wolf (1989).

| | |
|---|---|
| LMC | HDE 268835 (R66), HDE 269217 (R82), HD 37974 (R126), Hen S12, HD 34464 (Hen S22), HDE 269599s (Hen S111), HD 38489 (Hen S134) |
| SMC | R4, R50, Hen S18, (AZV172) |
| Milky Way | MWC300, MWC349, CD-24°5721, CPD-57°2874, CPD-52°9243 |

Table 2  Stellar parameters of B[e] supergiants

| star | Sp. type | $T_{eff}[10^3K]$ | $M_{bol}$ | $R[R_\odot]$ | $M[M_\odot]$ | $\dot{M}_{tot}[M_\odot \ yr^{-1}]$ |
|---|---|---|---|---|---|---|
| R4 | B0-0.5 | 23-26 | -8.8...-9.3 | 32 | 30-40 | $3 \ 10^{-5}...7 \ 10^{-5}$ |
| R50 | B2-3 | 17 | -9.5 | 81 | 40-50 | $1 \ 10^{-4}...2 \ 10^{-4}$ |
| S18 | B0 | 25 | -9.3 | 35 | 35-40 | $4 \ 10^{-5}...1 \ 10^{-4}$ |
| R82 | B2-3 | 18.5 | -8.8 | 50 | 30 | $4 \ 10^{-5}...1 \ 10^{-4}$ |
| S12 | B0.5 | 23 | -8.6 | 30 | 25-30 | $3 \ 10^{-5}...7 \ 10^{-5}$ |
| S22 | B0-0.5 | 23-26 | -9.7...-10.2 | 49 | 50-55 | $1 \ 10^{-4}...3 \ 10^{-4}$ |
| S134 | B0 | 26 | -10 | 45 | 60 | $8 \ 10^{-5}...2 \ 10^{-4}$ |
| R126 | B0.5 | 22.5 | -10.5 | 72 | 70-80 | $9 \ 10^{-5}...2 \ 10^{-4}$ |
| R66 | B8 | 12 | -8.9 | 125 | 30 | $3 \ 10^{-5}...1 \ 10^{-4}$ |

The position of the B[e] supergiants in the HR diagram led Zickgraf et al. (1986) to the conclusion that like the LBVs they represent a post-main sequence stage of evolution. It is, however, not clear whether there is an evolutionary connection between LBVs and B[e] supergiants.

In the following I will discuss some of the properties of the B[e] supergiants in more detail and, in particular, compare them with the LBVs.

## 3.1. Variability

An important difference between the two groups of stars is the photometric behaviour of the B[e] supergiants at least during the limited time interval over which observations are available, i.e. typically 20 to 30 years. Neither photometric nor spectroscopic variations similar to the LBV outbursts have been found for the B[e] supergiants in the MCs. A summary of photometric observations of the MC objects can be found in Zickgraf et al. (1986). The amplitudes of variations if detected at all are usually less than 0.1 mag. This is of the same order of magnitude as found in normal OB supergiants. Possibly R4 showed some larger variations but also only of about 0.1 to 0.2 mag. CPD-52°9243, however, could be an example that on longer times scales variations with larger amplitudes might occur. This star apparently was about 2 mag brighter when the CPD-catalogue was compiled than today (cf. Swings, 1981). However, the reality and possible nature of the indicated variation is not yet clear. The situation for the galactic objects is particularly bad, because only very few photometric and spectroscopic observations are available.

The lack of variability exceeding the normal small scale variations of OB stars is not only found in the visual wavelength region but also in the IR. The constant infrared fluxes were interpreted by Zickgraf et al. (1986) as evidence for ongoing dust formation around the B[e] supergiants.

The line spectra also do not show strong variations over the years of observations. Some smaller variations, however, seem to have occured in a few cases as e.g. Hen S22 (cf. Muratio 1978, and Zickgraf et al. 1986). A special case is Hen S18 which shows a strongly variable HeII4686 emission line. Shore et al. (1987) and Zickgraf et al. (1989) concluded that a (possibly main-sequence) companion accreting mass from the stellar wind of the primary could be the reason for this exceptional variation.

## 3.2 Spectroscopic Appearance

Despite the different photometric behaviour spectroscopically the B[e] supergiants show a remarkable overall resemblance with LBVs during (normal) outbursts. Both types of stars show strong Balmer emission lines. The most obvious similarity is the low-excitation character of the spectra. They exhibit narrow emission lines of singly ionized metals (particularly of FeII) indicating low wind velocities, typically of the order of 50 to 100 km s$^{-1}$ for the B[e] supergiants and 100 to 200 km s$^{-1}$ for the LBVs. [FeII] is usually strong in B[e] spectra but normally found only in minimum spectra of LBVs. The similarity was e.g. shown by Stahl et al. (1983b) for R66 whose spectrum closely resembles that of S Doradus during maximum. They proposed that R66 which apparently is photometrically very stable might be in an "extended stand still" phase of an S Doradus outburst.

There are, however, also some interesting differences between the two types of spectra. [NII] e.g. usually is weak or absent in many B[e] spectra but rather pronounced in LBVs (see above). On the other hand [OI] is always strong in B[e] spectra whereas it is not found in

252

LBVs. Whether this reflects a difference in the chemical composition
is not clear. If it does it would mean that the B[e] supergiants are
possibly not as evolved as LBVs. It could, however, also be due to
different excitation in the circumstellar matter around B[e]
supergiants and LBVs.

The most important spectroscopic difference between both classes
of stars is the hybrid nature of the B[e] spectra which is not found
in LBVs. Hybrid means that the narrow low-excitation lines of the
singly ionized metals are contrasted by the simultaneous presence of
broad resonance absorption features of high ionization species like
SiIV and CIV in the UV and/or broad P Cygni absorption components of
hydrogen and neutral helium (cf. Zickgraf et al. 1985, 1986, 1989).
These absorptions indicate the presence of a hot fast expanding wind
in addition to the slowly expanding cool wind as evidenced by the
singly ionized metals. The hybrid character of the B[e] spectra will
be discussed further in section 4.1. because it led to the two-
component wind model for the B[e] supergiants.

### 3.3 Infrared radiation: mass loss and dust

The strong infrared excesses are certainly one of obvious classification criteria of the B[e] supergiants. The presence of the IR excess
is made clearer in Fig. 1 which shows a (J-H)-(H-K) diagram for the MC
objects. They form a clearly distinguishable group above H-K = 0.7 to
0.8. The excess in J is mainly due to (f-f)-(f-b) radiation whereas
longward of about 2μm thermal radiation of dust dominates. The
excesses in J have been used to estimate the mass-loss rates of the
B[e] supergiants (see Table 2). They were found to be of the order of
log Ṁ = -4.5 to -4 (cf. Zickgraf 1989, Zickgraf et al. 1989). Similar
rates have been found by Stahl et al. (1983b), Shore and Sanduleak
(1983) and Friedjung and Muratorio (1988).

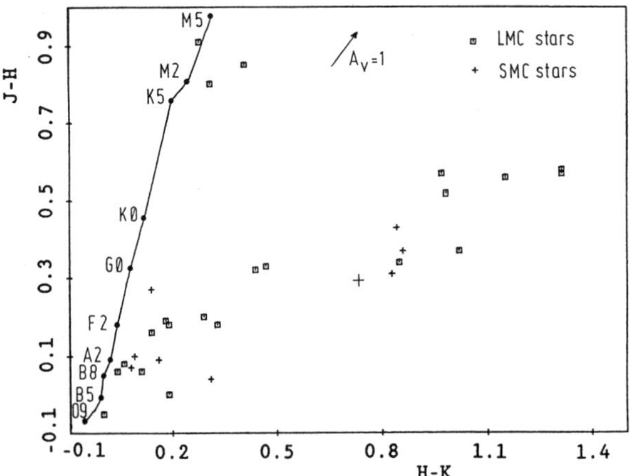


Fig. 1 (J-H)-(H-K)-diagram of MC stars. the B[e] supergiants form a
distinct group above H-K=0.7 to 0.8. The IR above about 2μm is due to
thermal radiation of circumstellar dust. The LBVs are among the stars
in the lower left corner.

The energy distributions were discussed by Zickgraf et al. (1986). They show the presence of a hot dust component with a typical temperature around T(dust) = 1000K. The black-body radius of the dust is typically R = 300 to 500 $R_*$. Note that the cool dust envelopes found around some LBVs are different.  They appear not to be related to the present mass loss but rather to shell ejection events some $10^2$ to $10^3$ years in the past. The dust around B[e] supergiants, however, is obviously originating in the present mass outflow which in certain regions must be dense and cool enough to enable ongoing grain condensation. For an assumed expansion velocity of 100 km s$^{-1}$ (see below) the flow time to reach the dust radius is of the order of 5 years.

## 4. OBSERVATIONAL EVIDENCE FOR AXIAL SYMMETRY IN B[e] SUPERGIANTS AND LBVs

In the previous sections the peculiar characteristics of the LBVs and B[e] supergiants were described. Are they related to axial symmetry or rotation? This question can probably be answered with yes in the case of the B[e] supergiants as will be shown in the following section. For the LBVs an answer cannot yet be given. However, as will be discussed below several observations give clear evidence for the presence of axial symmetry at least in two of these stars.

### 4.1. B[e] Supergiants

4.1.1 "Hybrid" spectra: The two-component wind model.  Strong evidence for the existence of a non-spherical distribution of matter around the B[e] supergiants comes from the observation of the hybrid character of most B[e] spectra. Zickgraf et al. (1985) studied in detail the case of R126 which became something like the proto-type of the B[e] supergiants. The continuous energy distribution and some presumably photospheric absorption lines in the satellite UV showed that the star is a B0.5Ia$^+$ hypergiant.  Its optical spectrum is characterized by extremely narrow metallic emission lines of FeII, [FeII] and [OI] with FWHM of about 10 km s$^{-1}$ (cf. Zickgraf, 1988) indicating a slowly expanding cool wind. In addition a split weak emission line of [NII] was found.  On the other hand, the IUE spectrum in the short wavelength region ($\lambda < 2000$Å) surprisingly was found to be dominated by broad absorption features of SiIV, CIV, and NV indicating the presence of a hot stellar wind with an expansion velocity of about 1800 km s$^{-1}$. This would be a velocity expected for an early-type B supergiant. Zickgraf et al. discussed several models to explain these observations, e.g. a deccelarated wind, generation of the forbidden and low-excitation lines in  a shock zone between the fast stellar wind and the surrounding interstellar material. However, the discussion showed that obviously two different environments of line formation are required to produce such spectra.  The "two-component wind model" suggested by Zickgraf et al. (1985) no longer assumes spherical symmetry but consists of a hot, fast line-driven wind in the polar region (CAK, Castor et al. 1975) as usually observed in OB stars and a dense, cool and slowly expanding wind in the equatorial zone forming a disk-like configuration. Zickgraf et al. (1989) suggested the term "excretion disk" for this structure in order to distinguish

it from the common accretion disks. Figure 2 shows a sketch of the
proposed geometry.

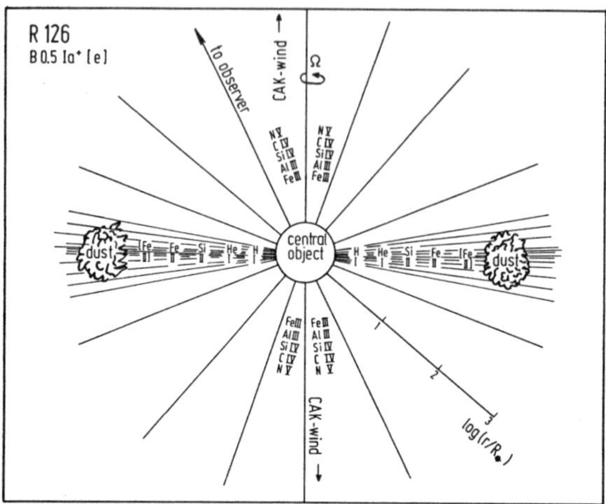


Fig. 2. Proposed geometry of the two-component wind model. The hot
high velocity wind emerges from the polar region, whereas in the
equatorial zone a dense, cool and slowly expanding wind originates,
which forms an excretion disk.  Here the formation of low-excitation
lines, molecular emission and dust condensation takes place.  (From
Zickgraf et al. 1985)

The discovery of molecules of CO and TiO around B[e] supergiants
by McGregor et al. (1988b,c) and Zickgraf et al. (1989) also proved
the presence of a cool and dense region around these stars.  McGregor
et al. estimated the densities in the CO emitting regions to be of the
order of $n_H = 10^{10}$ cm$^{-3}$ with a temperature of a few $10^3$ K.  They also
found band emission from $^{13}$CO in some objects with a strength which
(making some assumptions) would argue for the presence of CNO
processed material in the circumstellar matter and hence for an
evolved evolutionary stage of these stars.
The two-component model was suggested to be valid also for the
remaining B[e] supergiants in the Magellanic Clouds (Zickgraf et al.,
1986) and the related objects in our Milky Way (e.g. Winkler and Wolf,
1989).
The equatorial disk is supposed to be the site of dust
condensation and the formation of the low excitation emission lines.
It also could represent the region dense and cool enough to be
responsible for the observed molecular band emission of CO and TiO
(see above). An analysis of FeII, TiII and [FeII] lines of various
B[e] supergiants yielded excitation temperatures of 5000 to 10000 K
for the line emitting region with an indication of an outward drop of
$T_{ex}$; electron temperatures as low as 4200 K were derived from the broad
Thomson scattering wings of the Balmer lines (Friedjung and Muratorio
1980, Muratorio and Friedjung 1988, Zickgraf et al. 1986).
Individual characteristics of the stars are explained by taking
into account different angles of aspect. Whereas R126 is a example for
a pole-on seen object, R50 in the SMC represents the edge-on case

showing no evidence for a high velocity polar wind component. Rather its line spectrum resembles shell type stars (in the sense of classical Be stars) with central absorption features in the Balmer and FeII emission lines indicating very low expansion velocities of the order of 10 km s$^{-1}$. The maximum wind velocity from the blue edge of the Balmer lines is about 90 km s$^{-1}$ (Zickgraf et al. 1986).

The fast wind component is visible in most B[e] supergiants in the MCs. Only two stars show no evidence for the polar wind. With the assumption of the same overall geometry for all B[e] supergiants and random orientation of the inclination axis this small statistics allows to estimate the "opening" angle of the disk to about 20°. Zickgraf et al. (1989) used the emission measures from the IR excess in J and Hα to derive densities in the disks and to estimate the density and mass flux ratio between disk and polar wind. It was found that the disks are about 10$^2$ to 10$^3$ times denser than the polar wind regions. The mass flux ratio f(disk)/f(pole) was found to be of the order of 50. The total mass loss, typically of the order of 10$^{-4}$ M$_\circ$ yr$^{-1}$, is practically given by the dense and massive outflow in the excretion disk (see Table 2) (cf. also Zickgraf, 1989).

Zickgraf et al. (1985) assumed that rotation might play an important role in the formation of the two-component structure of R126 possibly supported by other forces yet to confirm. Unfortunately, photospheric absorption lines are weak or even absent in most B[e] supergiants. If they are visible they are usually contaminated by contributions from the stellar wind. This makes it practically impossible to determine the rotation velocities directly. There is, however, observational evidence e.g. from emission line profiles that rotation might in fact be present.

4.1.2. Emission line profiles. The low excitation emission lines of forbidden transitions are expected to originate at a distance from the star so large that the assumption of constant expansion velocity is reasonable. The line profiles should therefore be rectangular if the emitting region is a sphere expanding radially at constant velocity. High resolution spectroscopic observations, however, showed that in several stars split or asymmetric line profiles of [FeII] and [OI] are present which very likely originate in a disk-like structure (cf. Zickgraf, 1988). Winkler and Wolf (1989) observed double-peaked FeII lines in CPD-52°9243 which they attributed to a rotating disk. Fig. 3 shows forbidden line profiles of the galactic star MWC349. They clearly display the characteristics expected for a rotating disk. The line widths decrease with decreasing excitation of the lines, possibly reflecting an outwards decreasing rotational velocity. The observations particularly of the forbidden lines are therfore clearly in favor of a disk-like geometry (see also Hamann and Simon 1988). Also in R126 indication for rotation was found in the dependence of the emission line widths on the excitation potential and in the split [NII] line profile (Zickgraf et al. 1985).

4.1.3. Polarization. Polarization measurements provide a tool to investigate the spatial structure of otherwise unresolved light sources. Unfortunately no polarization data of the MC B[e] supergiants are available. However, some work has been done on galactic stars as e.g. CPD-52°9243 and MWC349 (Swings 1981, and Elvius 1974). Both stars possess an intrinsic polarization component showing that the scattering particles have a non-spherical distribution. Recently a

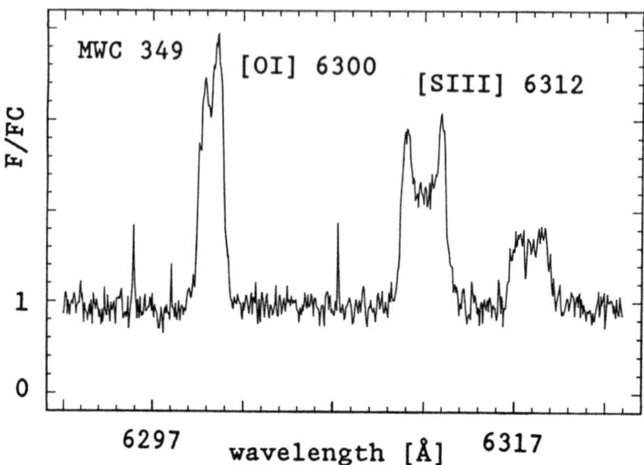


Fig. 3 Forbidden emission lines of MWC349. The line profiles have been observed by the author in 1987. The double peak profiles indicate that the lines originate in a rotating disk. The separation of the peaks is 32 and 95 km s$^{-1}$ indicating rotation velocities of 16 and 47 km s$^{-1}$ for [OI] and [SIII], respectively.

larger sample of galactic B[e] stars was studied by Zickgraf and Schulte-Ladbeck (1989). They found the polarization characteristics of the investigated stars in agreement with the two-component wind model.

4.1.4 MWC349: spatially resolved. The most obvious evidence for a two-component or bipolar structure of B[e] supergiants is found for the intrinsically polarized B[e] supergiant MWC349. This strong radio source has been spatially resolved with the VLA. The radio map of White and Becker (1985) shows a bipolar structure with polar lobes approximately in N-S direction and a perpendicularly E-W oriented equatorial dusty disk. White and Becker interprete the observations as being compatible with the presence of an equatorial disk with an opening angle of 30°. This is practically the same value as estimated above from a statistical argument.

The dusty disk is also confirmed by IR speckle interferometry obtained by Leinert (1986). He was able to resolve the source in E-W i.e. perpendicular to the radio lobes, but not in N-S direction. The IR source has a diameter of 85±19 marcs in the resolved E-W axis. Assuming the stellar parameters given by Hartmann et al. (1980) for MWC349 this yields a disk radius of about 500 $R_*$. Note that this is approximately the size of the dust component as estimated from the IR fluxes (see above).

The described observations are obviously all in favour of a non-spherical structure of the circumstellar environment of the B[e] supergiants, which presumably has a disk-like geometry as suggested by the two-component wind model.

## 4.2. Axial Symmetry in LBVs

Axial symmetric structures have been found in two cases, $\eta$ Car and AG Car.

4.2.1 Eta Car. The central source of $\eta$ Car which appears as a slightly extended (2") core is embedded in a dusty nebular structure called homunculus (Gaviola 1950). This nebula consists of matter ejected during the great eruption in the 1840s (Davidson, 1971, Walborn, 1976). The shape of the homunculus is elongated in NW-SE direction (position angle P.A. = 132°) and has an extension along the major axis of about 10 arcsec. The whole structure is surrounded by nebular condensations. From proper motion measurements of these features which show an outward directed and nearly radial motion with tangential velocities of up to 1000 km s$^{-1}$ Walborn et al. (1978) concluded that some of the condensations have been ejected in 1889, others during the eruption in the 1830s and 1840s and several condensations even earlier in the 15th century. A schematical sketch of the outer structure of $\eta$ Car can e.g. be found in Walborn et al. (1978).
 The central object was observed recently using speckle interferometric methods e.g. by Weigelt and Ebersberger (1986), Meaburn et al. (1983), Chelli et al. (1983) and Hofmann and Weigelt (1988). In particular the observations of Weigelt and Ebersberger and of Hofmann and Weigelt revealed the presence of three 12-times fainter companions separated by about 0.1" to 0.2" from the dominating primary object. None of the individual component is more extended than 0.03" (which was the diffraction limit of the observations). With the assumption that the companions are stars and not nebular knots of gas and dust Davidson (1989) concluded that these observations are consistent with the scenario of a very massive primary object with a bolometric magnitude of $M_{bol}$ = -11.2 and an initial mass of 120 $M_\odot$ (see also Davidson and Humphreys, 1986). The companions are probably O stars with masses around 30 to 60 $M_\odot$. The effective temperature of $\eta$ Car has been determined by Davidson et al. (1986) to be in the range between 24000 and 33000 K.
 The internal structure of the homunculus was studied during the past decade with various observational methods, in particular polarimetry, infrared mapping with high spatial resolution and long-slit spectroscopy. The main result of these observations was the detection of a bipolar structure with a circumstellar disk around $\eta$ Car.

4.2.1.1 Polarization. The polarisation map of Warren-Smith et al. (1979) showed that the edges of the homunculus are highly polarized with polarization degrees of up to 38%. The polarization drops towards the central object and is higher in the NW edge (i.e. in the "head" of the homunculus) than in the SE. In the high-polarization areas little light emission occurs (which would be unpolarized). The morphology of the polarization map lead Warren-Smith et al. to the conclusion that the illuminating source of the homunculus must be more complex than a single central source. The comparison of the polarisation patterns of $\eta$ Car and the bipolar nebula M1-92 (see e.g. Schmidt et al. 1978) showed a remarkable resemblance of both. This resulted in the model of a circumstellar disk in the inner 2" of the homunculus surrounding the central star. However, in contrast to M1-92 the disk of $\eta$ Car is bright rather than dark. It is illuminated by the central star. Both, disk and central star illuminate the matter in the homunculus which must contain small (ca. 0.1μm) dust grains in order to explain the polarization of the scattered light. The observed polarization pattern is in agreement with this model if the head of the homunculus is inclined away from the observer by about 20°.

4.2.1.2 Infrared mapping. Infrared mapping was carried out by numerous groups, e.g. Hyland et al. (1979), Mitchell et al. (1983), Hackwell et al. (1986) (more references of IR observations of η Car can be found in these papers). Hyland et al. and Mitchell et al. obtained maps at 3.6, 8.4, 10.2, 11.2μm and 2.2μm, respectively, with a spatial resolution of 1". These observations revealed the existence of two emission maxima in the inner 3" of the object. The line joining the two sources is perpendicular to the major axis of the homunculus. The separation of the sources is wavelength dependent and increases from 0.7" at 2.2μm to about 2.2" at 11.2μm. Furthermore, the intensity ratio of both peaks increases from 3.6 to 11.2μm. The FWHM of both sources also behave different. Whereas the width of the primary component increases with wavelength it is essentially constant for the secondary. According to Hyland et al. these observations which revealed such fine details of the structure in the inner 3 or 4" rule out a model with two distinct sources of luminosity illuminating a common envelope. They were, however, found to be compatible with a model involving a single luminosity source surrounded by a dust shell. The shell is enhanced around the equatorial plane by a dusty non-uniform disk at a position angle perdendicular to the major axis of the homunculus. It is supposed to be seen edge-on. This model agrees well with the results of the polarization measurements of Warren-Smith et al. (1979).

Further support for a bipolar structure of η Car came from IR mapping by Hackwell et al. (1986) in 6 wavelength bands of about 1μm width between 8.1 and 13.1μm, i.e. around the 10μm silicate feature. The spatial resolution was about 1.7", which could, however be enhanced by applying the maximum entropy statistical method (cf. Hackwell et al. 1986). They fitted a silicate emission model to the observations and found clear evidence for a bipolar ("horned") structure in the inner 5" of the homunculus. This is most obvious at longer wavelengths. The symmetry axis of the bipolar horns appears to be parallel to the major axis of the homunculus. The interpretation of the data further yielded some indication for a clumpy distribution of the dust the clumps being approximately 0.8" in size.

4.2.1.3 Longslit spectroscopy and spectropolarimetry. Meaburn et al. (1987) studied the structure of η Car with spatially resolved spectroscopy of Hα and [NII]. Additionally spectropolarimetric observations of Hα and of scattered continuum light were obtained. In the vincinity of η Car the spectra showed a broad Hα component centered on the systemic velocity with a narrow emission "spike". In the NW quadrant, i.e., in the region of the head of the homunculus this component extended out to $+1200$ km s$^{-1}$, and in the SE, i.e. on the side opposite of the head, the broad Hα-component is shift by $+200$ km s$^{-1}$. The broad components in the NW and SE were found to be highly polarized (25 to 40%). In the NW a narrow unpolarized spike in Hα at about systemic velocity was detected in addition to the broad polarized feature. In the model they suggest this spike is due to intrinsic emission from a radiatively ionized surface of a dusty disk surrounding $\eta$ Car. They conclude, that the broad unshifted H$\alpha$ feature must originate in the atmosphere of $\eta$ Car or in an outflow very close to $\eta$ Car. Dust particles flowing out in the direction of the homunculus scatter these photons thus creating the broad positively shifted and strongly polarized H$\alpha$ emission in the NW and SE corner of the homunculus. Evaluation of the involved velocities showed that at the edge of the homunculus the matter flows out with a velocity of about 800 km s$^{-1}$ in a cone tilted away from the observer in the NW by 33° and inclined towards the observer in the SE by 48°. Note that this is essentially consistent with the model of Warren-Smith et al. and with the results of the IR mapping particularly of Hackwell et al. which showed the horned bipolar structure. Meaburn et al. find even evidence for bipolarity outside the homunculus. This shows up as a jet-like feature and a knot of gas in the SE which could be part of an expanding lobe. A counter feature is also indicated by a filamentary arc in the NW. The outer bipolar structure appears to be consistent with the inner structure in the homunculus.

The described observations all show clear evidence for bipolarity or axial symmetry in this unique object.

4.2.2. AG Car.  The ring nebula around AG Car was first discovered by Thackeray (1950), who later also studied the velocity field in the nebula (Thackeray 1977).  A recent study of the morphology and mass-loss history of AG Car was published by Stahl (1987).  The shape of the nebula is roughly that of an elliptical ring. The size measured from the emission maxima of H$\alpha$ and [NII] was determined by Stahl (1987) to 20"x28". The total size is larger, namely 30"x39". The position angle of the major axis is P.A. = 150° (Thackeray 1950).  The narrow band [NII] image of Stahl (1987) showed clearly an anisotropic distribution of the emitting gas. The emission is strongest in the SW edge; a second weaker maximum is located in the NE edge of the nebula, i.e. the connecting line of the maxima is oriented perpendicular to the major axis.

A new approach to resolve the structure in the inner part of the nebula yielding a spectacular result was made by Nota and Paresce (1989).  They were able to overcome the problem of the high intensity ratio between central star and nebular emission by employing the technique of stellar coronography. A description of the equipment is given in their paper (Nota and Paresce 1989). The working principle is to occult the central star with a movable wedge allowing the adjustment of the occultation to the prevailing seeing conditions. They obtained narrow band images in emission lines and broad band images in the continuum.  The surprising result was the detection of a jet-like structure in the broad band continuum images extending from about 5 arcsec from the star in SW direction out to about 15 arcsec. The jet is not visible in the narrow band emission images.  Note that the nebular ring shows an emission maximum in the surrounding of the outer end of the jet. The jet itself has a helical structure. It consists of two outward spiralling twisted filaments. In the opposite direction an extended emission feature is visible at about 9.5 arcsec separated from the star. It could possibly be part of an otherwise invisible counter-jet.  The jet suggests that in the nebula a symmetry axis exists which is oriented perpendicularly to the major axis of the elongated nebula.  The jet is not visible in the narrow band emission-line images. The presence of dust in the nebula of AG Car has already been mentioned above. Nota and Paresce compared the optical with the infrared structure observed by McGregor et al. (1988a) and found a convincing coincidence of the brightest jet features with the brightest regions in the IR located in the SW and NE.  They furthermore found the broad-band (V-I) colors of the jet and the central star to be compatible with the assumption that the light seen from the jet is stellar light scattered by large dust particles not yet swept up by the stellar wind.


## 5. CONCLUSIONS


It was shown that various observations obviously favour a non-spherical model for the circumstellar environment of the B[e] supergiants. The geometry is very likely bipolar with a dusty equatorial disk as suggested by the two-component wind model.  The origin of the disk is not yet clear.  The apparent lack of photospheric absorption lines in most B[e] supergiant spectra makes it very difficult to determine v sin i. However, particularly the line profiles of optically thin emission lines give some indication that rotation in fact plays a role in the formation of the disks. Also magnetic fields could possibly be important (see e.g. the model calculations of Cassinelli et al. 1989).

Though there is not much evidence for binary systems among the B[e] supergiants this possibility can presently not be ruled out.  An alternative scenario for LBVs and possibly also for B[e] supergiants with close binary systems was discussed by Gallagher (1989).  However, it appears unlikely that accretion as suggested e.g. by Bensammar et al. (1983) for the star Hen S22 is the cause for the disk formation.

The spatially resolved observations of $\eta$ Car clearly show that this unique LBV can be described by a bipolar geometry with an equatorial disk. Also the surprising observation of the jet-like structure in the inner nebula of AG Car can certainly be regarded as strong evidence for dynamical processes involving angular momentum. Even though it is not yet clear from these observations whether rotation is an important parameter for the mass loss of AG Car and $\eta$ Car the structure of the nebulae themselves proves that one has to be very careful with the assumption of spherical symmetry not only in the case of the unique object $\eta$ Car but also in the case of "normal" LBVs.

It is not known whether a relation exists between the observed instabilities of LBVs and the bipolar structure of their nebulae. It is, however, clear that rotation would have strong effects on the internal structure.  The presence of e.g. differential rotation in these massive evolved stars cause shear forces. As discussed e.g. by Sreenivasan and Wilson (1989) they generate turbulence and in particular mixing. These effects should be described with hydrodynamical models which are not yet available.

ACKNOWLEDGEMENTS.  I wish to thank the Alexander-von-Humboldt Foundation, West Germany, for granting a Feodor-Lynen-Fellowship. Financial support by the NSF AST87-14579 is also gratefully acknowledged.

REFERENCES

Allen, D.A. (1973), Mon. Not. Roy. Astron. Soc. 161, 145
Allen, D.A., Glass, I.S. (1976), Astrophys. J. 210, 666
Allen, D.A., Swings, J.P. (1976), Astron. Astrophys. 47, 293
Appenzeller, I. (1986), IAU Symposium 116, "Luminous Stars and
  Associations in Galaxies," p. 139
Appenzeller, I. (1989), IAU Colloqium 113, "Physics of Luminous Blue
  Variables," p. 195
Azzopardi, M., Vigneau, J. (1982), Astron. Astrophys. Suppl. 50 291
Bensammar, S., Friedjung, M., Muratorio, G., Viotti, R. (1983),
  Astron. Astrophys. 126, 427
Cassinelli, J.P., Schulte-Ladbeck, R.E., Poe, C.H., Abbott, M. (1989),
  IAU Colloquium 113, "Physics of Luminous Blue Variables," p. 121
Castor, J.I., Abbott, D.C., Klein, R.I. (1975), Astrophys. J. 195,
  157
Chelli, A., Perrier, C., Biraud, Y.G. (1983), Astron. Astrophys. 117,
  199
Cohen, M., Bieging, J.H., Dreher, J.W., Welch, W.J. (1985), Astrophys.
  J. 292, 249
Conti, P.S. (1984), IAU Symposium 105, "Observational Tests of Stellar
  Evolution Theory," p. 233
Davidson, K. (1971), Mon. Not. Roy. Astron. Soc. 154, 415
Davidson, K. (1987), Astrophys. J. 317, 760
Davidson, K., Humphreys, R.M. (1986), Astron. Astrophys. 164, L7
Davidson, K. (1989), IAU Colloquium 113, "Physics of Luminous Blue
  Variables," p. 101
Davidson, K., Walborn, N.R., Gull, T.R. (1982), Astrophys. J. Lett.
  254, L47
Davidson, K., Dufour, R.J., Walborn, N.R., Gull, T.R. (1986),
  Astrophys. J. 305, 867
de Jager, C. (1980), The Brightest Stars, Reidel, Dordrecht
de Jager, C. (1984), Astron. Astrophys. 138, 246
Dufour, R.J., Mitra, P. (1987), Bull. A.A.S. 19, 1090
Elvius, A. (1974), Astron. Astrophys. 34, 371
Feast, M.W.M., Thackeray, A.D., Wesselink, A.J. (1960), Mon. Not. Roy.
  Astron. Soc. 121, 344
Friedjung, M., Muratorio, G. (1980), Astron. Astrophys. 85, 233
Gaviola, E. (1950), Astrophys. J. 111, 408
Geisel, S.L. (1970), Astrophys. J. Lett. 161, L105
Gallagher, J.S. (1989), IAU Colloquium 113, "Physics of Luminous Blue
  Variables," p. 185
Glass, I.S. (1977), Mon. Not. Roy. Astron. Soc. 178, 9P
Hackwell, J.A., Gehrz, R.D., Grasdalen, G.L. (1986), Astrophys. J.
  311, 380
Hamann, F., Simon, M. (1988), Astrophys. J. 327, 876
Hartmann, L., Jaffee, D., Huchra, J.P. (1980), Astrophys. J. 239, 905
Henize, K. (1956), Astrophys. J. Suppl. 2, 315
Hofmann, K.-H., Weigelt, G. (1988), Astron. Astrophys. 203, L21
Humphreys, R.M. (1987), in "Instabilities in Luminous Early Type
  Stars," H. Lamers and C. de Loore (eds.), Reidel Publishing, p. 1
Humphreys, R.M. (1989), IAU Colloquium 113, "Physics of Luminous Blue
  Variables," p. 3
Humphreys, R.M., Davidson, K. (1979), Astrophys. J. 232, 409
Humphreys, R.M., Davidson, K. (1984), Science 223, 243

Humphreys, R.M., Leitherer, C., Stahl, O., Wolf, B., Zickgraf, F.-J. (1988), Astron. Astrophys. 203, 306
Hyland, A.R., Robinson, G., Mitchell, R.M., Thomas, J.A., Becklin, E.E. (1979), Astrophys. J. 233, 145
Kudritzki, R.P., Gabler, A., Gabler, R., Groth, H.G., Pauldrach, A.W.A, Puls, J. (1989), IAU Colloquium 113, "Physics of Luminous Blue Variables," p. 67
Lamers, H.J.G.L.M. (1986), IAU Symposium 116, "Luminous Stars and Associations in Galaxies," p. 157
Lamers, H.J.G.L.M. (1986), in "Instablilities in Luminous Early Type Stars," H. Lamers and C. de Loore (eds.), Reidel Publishing, p. 99
Lamers, H.J.G.L.M. (1989), IAU Colloquium 113, "Physics of Luminous Blue Variables," p. 135
Lamers, H.J.G.L.M., Fitzpatrick, E. (1988), Astrophys. J. 324, 279
Leinert, C. (1986), Astron. Astrophys. 155, L6
Leitherer, C., Appenzeller, I., Klare, G., Lamers, H.J.G.L.M., Stahl, O., Waters, L.B.F.M., Wolf, B. (1985), Astron. Astrophys. 153, 168
Maeder, A. (1983), Astron. Astrophys. 120, 113
Maeder, A. (1989), IAU Colloquium 113, "Physics of Luminous Blue Variables," p. 15
Maeder, A., Meynet, G. (1987), Astron. Astrophys. 182, 243
McGregor, P.J., Finlayson,K., Hyland, A.R., Joy, M., Harvey, P.M., Lester, D.F. (1988a), Astrophys. J. 329, 874
McGregor, P.J., Hyland, A.R., Hillier, D.J. (1988b), Astrophys. J. 324, 1071
McGregor, P.J., Hillier, D.J., Hyland, A.R. (1988c), Astrophys. J. 334, 639
Meaburn, J., Walsh, J.R., Hebden, J.C., Morgan, B.L., Vine, H. (1983), Mon. Not. Roy. Astron. Soc. 204, 41P
Meaburn, J., Wolstencroft, R.D., Walsh, J.R. (1987), Astron. Astrophys. 181, 333
Mitchell, R.M., Robinson, G., Hyland, A.R., Jones, T.J. (1983), Astrophys. J. 271, 133
Muratorio, G. (1978), Astron. Astrophys. Suppl. 33, 125
Muratorio, G. (1981), Astron. Astrophys. Suppl. 43, 111
Muratorio, G., Friedjung, M. (1988), Astron. Astrophys. 190, 103
Nota, A., Paresce, F. (1989), IAU Colloquium 113, "Physics of Luminous Blue Variables", p. 159
Pauldrach, A.W.A., Puls, J., Kudritzki, R.P. (1989), IAU Colloquium 113, "Physics of Luminous Blue Variables", p. 261
Poe, C.H., Friend, D.B., Cassinelli, J.P. (1989), Astrophys. J. 337, 888
Schmidt, G.D., Angel, J.R.P., Beaver, E.A. (1978), Astrophys. J. 219, 477
Shore, S.N., Sanduleak, N. (1983), Astrophys. J. 273, 177
Shore, S.N., Sanduleak, N., Allen, D.A. (1987), Astron. Astrophys. 176, 59
Sreenivasan, S.R., Wilson, W.J.F. (1989), IAU Colloquium 113, "Physics of Luminous Blue Variables," p. 205
Stahl, O. (1987), Astron. Astrophys. 182, 229
Stahl, O. (1989), IAU Colloquium 113, "Physics of Luminous Blue Variables," p. 149
Stahl, O. Wolf, B. (1989), Astron. Astrophys. (submitted)
Stahl, O., Wolf, B., Klare, G., Cassatella, A., Krautter, J., Persi, P., Ferrari-Toniolo, M. (1983a), Astron. Astrophys. 127, 49

Stahl, O., Wolf, B., Zickgraf, F.-J., Bastian, U., de Groot, M.J.H.,
  Leitherer, C. (1983b), Astron. Astrophys. 120, 287
Stahl, O., Leitherer, C., Wolf, B., Zickgraf, F.-J. (1984), Astron.
  Astrophys. 131, L5
Stahl, O., Smolinski, J., Wolf, B., Zickgraf, F.-J. (1989), IAU
  Colloquium 113, "Physics of Luminous Blue Variables," p. 295
Swings, J.P. (1981), Astron. Astrophys. 98, 112
Tammann, G.A., Sandage, A. (1968), Astrophys. J. 151, 831
Thackeray, A.D. (1950), Mon. Not. Roy. Astron. Soc. 110, 524
Thackeray, A.D. (1977), Mon. Not. Roy. Astron. Soc. 180, 95
van Genderen, A.M., Steemers, W.J.G., Feldbrugge, P.T.M., de Groot,
  M., Damin, E., van den Boogaart, A.K. (1985), Astron. Astrophys.
  153, 163
Walborn, N.R. (1976), Astrophys. J. Lett. 204, L17
Walborn, N.R. (1977), Astrophys. J. 215, 53
Walborn, N.R. (1982), Astrophys. J. 256, 452
Walborn, N.R. (1989), IAU Colloquium 113, "Physics of Luminous Blue
  Variables," p. 27
Walborn, N.R., Blanco, B.M., Thackeray, A.D.: 1978, Astrophys. J. 219,
  498
Warren-Smith, R.F., Scarrott, S.M., Murdin, P., Bingham, R.G. (1979),
  Mon. Not. Roy. Astron. Soc. 187, 761
Waters, L.B.F.M., Wesselius, P.R. (1986), Astron. Astrophys. 155, 104
Weigelt, G., Ebersberger, J. (1986), Astron. Astrophys. 163, L5
White, R.L., Becker, R.H. (1985), Astrophys. J. 297, 677
Winkler, H. , Wolf, B. (1989), Astron. Astrophys. 219, 151
Wolf, B. (1986), IAU Symposium 122, p. 409
Wolf, B. (1989a), IAU Colloquium 113, "Physics of Luminous Blue
  Variables," p. 91
Wolf, B. (1989b), Astron. Astrophys. 217, 87
Wolf, B., Zickgraf, F.-J. (1986), Astron. Astrophys. 164, 435
Wolf, B., Stahl, O., de Groot, M., Sterken, C. (1981), Astron.
  Astrophys. 99, 351
Zickgraf, F.-J. (1988), in "Mass Outflow From Stars and Galactic
  Nuclei," L. Bianchi and R. Gilmozzi (eds.), Kluwer Academic
  Publishing, p. 211
Zickgraf, F.-J. (1989), IAU Colloquium 113, "Physics of Luminous Blue
  Variables," p. 117
Zickgraf, F.-J., Schulte-Ladbeck, R.E. (1989), Astron. Astrophys. 214,
  274
Zickgraf, F.-J., Wolf, B., Stahl, O., Leitherer, C., Klare, G. (1985),
  Astron. Astrophys. 143, 421
Zickgraf, F.-J., Wolf, B., Stahl, O., Leitherer, C., Appenzeller, I.
  (1986), Astron. Astrophys. 163, 119
Zickgraf, F.-J., Wolf, B., Stahl., O., Humphreys, R.M. (1989), Astron.
  Astrophys. 220, 206

# EFFECTS OF MASS LOSS ON LATE STAGES OF MASSIVE STAR EVOLUTION


N. LANGER[*]
*Lick Observatory*
*University of California at Santa Cruz*
*Santa Cruz, CA 95064, U.S.A.*



ABSTRACT. The current status of mass loss rates of massive stars ($M_{ZAMS} \gtrsim 15\,M_\odot$) and their effects on stellar evolution is reviewed with emphasis on observable properties of post main sequence stars. The different stages of massive star evolution are discussed, i.e. main sequence stars, supergiants, Luminous Blue Variables, and Wolf-Rayet stars, as well as the supernova explosion. A summary of the conclusions is given in the last section.


## 1. Introduction

Stellar mass loss is a very common phenomenon, being known as to affect the evolution of stars of nearly any mass. The present paper deals with stars more massive than $\sim 15\,M_\odot$. Since their mass may be significantly reduced during the main sequence phase, we refer to the initial mass as the zero age main sequence (ZAMS) mass. Note that the actual mass of 'massive stars' may decrease much below $15\,M_\odot$ during their evolution (see below).

We focus on the effects of mass loss on massive stars from the point of view of theoretical stellar models, but we restrict our discussion to observable evolutionary phases. Due to the enhanced speed of the late burning stages in the stellar interior, this means we concentrate mainly on hydrogen and helium burning and — a certainly observable phase in spite of its short duration — the supernova event. Though there are only two burning phases involved, a comparison with observations of massive (i.e. luminous) stars has to deal with a large number of different classes of spectral types. To attach a certain spectral type to a given stellar model (or vice versa) is a general problem. In principle it requires the construction of a model atmosphere including the spectral line information necessary for a spectral classification. This would require much computer time, and furthermore theoretical concepts for stars with the highest mass loss rates still have to be developed in this respect. For this reason only a rough correspondence between spectral types and stellar models can be established in most cases, mainly on the basis of the location of the model in the HR diagram. To attach an interior model to a certain observed star is additionally complicated by the fact that one stellar track may cross the same point in the HR diagram twice, or that one point may be crossed by two different tracks belonging to

---


*) On leave of absence from Universitäts-Sternwarte Göttingen, F.R.G.





*L. A. Willson and R. Stalio (eds.), Angular Momentum and Mass Loss for Hot Stars, 265–278.*
© 1990 *Kluwer Academic Publishers. Printed in the Netherlands.*

266

different ZAMS-masses. For example, the effective temperatures of late WN stars overlap considerably with the main sequence band (Smith and Willis, 1983; Schmutz et al., 1989), or blue supergiants (BSGs) on their way to the Hayashi line have the same HRD position as such BSGs, which perform a loop from the red supergiant (RSG) stage (cf. e.g. Langer et al., 1989). Often, the additional information of surface abundances can lead to an unambiguous correspondence of observed star and stellar model. In the following we shall discuss in different chapters the main evolutionary stages of massive stars: the main sequence (MS) phase, supergiants, the brief but important phase as Luminous Blue Variable (LBV), the Wolf-Rayet (WR) phase, and finally the supernova (SN) event. At the beginning of each chapter we will briefly define which sort of stellar model we mean when we use the above expressions for the different phases.

Mass loss is one of the largest problems for the theory of massive star evolution, because — with the exception of hot hydrogenrich stars — practically no quantitative mass loss theories exist, and empirical mass loss rates often are only order of magnitude estimates, e.g. for RSGs, LBVs, and WRs. This is a particular difficult situation, since a difference in mass loss rates of a factor of 2 often makes a huge difference for the evolution of a massive star, which is evident considering the fact that many massive stars are presumably losing half of their initial mass or more during evolution. Therefore it is often the only way in stellar evolution, to treat the mass loss rate of stars of a certain class as a free parameter to be varied and check for observable consequences, thereby restricting the mass loss rate semiempirically. However, this task would be much more easy if there were not other major uncertainties involved in massive star models, which we will therefore briefly mention here.

A problem for the theory of massive stars of at least equal importance as mass loss is convection theory. One item is the question whether convective regions are restricted to the part of the star where the local criterion for convective stability is violated, or if the inertia of convective blobs leads to an extension of the so defined convective regions ("overshooting"). Current convection theories cannot answer this general question (cf. Langer, 1986; Renzini, 1987), and observational evidence is also poor, partly because the large number of qualitatively different convection zones which do occur in massive stars complicate the situation (cf. Langer et al., 1985). A second and perhaps even more profound issue is the question for the correct local criterion for convection. It has been widely discussed in the last decades, whether the Schwarzschild criterion or the stricter Ledoux criterion should be adopted in stellar evolution calculations (cf. Stothers and Chin, 1976; 1977), and the Schwarzschild criterion has been preferred in most recent calculations (cf. references in Chiosi and Maeder, 1986). However, there is new strong evidence from observations of SN 1987A that to apply the Ledoux criterion may be more correct, while stellar layers which are unstable according to the Schwarzschild criterion but stable according to the Ledoux criterion should be mixed on a thermal time scale rather than on a dynamical time scale ("semiconvection"; cf. Langer et al., 1983).

More problems are involved in the modeling of massive stars, but most of them are not of the same global importance as mass loss and convection. We should mention the large uncertainty in the $^{12}C(\alpha,\gamma)^{16}O$ nuclear reaction rate (cf. Caughlan and Fowler, 1988), which determines whether the main fraction of helium is transformed into carbon or rather into oxygen during central helium burning, and thereby has large influence on the helium burning time scale and the post-helium burning evolution. Furthermore, stellar rotation (see Tassoul, this volume) may have some impact on the evolution, which is, however, largely unknown, and in some cases also magnetic fields or binarity may be of relevance.

However, in the present discussion we will basically discuss the consequences of mass loss on non-rotating, non-magnetic massive single stars, which turns out to be already rather complex. Furthermore, we will mainly consider models obtained with a convection theory which leads to a selfconsistent simulation of the SN 1987A progenitor (cf. Sect. 6.1). However, we will mention at which places this assumption is important for our conclusions and sometimes also which conclusions might emerge in case the Schwarzschild criterion for convection would have been assumed.

## 2. The Main Sequence Phase

We will refer to stars which have not yet exhausted their hydrogen in the center as of being in the main sequence phase. Due to this definition of the main sequence phase it is well possible, that not only luminosity class V stars ("dwarfs") belong to the main sequence, but possibly even stars of luminosity class I. E.g. a $100\,M_\odot$ star may increase its radius from $13\,R_\odot$ to $53\,R_\odot$ during main sequence evolution, indicating a decrease of surface gravity by almost two orders of magnitude. Note therefore, that we will designate only models beyond core hydrogen exhaustion as supergiants in this paper.

The MS phase is covered here since MS mass loss has a large influence on the post main sequence evolution of massive stars, as has been found in numerous studies of this topic (see Chiosi and Maeder, 1986). Besides the effect of a widening of the main sequence band, a decrease of the mass-luminosity ratio M/L, and a prolongation of the hydrogen burning lifetime $\tau_H$, MS mass loss affects the post MS evolution by an increase of the tendency of a massive star to evolve toward the Hayashi line right after core hydrogen exhaustion (cf. Stothers and Chin, 1979; Maeder, 1981); (as long as the surface hydrogen mass fraction is not greatly reduced by the mass loss).

In contrast to later evolutionary phases, we have a relatively good idea about the dependence of the mass loss rate on the stellar parameters as well as of the quantitative amount of mass loss on the main sequence. A couple of mass loss formulae for main sequence O and B stars have been derived from observations, which yield rather similar results for the well populated part of the main sequence, though a factor of $2-3$ uncertainty may still be involved (cf. Chiosi and Maeder, 1986). However, empirical mass loss formulae exist only for solar metallicity, though there is observational and theoretical evidence for a strong metallicity dependence of the MS mass loss rate. Furthermore, the observational data for the most massive stars is very rare, which means that empirical mass loss rates are especially uncertain in this regime.

However, also quantitative theoretical mass loss rates, based on the radiation driven wind theory of Castor et al. (1975), are meanwhile available (e.g. Kudritzki et al., 1989), and Langer and El Eid (in prep.) have utilized them for a grid of main sequence evolutionary calculations for stars in the range $20\,M_\odot \lesssim M_{ZAMS} \lesssim 200\,M_\odot$ and different metallicities. An interesting result of these calculations is the fact that for solar metallicity the total amount of mass lost during the MS evolution turned out to be a factor of $2-3$ smaller compared to the case when empirical rates (e.g. that of Lamers, 1981; or de Jager et al., 1986) have been adopted. For metallicities below solar, where scaling relations of the form $\dot{M} \sim (Z/Z_\odot)^\alpha$ with $\alpha = 0.5$ (Kudritzki et al., 1987) or $\alpha = 0.75$ (Pauldrach, priv. comm.) have been used, the MS mass loss is reduced by additional factors.

It is beyond the scope of the present paper to discuss whether the theoretical mass loss rates are better than the empirical mass loss formulae or not. However, if they turn out to be correct, it would imply that MS mass loss is much less efficient than previously assumed. For low metallicity stars (e.g. in the SMC, where $Z \simeq Z_\odot/10$), MS mass loss could be almost completely neglected even for the most massive stars (we found $\Delta M/M \lesssim 3\%$), and even for galactic stars it would lead to evolutionary consequences only for $M_{ZAMS} \gtrsim 40\, M_\odot$: for $Z = Z_\odot$ we found total main sequence mass losses of 0.8, 2.5, 4.5, 8.2, and 35.0 $M_\odot$ for $M_{ZAMS} =$25, 40, 60, 85, and 200 $M_\odot$, respectively. For comparison, Maeder and Meynet (1987), utilizing the mass loss formula of de Jager et al. (1986), found values of 2.2, 7.7, 17.0, and 27.4 $M_\odot$ for $M_{ZAMS} =$25, 40, 60, and 85 $M_\odot$ (note, however, that Maeder and Meynet had somewhat increased main sequence lifetimes due to their assumption of overshooting).

The utilization of theoretical mass loss rates has therefore two main implications for the post MS evolution. One is, that evidently the mass of the remaining hydrogenrich envelope at core hydrogen exhaustion is largely increased. E.g. the so called Conti-scenario for WR star formation, i.e. forming a WR star on the main sequence due to strong single star mass loss (cf. Maeder, 1982), would not work, even for the most massive stars, and even for optimistic assumptions on convective overshooting. Consequently, the amount of hydrogenrich matter left to be lost in other evolutionary phases in order to reach the WR stage is correspondingly increased (see below). Secondly, a smaller MS mass loss reduces the trend towards the RSG stage at core H-exhaustion and leads to evolutionary tracks which stay in the BSG region during early core He-burning for models which imply the Schwarzschild criterion for convection (cf. e.g. Lamb et al., 1977; Stothers and Chin, 1979; Maeder, 1981). However, both, semiconvection and convective overshooting during MS evolution, also lead to RSGs at core He-ignition, even for non-mass loosing stars (Stothers and Chin, 1985; Langer et al., 1985), which means that this second effect is only important when semiconvection and overshooting are not adopted. Since the occurrence of semiconvection in massive stars is strongly indicated by the fact that the SN 1987A progenitor was a BSG (see Sect. 6), the second effect may be unimportant in nature. Note, however, that most evolutionary calculations in the recent decade have been performed with the Schwarzschild criterion for convection.

## 3. Supergiants

We refer to supergiants as post main sequence stars (i.e. $X_c = 0$) which are neither Wolf-Rayet stars nor Luminous Blue Variables (see below). Since stars with initial masses above a certain limit ($M_{ZAMS} \gtrsim 45\, M_\odot$, cf. Sect. 4) are supposed to evolve into WR stars after a short LBV phase at the beginning of core helium burning, our discussion of the supergiant phase is restricted to the mass range $15\, M_\odot \lesssim M_{ZAMS} \lesssim 45\, M_\odot$. Note that the evolution of stars with initial masses close to the lower limit of this mass range are hardly affected by mass loss at all, since not only on the main sequence but also in the supergiant stage mass loss rates are not large enough to remove a substantial part of the total stellar mass. In fact, the mass range $10\, M_\odot \lesssim M_{ZAMS} \lesssim 15\, M_\odot$ may be the only one in which mass loss can be neglected completely.

Mass loss rates of blue supergiants are relatively well known. The radiation driven wind theory is still applicable for the main part of the BSGs and yields mass loss rates well in agreement with observatioanlly derived values (see Kudritzki, 1988), which can often be achieved by radio measurements in this case, which is the most reliable observational technique for mass loss determinations. RSG mass loss rates, an the other side, are almost completely unknown, since no theory is available to quantitatively predict RSG mass loss rates, and observatioanlly derived values are rare and often very uncertain (cf. Reimers, 1975; de Jager et al., 1986; 1988).

The post main sequence evolution of stars in the upper part of the considered mass range in the HR diagram is extremely uncertain due to the sensitivity of the stellar tracks to MS and post-MS mass loss rates, convection theory, and the $^{12}C(\alpha,\gamma)^{16}O$ cross section, i.e. the parameter space to be explored by theoretical models is simply too large. Basic evolutionary connections between supergiants and other types of stars are therefore mainly unknown, like the question whether the most luminous RSGs evolve into WR stars (Chiosi and Maeder, 1986), or whether some massive stars encounter an LBV phase prior to or after the supergiant phase (Langer, 1989).

Though some stars within the considered mass range spend some time of their post-MS evolution in different stages, the supergiant stage is certainly the one they spend the largest part of it. From the theoretical point of view, the supergiant stage can be divided into a red and a blue phase, corresponding to surface temperatures of the Hayashi-line (i.e. $T_{eff} \simeq 3000 - 4000\,K$, depending on metallicity and mass) and values of $T_{eff} \gtrsim 10\,000\,K$. The gap in between is crossed only on thermal time scales (some $10^4\,yr$; cf. e.g. Langer et al., 1989) and is observatioanlly confirmed (Humphreys and McElroy, 1984; Fitzpatrick and Garmany, 1989). Therefore, the number ratio of blue to red supergiants is a very important observational constraint to theoretical post-MS evolution of stars of the considered mass range. HR diagram positions of supergiants in the Milky Way and also in the LMC and SMC (Humphreys and McElroy, 1984) suggest, that part of their post hydrogen burning evolution should be spent in either of the supergiant stages. However, due to the possibility of a non-monotonous surface temperature evolution, this constraint does not rule out models which turn to the Hayashi-line at core helium ignition; blue loops during core helium burning may lead to a significant duration of the BSG stage in this case (cf. Langer et al., 1989), a scenario which seems to be favored by observations at least in the LMC (Fitzpatrick and Garmany, 1989). However, models which allow only for one of the two supergiant stages in this mass range have to be considered less likely, as e.g. the case of the Schwarzschild criterion for convection and no mass loss at all, where no RSG phase occurs (cf. Lamb et al., 1977; Langer et al., 1985), or models with strong mass loss and convective core overshooting, which basically have no BSG phase (e.g. Maeder and Meynet, 1987).

## 4. Luminous Blue Variables

For stars above a certain mass limit ($M_{ZAMS} \gtrsim 45\,M_\odot$) the evolution toward the RSG stage is inhibited, simply since no RSGs above that mass (i.e. more luminous than $\sim 10^{5.7}\,L_\odot$) are found in nature (Humphreys and McElroy, 1984). On the other side, it is known from WR stars in clusters and associations, that stars initially more massive than $\sim 45\,M_\odot$ eventually evolve into WR stars (Schild and Maeder, 1984; Humphreys et al., 1985), i.e. objects which are supposed to have none or almost no hydrogen envelope left. The simplest explanation of both observations is to assume that those massive stars loose their H-rich envelope very early in their post-MS evolution, which prevents them from evolving to cool surface temperatures and at the same time creates WR stars (cf. Maeder, 1983). The evolutionary phase in which the main part of the H-rich envelope, which remained after central H-burning, is lost, is required to be smaller than some $10^4\,yr$, which is the time in which massive star models would cross the HR diagram toward the RSG branch if no mass loss would prevent that. The amount of mass, which has to be lost in order to make the stars evolve blueward rather than redward is of the order of $10\,M_\odot$, depending on the initial mass, on the MS mass loss, and on assumptions about overshooting. The required mass loss rate for this phase is therefore $\sim 10^{-3} \, M_\odot \, yr^{-1}$.

It turned out that the so called Luminous Blue Variables may be the observational counterpart which corresponds to this brief heavy mass loss stage. Though this class is very heterogeneous (see Humphreys, 1989, for a summary of LBV-defining properties) their number is of the right order (i.e. very small), their highly variable mass loss rates have been estimated to be — time-averaged — of the order of $10^{-4} - 10^{-2} \, M_\odot \, yr^{-1}$ (Lamers, 1989), and also their location in the HR diagram fits into this general picture (Humphreys, 1984).

However, up to now no general understanding of the physical cause of the very high mass loss rates could be achieved. Several mass loss mechanisms have been proposed (cf. Davidson and Moffat, 1989, for details), but none of them can cope with the overwhelming variety of observed features.

The simplest approach to simulate the LBV phase in stellar evolution calculations is to apply a very high rate of mass loss at the time when the stellar track turns toward the Hayashi line (e.g. Maeder and Meynet, 1987; Langer, 1990). Though in reality there may be several mass loss episodes of largely different characteristics, the simplification of assuming a smooth time-averaged mass loss rate has no consequences for the post-LBV evolution, since once a WR star is formed, its structure is basically determined by its mass and is therefore independent of the formation history (Langer, 1989a). Furthermore, the rate of mass loss, which has to be applied to a given stellar model in order to stop its redward evolution, does not depend on the specific functional form of the mass loss rate but is rather determined by the condition $\frac{d}{dt}T_{eff} = 0$ (see below). The internal trend to cool temperatures is stopped only when the surface hydrogen mass fraction has been reduced to values of the order of $X_s \simeq 0.25$ (cf. Langer and El Eid, 1986; Maeder and Meynet, 1987; Langer, 1990), due to the related reduction of the opacity and increase of the mean molecular weight in the remaining envelope. At this point the star may certainly be considered a WR star (see next section).

As mentioned above, the LBV mass loss rate in evolutionary calculations is determined by the condition $\dot{T}_{eff} = 0$, or equivalently, $\dot{R} = 0$, which can be understood as follows: in absence of mass loss, the star would evolve to cool temperatures, thereby increasing its radius at a certain rate $\dot{R}_1 > 0$. The mass loss rate, then, has to be adjusted in order to yield a compensating radius decrease per unit time $\dot{R}_2(\dot{M}) = -\dot{R}_1$, i.e. resulting to $\dot{R} = 0$. In numerical calculations, this can be achieved by adopting a mass loss formula as e.g. $\dot{M} = \dot{M}_0(T_0/T_{eff})^\alpha$ for $T_{eff} > T_0$, $\dot{M}_0$ being a standard mass loss rate (e.g. that of de Jager et al., 1986), $T_0$ some limiting effective temperature, and $\alpha$ a large power (e.g. $\alpha = 5$).

The problem is that the rate of radius increase due to the internal structure $\dot{R}_1$ depends greatly on the previous evolution, i.e. on assumptions on MS mass loss, semiconvection, and overshooting. For example, as mentioned in Sect. 2, small MS mass loss rates in connection with the Schwarzschild criterion for convection lead to a post-MS HRD crossing time of the order of the nuclear time scale, i.e. $R/\dot{R}_1 \simeq \tau_{He}$. In a recent calculation of that kind for $M_{ZAMS} = 100 \, M_\odot$ and $Z = Z_\odot$ (Langer and El Eid, in prep.), using the small theoretical MS mass loss rates (cf. Sect. 2) and the Schwarzschild criterion, the LBV mass loss rate was actually found to be so much decreased that the star was still in the LBV phase at the end of its evolution. However, when semiconvection is taken into account, the duration of the LBV phase is found to be only $3 \cdot 10^4 \, yr$, which agrees much better with the observed small number of LBVs.

Finally, we want to point out that the duration of the LBV phase $\tau_{LBV}$ depends greatly on metallicity due to two consequences of the metallicity dependence of the MS mass loss (cf. Sect. 2). The LBV time scale is simply $\tau_{LBV} = \Delta M_{LBV}/\dot{M}_{LBV}$, $\Delta M_{LBV}$ being the amount of mass to be lost in the LBV phase to sufficiently reduce the surface hydrogen mass fraction in order to stop the redward trend, and $\dot{M}_{LBV}$ is the time-averaged LBV mass loss rate, established by the condition $\dot{R} = 0$. First, of course, a lower MS mass loss rate increases $\Delta M_{LBV}$ and therefore $\tau_{LBV}$, and second, it decreases $\dot{R}_1$, thereby also decreasing $-\dot{R}_2(\dot{M})$ which leads to smaller LBV mass loss rates. Therefore, a relatively larger number of LBVs should be expected e.g. in the LMC as compared to the Milky Way.

## 5. Wolf-Rayet stars

Already in Sect. 1 we mentioned, that WR stars cannot be discriminated from other stars on the basis of their HRD position alone, since e.g. WNL stars (see below) partly have the same position as main sequence stars. Moreover, neither observed WR stars nor theoretical WR models can yet be accurately placed into the HR diagram (cf. Langer, 1989a). However, in recent years it became evident, that the surface abundances of WR stars are so extreme, that they can be almost unambiguously defined through them. Here, we will designate such objects as WR stars, which are hot (i.e. $T_{eff} \gtrsim 20\,000\,K$) and have substantially hydrogen depleted envelopes ($X_s \lesssim 0.4$). Note that the exact value of the limiting hydrogen abundance is of minor importance in the context of stellar modeling, since when a stellar model evolves blueward toward the WR stage, its surface H-mass fraction will be smaller than 0.4 anyway (see above). However, the transition from e.g. the LBV phase to the WNL stage is continuous, and the limiting value of the effective temperature is somewhat arbitrary.

Spectroscopically, WR stars are subclassified into several groups, e.g. late WN (WNL), early WN (WNE) and WC stars (cf. van der Hucht et al., 1981), and obviously this distinction corresponds to differences in the surface composition of stars of each group: envelopes of WNL stars are composed of products of partial hydrogen burning (e.g. H, He, and N), WNE stars show products of complete H-burning (He and N), and WC stars show partial He-burning products (He, C, and O); cf. Willis (1982); Smith and Hummer (1988); Torres (1988). Since it is very important for the stellar structure whether hydrogen is present or absent in a WR envelope, we split this section into two parts, first discussing the WNL stars and then the WNE and WC stars.

### 5.1 WOLF-RAYET STARS WHICH CONTAIN HYDROGEN

Though the discrimination between hydrogen containing and hydrogenless WN stars as WNL and WNE stars is not in general agreement with observations (Hamann, 1990) it seems to cover most cases and is therefore maintained here for simplicity.

Hydrogen containing WR models are very different from those of hydrogenless WRs in several respects. The presence of hydrogen implies the presence of a hydrogen burning shell, i.e. a second independent region of nuclear energy generation besides the helium burning central energy source. Due to this, and also due to the high opacity and the low mean molecular weight of ionized hydrogen compared to ionized helium, WNL models have much larger radii and correspondingly smaller surface temperatures compared to hydrogenless models (note that the effect of the WR wind on apparent radius and surface temperature is still neglected here). One important consequence is that WNL models are found to be vibrationally stable even if they are extremely luminous, which is not the case for WNE and WC stars (Maeder, 1985).

Also the observed properties of WNL stars are quite different from those of WNE and WC stars: WNL stars are by far more luminous and more massive (Smith and Willis, 1983; Niemela, 1983; Schmutz et al., 1989). Therefore, Langer (1989b) argues that it may be purely accidental that the observed mass loss rates are of the same order of magnitude for both groups, i.e. $\dot{M} \simeq 3 \cdot 10^{-5}\, M_\odot\, yr^{-1}$ (Barlow et al., 1981; Conti, 1986). E.g. the so called momentum problem for WR winds in the framework of the radiation driven wind theory, i.e. the fact that the wind momentum is much larger than the momentum of the emerging radiation for many WR stars, is one order of magnitude smaller for WNLs compared to WNE and WC stars, since the WNL luminosities are on average almost one order of magnitude larger (Lundström and Stenholm, 1984; Schmutz et al., 1989; Smith and Maeder, 1989) and their final wind velocities $v_\infty$ are smaller by a factor of $2 - 3$ (e.g. Schmutz et al., 1989). For example, the well studied WN6 star WR47, component of a WNL+O binary, is found to have a mass of $48\, M_\odot$ and a mass loss rate of $3 \cdot 10^{-5}\, M_\odot\, yr^{-1}$ (Moffat et al., 1989). From its mass, a lower limit on its luminosity of $L \gtrsim 10^6\, L_\odot$ can be derived, and the spectral type implies a final wind velocity of $v_\infty \simeq 1000\, km\, s^{-1}$. This leads to a momentum ratio of $\dot{M} v_\infty/(L/c) \lesssim 2$, i.e. of order unity, which therefore allows for the possibility that the mass loss is driven by radiation alone (cf. also Schmutz et al., 1989).

Looking qualitatively at the predictions of the radiation driven wind theory, a comparison of O-stars and WNL stars indicates that radiation pressure as cause for the WNL wind may be a promising possibility: consider a WNL star and an O-star at the same position in the HR diagram. They have the same luminosity and effective temperature and hence — in a first approximation — the same radiation field. They also have the same radius, but since the mass of the WNL stars must be smaller (WNL stars have much higher L/M ratios than O-stars; cf. e.g. Maeder and Meynet, 1987), it has a much smaller surface gravity and therefore a correspondingly larger mass loss rate. Due to the smaller surface gravity also the escape velocity is smaller leading to a smaller final wind velocity. Both effects go quite into the right direction. Of course, though the metal abundance in O-star and WNL star envelopes are comparable, the effect of the different electron density due to the difference in the hydrogen content has to be examined in more elaborated calculations.

The WNL mass loss rate is of special importance for the most massive stars $(M_{ZAMS} \gtrsim 100\, M_\odot)$, since the duration of the WNL phase may increase with increasing initial mass, offering the possibility that the WNL time scale becomes as large as the He-burning time scale (Langer, 1987) with the implication that the most massive stars would never loose their H-envelope completely. Consequences for possible related supernova events are mentioned in Sect. 6.

## 5.2 HYDROGENLESS WOLF-RAYET STARS

Once a massive stars has lost its hydrogen envelope completely, i.e. being either a WNE or a WC star, its internal structure becomes extremely simple and independent of its previous evolution. Langer (1989a) showed, that due to the overall dominance of radiation pressure, the whole structure of the star is determined by its total mass and its surface chemical composition, which allows to develop simple formulae for the dependence of the global structural quantities on those two parameters.

The basic missing information for a reliable description of the evolution of hydrogenless WR stars is — again — their mass loss rate e.g. as function of the stellar parameter. Observations indicate such high values that the momentum of the wind exceeds that of the radiation by more than one order of magnitude (cf. Schmutz et al., 1989), which makes it unlikely that those winds are radiatively driven. No theories are available for quantitative predictions (cf., however, Cassinelli, this volume; and Maeder, 1985). Langer (1989b) suggested, that due to the fact that mainly the mass determines the structure of WR stars — i.e. also their surface conditions — the mass loss rate should be primarily a function of their mass. By investigating the effect of different mass dependences of the form $\dot{M}_{WR} \sim M_{WR}^{\alpha}$ on IMF averaged properties of WNE and WC stars, which can be easily compared with observations, he found that only values of $\alpha > 1$ lead to an agreement with observations. Note that hitherto almost all evolutionary calculations for WR stars are performed with $\alpha = 0$, i.e. a constant WR mass loss rate (cf. however Maeder, this volume; Langer, 1990). The best agreement was achieved for a rate of the form $\dot{M}_{WR} = (0.6 - 1.0) \cdot 10^{-7} (M_{WR}/M_{\odot})^{2.5} \ [M_{\odot} \, yr^{-1}]$. A direct comparison of this equation with observed WR masses (or masses derived from observed luminosities via the mass-luminosity relation of Langer, 1989a) and mass loss rates (e.g. Schmutz et al., 1989) yields to no contradiction; however, the scatter of the observational data as well as their error bars are large. Note further that Abbott et al. (1986) already proposed a WR mass loss rate as $\dot{M} \sim M^{2.5}$ on a purely observational basis, which was, however, criticized later by Cassinelli and van der Hucht (1987).

The effect of mass dependent WNE and WC mass loss rates, as shown by Langer (1989b), is a very flat initial-final mass relation for stars in the mass range $45 \, M_{\odot} \lesssim M_{ZAMS} \lesssim 100 \, M_{\odot}$: the final mass is almost the same for the whole range, and it is of the order of $5 - 10 \, M_{\odot}$, depending on the exact value of the coefficients in the mass loss rate. This may have many consequences e.g. for galactic chemical evolution, which still have to be worked out. It may also lead to a rather homogeneous class of supernovae of Type I, which will be mentioned in the next section.

Finally, we note that due to the fact that the physical origin of the mass loss of hydrogenless WR stars is unknown, nothing can be said about its metallicity dependence.

## 6. Mass loss and supernovae from massive stars

The advanced burning stages of massive stars beyond core helium exhaustion proceed so rapidly, that the number of stars to be observed in those phases is too small to allow a useful comparison between theory and observations. Only at the very end of the evolution, when massive stars explode as a supernova, the observational statistic is sufficiently improved due to the large brightness of these events. Supernova observations are capable of yielding significant information about the mass loss history of their progenitor stars: their spectra contain clues about the chemical composition of the surface layers, and their light curves can be used, together with the velocity information from the spectra, to find the total mass and/or the mass of the hydrogen envelope of the progenitor. Furthermore, for supernova observations there is, unlike for other stellar observations, an evident correspondence of the observation and the evolutionary state of the observed object. Therefore, SN observations are capable of yielding strongest constraints to the theory of massive star evolution and thereby also to mass loss rates of massive stars.

### 6.1 TYPE II SUPERNOVAE

The SN 1987A in the LMC is a good example of the general statement above. From its light curve we know that at the time of explosion the progenitor star still had a hydrogenrich envelope of $\sim 8 - 13 \, M_{\odot}$ (cf. Woosley, 1988; Shigeyama et al., 1988). Since the progenitor star of this SN was known, its He-core mass could be constrained from its luminosity to

$\sim 5 - 7\,M_\odot$ (cf. Arnett et al., 1989), which leads to a ZAMS-mass of the order of $20\,M_\odot$. One can immediately conclude, that e.g. $M_{ZAMS} = 20\,M_\odot$ is still far from the lower ZAMS mass limit for WR star formation in the LMC.

Furthermore, we should mention that SN 1987A observations strongly indicate, that semiconvection may not be negligible in massive stars, since only when it is taken into account the blue-red-blue evolution of the SN 1987A progenitor (cf. Fransson et al., 1989) is obtained selfconsistently (Langer et al., 1989). This has far reaching consequences for both, the internal evolution and nucleosynthesis and the HRD track of massive stars. E.g., it means that all massive stars would evolve redward after core hydrogen exhaustion on a thermal time scale, independent of the main sequence mass loss (cf. Sect. 2), and that almost all blue supergiants were in the red supergiant stage before (and should therefore be nitrogen enhanced). Also, relatively small main sequence mass loss rates cannot be ruled out with the argument of a short LBV time scale (cf. Sect. 4).

Type II supernovae in general do not perform a homogeneous group, and the classification of their light curves is difficult (cf. e.g. Dogget and Branch, 1985). Since all of them indicate — per definition — the presence of hydrogen, they are certainly all related to massive stars. From the theoretical point of view, no homogeneous SN II class can be expected, since both, core- and envelope masses, may vary over wide ranges in the pre-supernova stage, depending on the initial stellar mass. Furthermore, not only RSGs may be Type II progenitors, but also BSGs (as in case of SN 1987A) and Wolf-Rayet stars. As mentioned in Sect. 5.1, WNL stars — the only WR stars which could be SN II progenitors — are mainly very massive objects, and possibly for the highest stellar masses ($M_{ZAMS} \gtrsim 100\,M_\odot$) they might be pre-SN configurations (then exploding due to the $e^{\pm}$-pair formation mechanism); cf. Langer, 1990a. However, due to the steep decline of the IMF, those might be very rare (cf. Herzig et al., 1990). However, close to the lower ZAMS mass limit for WR formation may also be a narrow mass range where the WNL phase is the final evolutionary stage. Though these low mass WNL stars may be only a very small fraction of all observed stars, since their progenitors would spend most of core helium burning in the supergiant phase, they may be statistically significant as pre-supernova configurations. No firm conclusions can yet be drawn from this field due to a lack of both, observational data and theoretical models, but it is well possible that this will change in the near future. Work on the theoretical side is in progress.

## 6.2 TYPE I SUPERNOVAE

Since a large fraction of massive stars looses their hydrogen envelope completely during the evolution, the question is justified, whether those give rise to type I SN events. In Sect. 5.2 we mentioned that the bulk of stars with $45\,M_\odot \lesssim M_{ZAMS} \lesssim 100\,M_\odot$ end their evolution as a low mass ($M \lesssim 10\,M_\odot$) WC star. Ensman and Woosley (1988) found, that light curves of exploding low mass hydrogenless WR stars look similar to observations of SNe of the recently classified subgroup of Type Ib (cf. Branch, 1986). For this reason, a relation between massive stars and Type Ib SNe is well possible. Cf. Langer (1990a) for a more thorough discussion of this topic.

## 7. Conclusions

It has been shown in the previous sections, that mass loss affects qualitatively the evolution of massive stars and is essential for an understanding of their observed properties. However, except for hot hydrogenrich stars, neither theories for quantitative predictions are available, nor do observed mass loss rates yield accurate mass loss relations e.g. as function of the stellar parameter. Due to further large uncertainties involved in the theoretical modeling of massive stars — e.g. convection theory — basic problems of massive star evolution are still to be solved. However, the incorporation of mass loss in stellar evolution calculations allowed for a large progress of our understanding of massive stars in the last decades (cf. Chiosi and Maeder, 1986). In the following we will briefly summarize conclusions of recent developments.

For **massive main sequence stars** (cf. Sect. 2) the radiation driven wind theory is able to predict mass loss rates for given stellar parameter, yielding excellent qualitative agreement with observations. Recent evolutionary calculations indicate a systematic difference of a factor $2-3$ between the theoretical rates and empirical mass loss formulae, the theoretical rates being smaller. Both seem to agree within the estimated uncertainty limits. Due to the probable importance of semiconvection in massive stars, indicated by the SN 1987A progenitor evolution, small main sequence mass loss rates do not lead to contradictions with observations of He-burning massive stars (cf. sections 3 and 4).

The radiation driven wind theory applies also for hot **supergiants** (see Sect. 3). However, the main mass loosing supergiant stage is that of red supergiants, for which mass loss rates are known only to an order of magnitude. Since probably all massive stars in the range $15\,M_\odot \lesssim M_{ZAMS} \lesssim 45\,M_\odot$ have a RSG phase, several basic questions are still open, e.g. whether the most massive RSGs evolve into Wolf-Rayet stars or not, or whether some massive stars have an LBV stage prior to or after a RSG phase.

Also for **Luminous Blue Variables**, time-averaged mass loss rates are only known to an order of magnitude, and no quantitative mass loss theories exist. However, time-averaged mass loss rates can be computed from stellar models by imposing the condition $\dot{R} = 0$ (cf. Sect. 4). The LBV mass loss rates obtained thereby depend greatly on the previous evolution of the star as well as on its metallicity.

The mass loss rates of **Wolf-Rayet stars** seem to depend much on whether hydrogen is present or not. In case hydrogen is present ("WNL"; cf. Sect. 5.1), the momentum problem of the wind is small, i.e. the radiation driven wind theory may apply, since also qualitative expectations agree with observations. This is not the case for hydrogenless WR stars, where the momentum problem is large, and the radiation driven wind theory is less likely to apply. However, from theoretical and observational arguments one finds a mass dependence of the mass loss rate of the form $\dot{M}_{WR} \sim M_{WR}^{2.5}$ for WNE and WC stars (cf. Sect. 5.2). The consequence is a very flat initial-final mass relation for a wide mass range $(45\,M_\odot \lesssim M_{ZAMS} \lesssim 100\,M_\odot)$ with final masses well below $10\,M_\odot$.

In Sect. 6 we outline the potential of **supernova** observations as tool for the analysis of the mass loss history of the progenitor stars. A confirmation of a relation of Type Ib SNe to WR stars would strongly support the concept of mass dependent WR mass loss rates. Furthermore, the variety of Type II SN light curves may be related to the large number of different possible pre-supernova configurations from massive stars, which do not lose their envelope completely during the evolution, like RSGs with thick or thin envelopes, BSGs, low mass WNLs and very massive WNLs. The SN 1987A provides an excellent example of what and how SN observations can teach us about physical processes relevant to the pre-supernova evolution, including mass loss.

*Acknowledgment.* I am grateful to S. Woosley for discussions and for his hospitality at Lick Observatory, and for discussions with R.P. Kudritzki, C. Leitherer, A. Maeder, and

A. Pauldrach. This work has been supported by the Deutsche Forschungsgemeinschaft (DFG) through grants La 587/1-2 and La 587/2-1, by NASA through grant NAGW-1273, and by the Astronomische Gesellschaft (AG) through the Ludwig-Biermann award 1989.

**References**

Abbott, D.C., Bieging, J.H., Churchwell, E., Torres, A.V.: 1986, Astrophys. J. **303**, 239

Arnett, W.D., Bahcall, J.N., Kirshner, R.P., Woosley, S.E.: 1989, Ann. Rev. Astron. Astrophys. **27**, 629

Barlow, M.J., Smith, L.J., Willis, A.J.: 1981, M.N.R.A.S. **196**, 191

Branch, D.: 1986, Astrophys. J. *Letters* **300**, L51

Cassinelli, J.P., van der Hucht, K.A.: 1987, in: *Instabilities in Luminous Early Type Stars*, eds. H. Lamers et al., Reidel, Dordrecht, p. 231

Castor, J.C., Abbott, D.C., Klein, R.I.: 1975, Astrophys. J. **195**, 157

Caughlan, G.R., Fowler, W.A.: 1988, Atomic Data and Nuclear Data Tables

Chiosi, C., Maeder, A.: 1986, Ann. Rev. Astron. Astrophys. **24**, 329

Conti, P.S.: 1986, IAU-Symp. **116**, 199

Davidson, K., Moffat, A.F.: 1989, eds. IAU-Colloq. **113**, in press

Dogget, J.B., Branch, D.: 1985, A. J. **90**, 2303

Ensman, L.M., Woosley, S.E.: 1988, Astrophys. J. **333**, 754

Fitzpatrick, E.L., Garmany, C.D.: 1989, Astrophys. J. , in press

Fransson, C., Cassatella, A., Gilmozzi, R., Panagia, N., Wamsteker, W., Kirshner, R.P., Sonneborn, G.: 1989, Astrophys. J. **336**, 429

Hamann, W.-R.: 1990, Proc. Hot Star Workshop, Boulder, C. Garmany, ed., in press

Herzig, K., El Eid, M.F., Fricke, K.J., Langer, N.: 1990, Astron. Astrophys. , submitted

van der Hucht, K.A., Conti, P.S., Lundström, I., Stenholm, B.: 1981, Space Sci. Rev. **28**, 227

Humphreys, R.M.: 1984, IAU-Symp. **105**, 279

Humphreys, R.M.: 1989, in: IAU-Colloq. **113**, in press

Humphreys, R.M., McElroy, D.B.: 1984, Astrophys. J. **284**, 565

Humphreys, R.M., Nichols, M., Massey, P.: 1985, Astron. J. **90**, 101

de Jager, C., Nieuwenhuijzen, H., van der Hucht, K.A.: 1986, IAU-Symp. **116**, 109

de Jager, C., Nieuwenhuijzen, H., van der Hucht, K.A.: 1988, Astron. Astrophys. Suppl. **72**, 259

Kudritzki, R.P.: 1988, in: 18$^{th}$ Saas-Fee course, Swiss Soc. of Astronomy and Astrophysics

Kudritzki, R.P., Pauldrach, A., Puls, J.: 1987, Astron. Astrophys. **173**, 293

Kudritzki, R.P., Pauldrach, A., Puls, J., Abbott, D.C.: 1989, Astron. Astrophys. **219**, 205

Langer, N.: 1986, Astron. Astrophys. **164**, 45

Langer, N.: 1987, Astron. Astrophys. *Letter* **171**, L1

Langer, N.: 1989, in: IAU-Colloq. **113**, in press

Langer, N.: 1989a, Astron. Astrophys. **210**, 93

Langer, N.: 1989b, Astron. Astrophys. **220**, 135

Langer, N.: 1990, in: *Hot Stars*, proc. 1$^{st}$ Boulder-Munich workshop, ed. C. Garmany, in press

Langer, N.: 1990a, in: *Supernovae*, proc. 10$^{th}$ Santa Cruz summer workshop, ed. S. Woosley, in press

Langer, N., Sugimoto, D., Fricke, K.J.: 1983, Astron. Astrophys. **126**, 207

Langer, N., El Eid, M.F., Fricke, K.J.: 1985, Astron. Astrophys. **145**, 179

Langer, N., El Eid, M.F.: 1986, Astron. Astrophys. **167**, 265

Langer, N., El Eid, M.F., Baraffe, I.: 1989, Astron. Astrophys. *Letters* **224**, L17

Lamb, S., Howard, W.M., Truran, J.W., Iben, I. Jr.: 1977, Astrophys. J. **217**, 213

Lamers, H.J.G.L.M.: 1981, Astrophys. J. **245**, 593

Lamers, H.J.G.L.M.: 1989, in: IAU-Colloq. **113**, in press

Lundström, I., Stenholm, B.: 1984, Astron. Astrophys. Suppl. **58**, 163

Maeder, A.: 1981, Astron. Astrophys. **102**, 401

Maeder, A.: 1982, Astron. Astrophys. **105**, 149

Maeder, A.: 1983, Astron. Astrophys. **120**, 113

Maeder, A.: 1985 Astron. Astrophys. **147**, 300

Maeder, A., Meynet, G.: 1987, Astron. Astrophys. **182**, 243

Moffat, A.F., Drissen, L., Robert, C., Lamontange, R., Coziol, R., Mousseau, N.: 1989, preprint

Niemela, V.S.: 1983, in: Proc. *Workshop on Wolf-Rayet stars*, Paris-Meudon, eds M.C. Lortet, A, Piltaut, p. III.3

Reimers, D.: 1975, Mem. Soc. Roy. Liege, 6th Ser. **8**, 369

Renzini, A.: 1987, Astron. Astrophys. **188**, 49

Schild, H., Maeder, A.: 1984, Astron. Astrophys. **136**, 237

Schmutz, W., Hamann, W.-R., Wessolowski, K.: 1989, Astron. Astrophys. **210**, 236

Shigeyama, T., Nomoto, K., Hashimoto, M.: 1988, Astron. Astrophys. **196**, 141

Smith, L.F., Hummer, D.G.: 1988, M.N.R.A.S. **230**, 511

Smith, L.F., Maeder, A.: 1989, Astron. Astrophys. **211**, 71

Smith, L.J., Willis, A.J.: 1983, Astron. Astrophys. Suppl. **54**, 229

Stothers, R., Chin, C.-W.: 1976, Astrophys. J. **204**, 472

Stothers, R., Chin, C.-W.: 1977, Astrophys. J. **211**, 189

Stothers, R., Chin, C.-W.: 1979, Astrophys. J. **233**, 267

Stothers, R., Chin, C.-W.: 1985, Astrophys. J. **292**, 222

Torres, A.V.: 1988, Astrophys. J. **325**, 759

Willis, A.J.: 1982, in: *Wolf-Rayet Stars: Observations, Physics, Evolution*, IAU-Symp. **99**, C. de Loore, A.J. Willis, eds., p. 87

Woosley, S.E.: 1988, Astrophys. J. **330**, 218

# V444 CYGNI AND CQ CEPHEI, KEY WOLF-RAYET BINARY STARS


Anne B. Underhill
Department of Geophysics and Astronomy
University of British Columbia
Vancouver, B. C., V6T 1W5, Canada



ABSTRACT. The fundamental properties of the Wolf-Rayet components of V444 Cygni and CQ Cephei such as mass, effective temperature, and luminosity are presented. These properties are consistent with these stars being young objects recently arrived on the main sequence. The properties of evolved, peeled-down model stars are not consistent with the observed properties of V444 Cygni and CQ Cephei.


## 1. Introduction

Radial-velocity and photometric observations of eclipsing binaries allow one to estimate the masses of the stars in such systems, their relative brightnesses, and sizes. If polarization variations are detected throughout the binary period, one may confirm the inclination of the system, see Rudy and Kemp (1978). Because all three types of information are available for V444 Cygni and CQ Cephei, these systems are important for determining the fundamental properties of Wolf-Rayet stars.

The spectral type of V444 Cygni is WN5 + O6. Features from the spectra of both stars are easily seen in the spectrum of the pair. In the visible spectral range, the light curve shows a broad primary minimum and a small secondary minimum. Underhill and Fahey (1987) have noted that the shape of the light curve, particularly near secondary minimum when the Wolf-Rayet star is eclipsed, is strongly distorted by scattered light from the O star.

The spectral type of CQ Cephei is WN7. No spectroscopic evidence of the companion star has been detected although it is known that two similar stars are present because the light curve shows two similar minima, see, for instance, Stickland et al. (1984). The Wolf-Rayet star is eclipsed at primary minimum.

## 2. Observations

Light curves in several wavelength bands and radial-velocity curves for several spectral lines are available for HD 193576 = V444 Cygni and for HD 214419 = CQ Cephei. References to the observations of V444 Cygni can be found in Underhill and Fahey (1987) and in Underhill, Yang, and Hill (1988a, 1988b). References for CQ Cephei may be found in Stickland et al. (1984) and in Underhill, Gilroy, and Hill (1990). Polarization observations of V444 Cygni are given by Rudy and Kemp (1978); polarization observations of CQ Cephei are given by Drissen et al. (1986). The parameters given in Table 1 for the binary systems V444 Cygni and CQ Cephei have been extracted from these references. The orbits are circular. The large polarization of CQ Cephei is mostly interstellar.




*L. A. Willson and R. Stalio (eds.), Angular Momentum and Mass Loss for Hot Stars, 279–282.*
© 1990 *Kluwer Academic Publishers. Printed in the Netherlands.*

The visible and ultraviolet spectrum of V444 Cygni presents spectroscopic evidence for a wind from each star and for streams of gas flowing between, and perhaps, around the two components. The visible and ultraviolet spectrum of CQ Cephei presents spectroscopic evidence for a wind from the Wolf-Rayet star and for streams of gas in the system. In the case of CQ Cephei, the line spectrum of the companion star has not been detected in any spectral range although the continuum of the companion star is detected. Possibly the absorption lines from the companion star are so broadened by electron scattering in the disk of material which generates the variable component of the polarization that they cannot be detected. Some of the emission lines are accompanied by shortward displaced absorption components. However, all of these features move with the Wolf-Rayet star.

## 3. Discussion

In the radial-velocity studies of V444 Cygni and CQ Cephei it is argued that the orbital motion of the Wolf-Rayet star is shown in each case by the wavelength changes of weak N IV emission lines. The strong emission lines of He I and He II are longward displaced by approximately 100 km s$^{-1}$. This is with respect to the systemic velocity shown by the O star in the case of V444 Cygni and it may be inferred from galactic rotation in the case of CQ Cephei. These displacements suggest that by means of radiation from the abundant element helium we are seeing infall from parts of the line-emitting region (LER) between the observer and the photosphere as well as orbital motion. The infall is not detected by means of the weak N IV lines. Apparently the N IV lines are formed in plasma closely attached to the Wolf-Rayet star. The He I and He II emission lines do show the orbital motion of the Wolf-Rayet component, but they are displaced longward. In the case of CQ Cephei, He I $\lambda$5876 is accompanied by several changing shortward displaced absorption components. In the case of V444 Cygni a shortward displaced He I $\lambda$5876 component is seen only at a few phases.

The effective temperatures, luminosities, radii, and log $g$ values of our stars, see Table 2, have been estimated by interpolating in the tables of properties of models provided by Maeder and Meynet (1988). I interpolated for stars having an age of 5 $\times 10^6$ years and the masses given in Table 1. These models have solar surface composition. The stellar properties have been estimated in this way because I believe that the stellar-evolution models for massive stars which are burning hydrogen are reliable, and that the masses given in Table 1 are reliable. The inferred radii easily fit into the circular orbits of the two binary systems.

I prefer not to use the radii for the components of V444 Cygni and CQ Cephei which have been derived by solving the light curves, because, in each case, the two stars appear to be buried in much plasma which scatters light from the companion star into the cone of sight and seriously distorts the light curve.

If one represents the Wolf-Rayet components of V444 Cygni and CQ Cephei using the properties of the highly evolved remnants of a massive star with an initial mass of 40 $M_\odot$ (Maeder and Meynet 1988), one finds that in both cases the brightness of the Wolf-Rayet star is more than 6 times that of the O star at 5500 Å, the effective temperatures of the two Wolf-Rayet stars are greater than $10^5$ K, the luminosities are of the order of 3 $\times 10^5 L_\odot$, and the radius of the Wolf-Rayet star in V444 Cygni is of the order of 0.3 $R_\odot$ while that of the Wolf-Rayet star in CQ Cephei is about 0.95 $R_\odot$. The ages of such remnants are less than 5.4 $\times 10^6$ years, which is a suitable age for early-type stars in Cygnus and Cepheus.

If one uses the models for remnants of stars of initial mass less than or equal to 25 $M_\odot$, the age of a remnant which has a mass like the observed masses for the above Wolf-Rayet stars is more than 8 $\times 10^6$ years, $T_{eff}$ is less than $10^4$ K, log $L/L_\odot$ is greater than 5.3, and the radius is greater than 700 $R_\odot$. Such model stars are not suitable for representing Wolf-Rayet stars.

The spectroscopic analyses of Bhatia and Underhill (1986, 1988, 1989) clearly indicate that the effective temperatures of Wolf-Rayet stars must be of the order of or less than 30,000 K. A radiation temperature corresponding to $T_{eff} \geq 10^5$ K will ionize the stellar LER to such an extent that spectral lines like what are observed will not be generated.

## 4. Conclusions

The highly evolved remnant model stars of Maeder and Meynet (1987, 1988) which show anomalous abundances on their surfaces do not have the properties of the Wolf-Rayet stars in V444 Cygni and CQ Cephei.

The hydrogen-burning models of Maeder and Meynet are satisfactory for representing the Wolf-Rayet stars in V444 Cygni and CQ Cephei. This conclusion strengthens the suggestion of Bhatia and Underhill (1986, 1988, 1989) that Wolf-Rayet stars are young stars still surrounded by large remnant disks.

It is chiefly the high electron temperature in the LER which causes the characterictic line patterns of WC and WN spectra to appear. A solar composition and careful modelling of dielectronic recombination will generate the line ratios observed in the spectra of Wolf-Rayet stars.

The disk-like LER is attached to the Wolf-Rayet star in a binary system and it rotates with and moves with the Wolf-Rayet star. Because much of the line emission as well as infrared and radio-flux emission may be generated by the remnant disk rather than by the star, there is no reason to infer that the winds of Wolf-Rayet stars are very dense. The available observations can be interpreted with $\dot{M} \leq 10^{-6} M_\odot$ yr$^{-1}$ and a remnant disk having the properties specified by Bhatia and Underhill.

## 5. References


Bhatia, A. K., and Underhill, A. B. 1986, *Ap. J. Suppl.*, **60**, 323.

Bhatia, A. K., and Underhill, A. B. 1988, *Ap. J. Suppl.*, **67**, 187.

Bhatia, A. K., and Underhill, A. B. 1989, *Ap. J.*, submitted.

Drissen, L., Moffat, A. F. J., Bastien, P., and Lamontagne, R. 1986, *Ap. J.*, **306**, 215.
  Maeder, A., and Meynet, G. 1987, *Astr. Ap.*, **182**, 243.

Maeder, A., and Meynet, G. 1988, *Astr. Ap. Suppl. Ser.*, **76**, 411.

Rudy, R. J., and Kemp, J. C. 1978, *Ap. J.*, **221**, 200.

Stickland, D. J., Bromage, G. E., Budding, E., Burton, W. M., Howarth, I. D., Jameson, R., Sherrington, M. R., and Willis, A. J. 1984, *Astr. Ap.*, **134**, 45.

Underhill, A. B., and Fahey, R. P. 1987, *Ap. J.*, **313**, 358.

Underhill, A. B., Gilroy, K. K., and Hill, G. M. 1990, *Ap. J.*, **351**, March 10.

Underhill, A. B., Yang, S., and Hill, G. M. 1988a, *Pub. Astr. Soc. Pac.*, **100**, 741.

Underhill, A. B., Yang, S., and Hill, G. M. 1988b, *Pub. Astr. Soc. Pac.*, **100**, 1256.

TABLE 1

Orbital Properties of V444 Cygni and CQ Cephei

| Quantity | V444 Cygni | CQ Cephei |
|---|---|---|
| Period (days) | 4.212424 | 1.6412436 |
| Polarization (%) | $\approx 0.5$ | $\approx 5.0$ |
| $i$ (degrees) | $78 \pm 4$ | $74 \pm 6$ |
| $a_{WR}$ $(R_\odot)$ | 27.4 | 9.7 |
| $a_{companion}$ $(R_\odot)$ | 9.1 | 8.2 |
| $M_{WR}$ $(M_\odot)$ | 9.8 | 13.6[a] |
| $M_{companion}$ $(M_\odot)$ | 29.6 | 16.0[a] |

[a]Assuming $M_{WR}/M_{companion} = 0.85$. This mass ratio is determined by matching the observed brightness ratio in the V band (Stickland et al. 1984) by a theoretical brightness ratio, see Underhill, Gilroy, and Hill (1989). Stickland et al. note that the observed brightness ratio of CQ Cephei in the V band is approximately 0.9.

TABLE 2

Interpolated Properties of the Stars in V444 Cygni and CQ Cephei[a]

| Quantity | V444 Cygni | CQ Cephei |
|---|---|---|
| WR Radius $(R_\odot)$ | 4.0 | 5.5 |
| WR $T_{eff}$ (K) | 27,200 | 29,000 |
| $\log (L_{WR}/L_\odot)$ | 3.91 | 4.27 |
| $\log g_{WR}$ | 4.22 | 4.09 |
| Surface abundances | solar | solar |
| Companion radius $(R_\odot)$ | 9.92 | 6.4 |
| Companion $T_{eff}$ (K) | 40,100 | 30,800 |
| $\log (L_{comp}/L_\odot)$ | 5.36 | 4.52 |
| $\log g_{comp}$ | 3.92 | 4.03 |
| Surface abundances | solar | solar |

[a]These are interpolated for an age of $5 \times 10^6$ yrs from the results of Maeder and Meynet (1988). The radii are estimated from the interpolated values for $\log L/L_\odot$ and $\log T_{eff}$. The results for the O star of V444 Cygni are quite uncertain because at an age of $5 \times 10^6$ yrs the properties of stars with initial masses of 25 and 40 $M_\odot$ differ greatly. Consequently interpolating between these masses is inexact.

# ROTATION OF HOT STARS AFTER THEY COOL OFF

David F. Gray
Department of Astronomy
University of Western Ontario
London, Ontario  N6A 3K7, Canada

ABSTRACT.  Rotation and convection are controllers of atmospheric phenomena.

## 1. INTRODUCTION

Some very interesting things happen to rotating stars when they evolve off the main sequence. Naturally they slow down with their increasing moment of inertia, but we also observe changes in convection, dramatic angular momentum loss, the turning on and off of magnetic coronae, and changes in mass loss (see the contributions in this volume by Antonello, Cassinelli, and Lamers). All of these phenomena are dynamically connected to rotation, either controlling rotation or being controlled by rotation.

There are several "boundaries" in the H-R diagram of relevance here. Let me start by explaining what the observations show.

## 2. THE BOUNDARIES

Line asymmetries are seen in most spectral lines. Many of these arise from line blends and as such are physically uninteresting. But unblended lines also are asymmetric. Spectral resolving power in excess of 100,000 and signal-to-noise ratios of a few hundred are needed to see the asymmetries. The line bisector is a convenient way to specify the asymmetries. We construct it as a locus of points connecting the midpoints of horizontal line segments bounded by the sides of the line profile. To first order, the asymmetries are the same in all lines of a given star, so we know we are not dealing with asymmetries in the atomic line broadening, but with a Doppler-shift distribution -- very much like the effects of rotational broadening. We believe these arise from the velocity fields of photospheric granulation. A remarkable and abrupt change in asymmetry is seen at boundary $\underline{G}$ in Fig. 1. Stars on the left show strong left-leaning bisectors; those on the right show right-leaning bisectors. The G boundary runs from the main sequence up to at least class Ib. (Line asymmetries have not been measured for more luminous stars.)

A cut across the G boundary at any luminosity shows the bisectors to change rapidly but continuously across the boundary. The boundary is well defined above class III, and less well defined toward the main sequence simply because of the difficulty of finding suitable stars.

The second boundary, labeled $\underline{R}$ in Fig. 1, is more dramatic. Stars on the left show widely ranging and large rotational line broadening. Those on the right show a single unique rotation rate for any given spot on the H-R diagram, and fast rotators appear only as anomalies.



*L. A. Willson and R. Stalio (eds.), Angular Momentum and Mass Loss for Hot Stars, 283–290.*

284

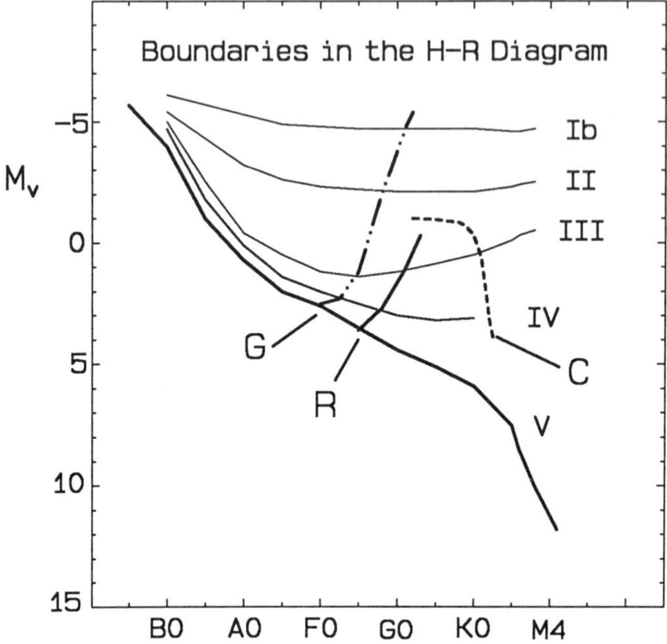


Fig. 1. The G̲, R̲, and C̲ boundaries are shown on this H-R diagram.

X-ray emission is seen for a sizeable fraction of stars to the left and under the boundary C̲ in Fig. 1. Hot coronal (magnetic) loops are the source of the x-rays for the sun, and assumed to be so for these stars. Only a few stars outside the C boundary show detectable x-ray emission. We don't know how sharp the C boundary is because only a few x-ray observations have been made.

These are the bare observations. Now let me tie things together with a few ideas.

## 3. G̲ IS FOR GRANULATION

Solar granulation gives us a ready interpretation of the line asymmetries on the cool side of the G boundary. They are there because the rising granules give more light than the falling cool material, leading to a skew distribution of the Doppler shifts shaping the lines. To be sure, significant differences are seen from star to star, with granulation loosing vigor with declining effective temperature and with lower luminosity class (Gray 1988). But the essential point is that granulation is the top of the convective envelope, and stars evolving from the main sequence all take up solar-type granulation upon crossing the G boundary, giving evidence that formation of deep convective envelopes has begun (Gray and Nagel 1989). The position of the granulation boundary agrees roughly with the predicted onset of convection in the stellar-interiors sense, i.e., significant energy transport. The average granule rise velocities amount to 1-2 km/s; velocities of fall are 3-4 times larger. A large fraction ($\approx$ 80%) of the photons come from rising material.

By contrast (Fig. 2), stars hotter than the granulation boundary show larger (reversed) asym-

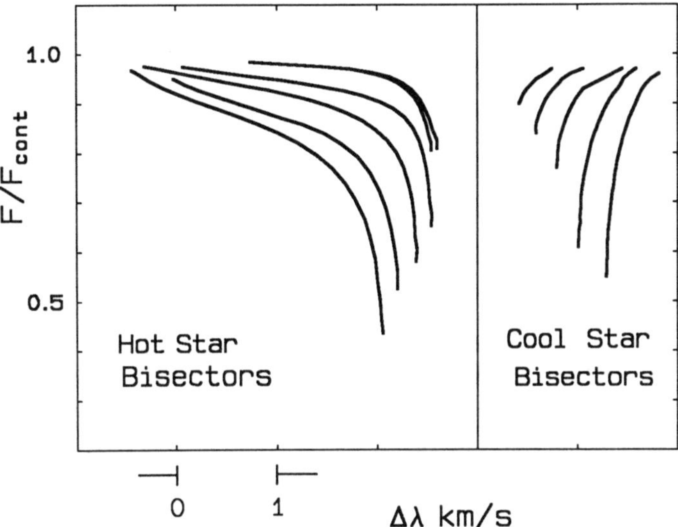



Fig. 2. The two types of spectral-line bisectors are illustrated here. The hot star on the left is 41 Cyg; the cool star on the right is α C Mi. The velocity scale is the same on both sides of the figure, and the several-times larger velocity span for the hot stars is typical. Adapted from Gray (1989a).

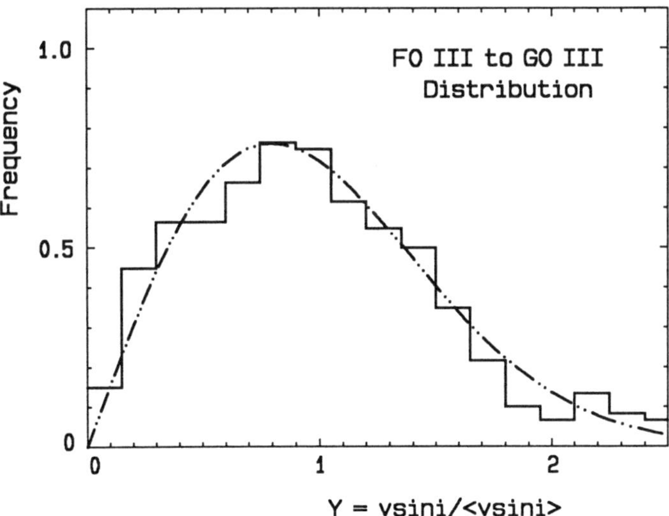



Fig. 3. Giants on the hot side of the R boundary have a Maxwell-Boltzmann distribution. Adapted from Gray (1989b).

metry, indicating an order of magnitude higher rise velocities ($\approx$ 15 km/s) but for a much smaller fraction of the photon flux ($\approx$ 10%). We appear to be seeing the characteristics of shallow convection zones about which very little is known. Shallow convection zones are probably common, perhaps universal, for early-type stars. From the available data, I cannot distinguish between 1) a small fraction of the surface ($\lesssim$ 10%) being involved in the rising stream and 2) a much larger fraction of the surface showing sporadic bubbling such that the time-average of rising material contributes $\approx$ 10% of the photons. But either way, velocities as large as 10-20 km/s will play a significant role in the atmospheric dynamics and the mass loss.

Naturally the granulation produces acoustic power, and it alone (without magnetic fields) can lead to significant heating of the upper atmosphere (Ulmschneider 1990). This in turn affects the mass-loss rates and the dissipation of angular momentum.

But convection in cool stars is much more important in the physical picture than a noise generator. It is instrumental in the manufacturing of magnetic fields, structuring them, and ultimately controlling rotation for many stars. Solar studies show how magnetic fields in the photosphere are concentrated by the horizontal flow patterns of the photosphere (Brandt et al. 1989, Scharmer 1989, Title et al. 1989). Velocity-field mapping also shows "drain" or sink areas with high downward velocities and strong vortex motion -- the kind needed to wind up magnetic field lines into flux tubes or ropes. This is one of the essential processes needed for a dynamo to function. It therefore comes as no surprise that the granulation boundary is also the "activity" boundary as seen in the onset of temperature inversions and chromospheres: stars on the hot side of the G boundary show no H & K line emission; those on the cool side do (Dravins 1981, Gray and Toner 1986).

The real dividing line defining "hot" and "cool" stars is the granulation boundary.

The second essential ingredient needed for dynamo activity is rotation, to which we now turn our attention.


## 4. R IS FOR ROTATION

Although different distributions of v sin i have been discussed for main-sequence stars, F giants show a Maxwell-Boltzmann distribution (Fig. 3). This distribution is destroyed as the giants evolve redward and strong magnetic braking dissipates the angular momentum of the convective envelope. The strong braking occurs as stars evolve across the rotation boundary, between G0 and G3 for class III giants (Fig. 4). The new v sin i distribution corresponds to rotation being a single-valued function of spectral type between G3 III and K2 III (Fig. 5) . How can this come about? In the first instance, we can understand it as a result of dynamo behavior. Rotation must exceed a certain limiting value to keep a dynamo wound up and ticking. Following the dimensional argument of Durney and Latour (1978), the limiting value is given by

$$v_{rot} = v_{conv}/(\ell/R), \tag{1}$$

where $v_{conv}$ is a characteristic convective velocity and $\ell/R$ is a characteristic convective length normalized to the stellar radius. The magnetic brake will slow surface rotation until the limiting value of the above equation is reached. Then the brake turns off. The turnoff depends on $v_{conv}$ and $\ell/R$, and so will be the same for all stars of the same mass and age, independent of their initial rotation rate. The observed value for G3 III giants is 6.1 km/s. From this point on, the behavior of rotation is essentially independent of what took place when these stars were hot rotators. Specifically, the initial rotational velocity distribution is erased.

Model envelope calculations of $v_{conv}$ and $\ell/R$ have been followed for evolution in this region of the H-R diagram. They give the relation shown in Fig. 6. The agreement is rather good, but remember the dimensional nature of the dynamo criterion expressed in eq. 1 allows vertical translation on this log v sin i plot. Nevertheless, the rapid decline in $v_{rot}$ according to the models occurs at the

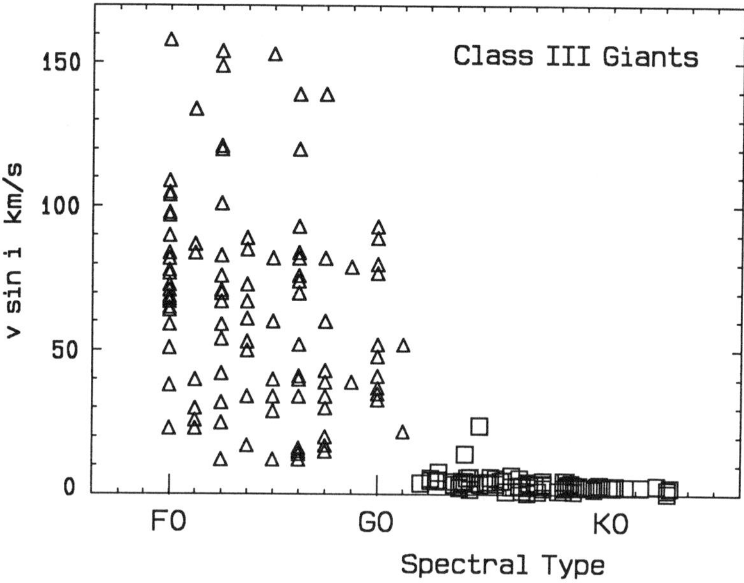



Fig. 4. Rotation drops abruptly between G0 III and G3 III. This is one point on the R boundary. Adapted form Gray (1989b).

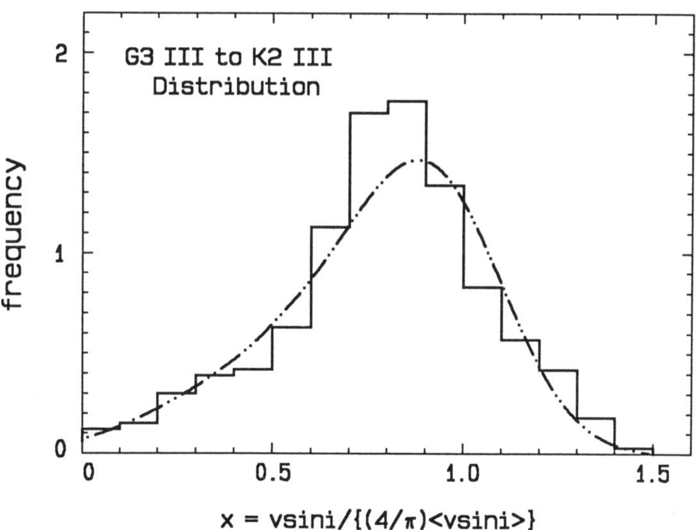



Fig. 5. The new distribution of rotation velocities (histogram) now shows a single value of rotation at each spectral type (dot-dashed curve). Adapted form Gray (1989b).

observed position of the sudden drop in rotation, and the slow decline from G3 III to K2 III also agrees within error. Why should the stars continue to rotate at the dynamo-criterion value? One way to understand it is through the rotostat hypothesis (Gray 1986, 1988). If the dynamo were turned off permanently at G3 III, angular momentum from the rapidly rotating core would be brought into the ever deepening convection zone and spin up the surface layers again. As the dynamo-criterion rate is exceeded (new $v_{conv}$ and $\ell/R$ now), the magnetic brake is reactivated. This reasoning leads me to believe that the dynamo brake flickers on and off repeatedly, maintaining the star's rotation close to the evolving value of $v_{rot}$.

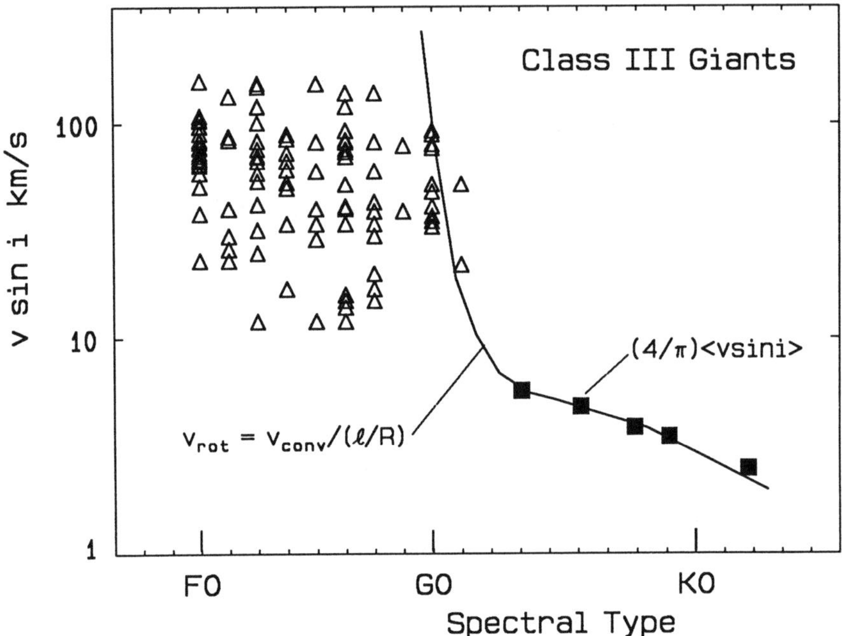


Fig. 6. The dynamo criterion (solid line) is compared to the observations. Actual rotation rates are plotted as boxes for the cooler giants. Adapted from Gray (1989).

Close binaries can show rapid rotation, enforced by tidal coupling and drawing on the orbital angular momentum, which apparently is too large to be rapidly dissipated by the dynamo braking. Such binaries, located between the rotation and coronal boundaries, are hyperactive, showing extra strong chromospheres and coronae, starspots, etc. They are hyperactive by single-star standards, but they are just showing us what is normal for rapidly rotating cool stars -- what all cool stars would look like without the rotational braking at the rotation boundary.

In summary, the convection is not only instrumental in generating magnetic fields and magnetic braking, it may well regulate and determine the rotation rates of stars to the right of the rotation boundary.

## 5. C IS FOR CORONA

Thermal x-ray emission is seen to come from hot coronal loops on the sun. All available evidence suggests the same interpretation of stellar x-ray emission for stars cooler than the granulation boundary. But not all stars to the right of the G boundary show x-ray emission. Those above and to the right of the C boundary (Fig. 1) are not detected x-ray emitters (Linsky and Haisch 1979, Haisch 1987). Explanations have been put forward involving the scale height of power dissipation (Bohm-Vitnese 1987) and the temperature of the plasma in coronal loops (Antiochos and Noci 1986). But I think the explanation is no more complicated than the natural dynamo response to stellar rotation. Namely, stars outside the coronal boundary do not rotate fast enough (eq. 1 again) to generate magnetic field. Without magnetic fields, there are no coronal loops to restrain the plasma and no magnetic mechanisms to heat the gas. With this interpretation, the coronal boundary is a magnetic boundary. Why is the rotation outside the boundary so low? Above class II, the natural increase in moment of inertia is sufficient to bring the rotation below the dynamo criterion. Observed v sin i rates show no evidence for angular momentum loss (Gray and Toner 1987). To the right of $\approx$ K2 III, evolution up the giant branch again ensures rapidly increasing moments of inertia and rotation rates depressed below the dynamo-criterion limit.

## 6. ROTATION, ROTATION EVERYWHERE

Chromospheres and coronae, and magnetic activity generally, result from the combined efforts of rotation and deep-envelope convection. Rotation contributes to its own demise with magnetic braking as stars cross the rotation boundary. Those stars rotating slowly enough, that is, less than $v_{rot}$ in eq. 1, bypass the dynamo-magnetic stage of evolution. Shallow-envelope convection, the kind seen for the stars on the hot side of the granulation boundary, is apparently not suitable for dynamo action, but may still be a controlling factor in the atmospheric dynamics.

As one who lives mainly among the cool stars, I dedicate to you "hot-star people" the following rhyme.

### A TWIST OF NATURE

Rotation, rotation everywhere you go.
It really puts on a beautiful show.
On the main sequence, stars spin like the devil,
but later on they are more calm and level.
If the rotation boundary they should cross,
expect plenty of angular momentum loss!
Magnetic fields and dynamos control the day,
except where convection dictates its say.
Are hot stars safe with rotation alone
from the antics and troubles to which cool stars are prone?
I really doubt it, but can't be quite sure
that the hot stars are really so terribly pure.
And maybe I should add with a grin and a smile
that Ap stars add some spice to that pile.
No! shouts a disenter from the front row.
It's the Be stars upon which we're stubbing our toe.
With Be's and LBV's and WR's doing their thing,
and the theorists and observers going around in a ring,

> well maybe it's not so dull when they're hot.
> You people seem to like then a lot.
> But each from the other we surely can learn,
> as hot stars become cool but continue to turn!

D.F.G.

## REFERENCES


Antiochos, S.K., and G. Noci   1986, Ap.J., <u>301</u>, 440.

Bohm-Vitense, E.   1987, Ap.J., <u>301</u>, 297.

Brandt, P.N., G.B. Scharmer, S.H. Ferguson, R.A. Shine, T.D. Tarbell, and A.M. Title 1989, <u>Solar and Stellar Granulation</u>, (Kluwer: Dordrecht), R.J. Rutten and G. Severino, eds., p. 305.

Dravins, D.   1981, Astron. Astrophys., <u>98</u>, 367.

Durney, B.R., and Latour, J.   1978, Geophys. Ap. Fluid Dyn., <u>9</u>, 241.

Gray, D.F.   1986, High lights in Astronomy, Reidel, Dordrecht, J.-P. Swings, ed., p. 411.

Gray, D.F.   1988, <u>Lectures on Spectral-Line Analysis: F, G, and K Stars</u>, (The Publisher: Arva, Ontario).

Gray, D.F.   1989a, Publ. Astron. Soc. Pacific <u>101</u>, 832.

Gray, D.F.   1989b, Ap.J., Dec. 15th issue.

Gray, D.F., and T. Nagel   1989, Ap.J., <u>341</u>, 421.

Gray, D.F., and C.G. Toner   1986, Publ. Astron. Soc. Pacific <u>98</u>, 499.

Gray, D.F., and C.G. Toner   1987 Ap.J., <u>322</u>, 360.

Haisch, B.M.   1987, <u>Fifth Cambridge Workshop on Cool Stars, Stellar Systems, and the Sun</u>, (Springer-Verlag: Berlin), J.L. Linsky and R.E. Stencel, eds., p. 269.

Linsky, J.L., and B. Haisch   1979, Ap.J. Letters, <u>229</u>, L27.

Scharmer, G.B.   1989, <u>Solar and Stellar Granulation</u>, (Kluwer: Dordrecht), R.J. Rutten and G. Severino, eds., p. 161.

Title, A.M., T.D. Tarbell, K.P. Topka, S.H. Ferguson, R.A. Shine, and the SOUP Team 1989, <u>Solar and Stellar Granulation</u>, (Kluwer: Dordrecht), R.J. Rutten and G. Severino, eds., p. 225.

Ulmschneider, P.   1990, <u>Sixth Cambridge Workshop on Cool Stars, Stellar Systems, and the Sun</u>, Publ. Astron. Soc. Pacific Conf. Series, G. Wallerstein, ed.

WINDS, MASS LOSS AND ROTATION IN
CENTRAL STARS OF PLANETARY NEBULAE

M. Perinotto
Dipartimento di Astronomia e Scienza dello Spazio
Università di Firenze
Largo E. Fermi, 5
50125 Firenze - Italy

ABSTRACT: The expected range of stars being the precursors of the central stars of Planetary Nebulae (CSPNs) in the main sequence, is first considered. This information is relevant in particular to establish the status of axial rotation of these objects before they evolve to the subsequent phases of their history.

Then the present knowledge of the different kind of winds at work in the various phases of the evolution of these stars is summarized. The effects of these winds for the evolution of the central stars and their ejecta is discussed. Finally I will speculate on the possible role of the rotation as a mechanism which can enhance the wind production in CSPNs and their precursors both in single stars and in double systems.

## 1. THE PRECURSORS OF CSPNs

The theory of stellar evolution tell us that all single stars with masses between 0.7-0.8 and 8-9 $M_\odot$, the precise values depending on the initial chemical composition, evolve from the main sequence (MS) through the planetary nebula (PN) stage up to the white dwarf degenerate status (cf. Iben and Renzini, 1983).

The upper limit should be lowered to about 6 $M_\odot$ if the overshooting process is fully active in the convective zones.

It is of some interest to try to examine to which extent the observations confirm this prediction. To do this, we may start from the individually observed PNe and look at their masses. By adding the masses of the associated central stars and considering also some mass which at the present time may be very tenuous and dispersed in the interstellar regions, corresponding to material ejected from the central stars during the previous evolutionary stages and not visible at the present time, we may come up with the mass of the progenitors while in the main sequence.

### 1.1 Mass of the nebulae

To have the mass of the nebulae, we must count the mass of the ionized component and that of the neutral component, if present. In both



L. A. Willson and R. Stalio (eds.), Angular Momentum and Mass Loss for Hot Stars, 291–305.

cases the distances of the objects must be known, something which is in general not easy in PNe. However in some cases the distances are relatively well known. It is so for the PN named K 648, the only one known to belong to a globular cluster (M15), for 35 galactic PNe having individual (not "statistical") distances (Mallik and Peimbert, 1988) and for 15 PNe near to the galactic center, studied by Kinmann et al. (1988).

The mass of the ionized component of K 648 is measured to be 0.02 $M_\odot$, while that of the above mentioned groups of stars was found to range in the intervals 0.4 - 0.004 and 0.3 - 0.003 $M_\odot$, respectively. To these PNe we may add the PNe studied in the Magellanic Clouds: 24 PNe by Wood et al. (1987) and 54 PNe by Meatheringham et al. (1988). The masses of the ionized components fall in the interval 0.56 - 0.02 and 0.35 - 0.01 $M_\odot$, respectively. Among the less massive objects in these samples, some might be optically thick. The authors give however arguments to support the view that some of these PNe are instead thin, so that the large spread in masses indicated by the above figures should be real. Actually in a few optically thick PNe, a neutral component is also observed with masses ranging between ~ 0.1 up to 1 $M_\odot$ and more (cf. Pottasch, 1984).

### 1.2 *Mass of the central stars*

In CSPNs there is yet not a single case known of a double line spectroscopic binary and eclipsing system from which to have a direct determination of the masses of the stars. We must then rely on indirect methods, namely on the comparison of the location of central stars in the HR diagram with the evolutionary tracks predicted for different values of the residual stellar mass after the ejection of the hydrogen rich envelope.

The location of the observed CSPNs in the HR diagram has its well known problems. However, from four samples considered by Pottasch (1984), Heap and Augensen (1987), Méndez et al. (1988), and Jacoby and Kaler (1989) the following intervals of masses: 0.54 - 0.9, 0.55 - 0.8, 0.55 - 0.9 and 0.55 - 0.8 are respectively indicated by the above analysis.

To an interval of 0.54 - 0.9 $M_\odot$ for the residual mass of the central star in the PN phase, the corresponding interval of mass of the progenitors in the MS is of ~ 0.8 up to 2-3 $M_\odot$, in case of solar chemical composition. The range of 2-3 $M_\odot$ for the upper limit reflects the uncertainty between 1/3 to 1 of the $\eta$ parameter in the adopted mass loss law, the Reimers (1975) law, during the RG and AGB phase of the evolution (cf. Iben and Renzini, 1983).

### 1.3 *Mass of the precursors of planetary nebulae*

Considering that the most massive precursors should likely originate the most massive nebulae, we conclude from the previous analysis that a range of masses of the progenitors of PNe of 0.8 to about 3.5 $M_\odot$ is indicated. This is less than the interval of 0.7 - 0.8 to 8 - 9 $M_\odot$ predicted from the theory of stellar evolution, also if the upper limit is lowered to about 6 $M_\odot$ because of the overshooting process (cf. Section 1). On the other hand, we cannot expect to easily detect the most massive CSPNs because their evolution in the PN phase is very rapid. Moreover the analysis in Sections 1.1 and 1.2 refers to samples of relatively better studied and therefore brighter objects, while the most massive central stars are expected to be associated with the presently fainter objects (Renzini, 1981).

We then consider the found interval of masses of 0.8 to about 3.5 $M_\odot$ as representative of the initial masses in the MS of the progenitors of the most "common" planetary nebulae.

Probably the best representation of the most "frequent" planetary nebula is an object a bit more massive than one solar mass, while in the MS, which comes to the PN stage with a residual mass of the central star of 0.6 $M_\odot$.

Considerations similar to that made in Sections 1.1 to 1.3 have been made by the author in a note for a conference on "Chemical and Dynamical Evolution of Galaxies" (Elba, September 1989). I have considered useful to repeat them here, for a matter of opportunity, because they are also relevant to the present paper.

## 2.   *EVOLUTION FROM THE MAIN SEQUENCE AND ASSOCIATED WINDS*

Because of the conclusions of Section 1.3, from now on we may refer to the evolution from the MS of 1.1 $M_\odot$ (e.g. Sweigart and Gross), followed by the evolution of a star of 0.598 $M_\odot$ in the PN phase (Schönberner, 1979; 1981) as representative of the evolutionary behaviour of a large fraction of "common" CSPNs. And we will consider the evolution from the MS of a 5.0 $M_\odot$ (e.g. Becker, 1981) followed by the evolution of a 0.8 $M_\odot$ (Paczynski, (1971) as representative of the high mass tail of the mass distribution of CSPNs.

Looking at the mentioned evolutionary tracks, we have the red giant branch (RG), terminated by the helium flash with the subsequent horizontal branch helium burning phase followed by the double shell asymtotic giant branch (AGB), in its early AGB and thermal pulsed AGB portions. During these phases the so called regular RG wind first occurs, with the star loosing mass at the empirically determined Reimers (1975) rate of $\dot{M}_R = - \eta \ 4 \cdot 10^{-13}$ LR/M (solar units, $yr^{-1}$) with low velocity (5–10 km s$^{-1}$) untill the star reaches the thermal pulsed AGB phase where a more conspicuous wind ($\dot{M} \cong 100 \ \dot{M}_R$) takes place, called superwind (cf. Iben, Renzini, 1983). This stronger wind is simply postulated in order to account at least for the well observed mass in the PN K648 (see Section 1.1). The mechanisms responsible for these winds, also in the case of the regular RG wind, are still matter of discussion (cf. Drinkwater and Wood, 1985; Sreenivasan and Wilson, 1985; Holzer, 1987; Fadeyev, 1988).

After the star gets rid of almost all its hydrogen rich envelope, it starts evolving to the left of the HR diagram at approximately constant luminosity. When its surface temperature becomes high enough to excite the previously ejected material, the object becomes finally visible as a regular planetary nebula. At this point we observe soon a fast wind, which remains present during the subsequent phases up to lower luminosity and higher gravity of the central stars when it starts to disappear.

294



## 3.    THE FAST WIND

The properties of these winds in CSPNs have been reviewed by M.
Perinotto (1989). We summarize here some relevant aspects and report
about new work which has been made meanwhile. As for the methods to
detect and study stellar winds, those based on the continuum free-free
or free-bound radiation, either in the radio, in the infrared or in
the optical domain, have not yet been applied to CSPNs. That's
because of the faintness of these objects and of the contamination by
the light of the nebula.

P Cygni profiles of hydrogen or helium lines have been examined
in one or two stars. Almost all the available information rests on P
Cygni profiles of lines of heavy ions observed in the UV range with
the IUE satellite ($\lambda$ 1200 - 3200 A), either at low (6 A) or high (0.15
A) spectral resolution. The lines in which P Cygni profiles have been
seen in CSPNs are essentially the same lines which display the P Cygni
phenomenon in hot stars of population I, i.e. the resonance lines of
NV $\lambda$ 1238.82, 1242.82; CIV 1548.20, 1550.77; Si IV 1393.73, 1402.73,
and the subordinate lines of OIV $\lambda$ 1338.60, 1342.98, 1343.51; OV
1371.29; NIV 1718.15; CIII 2297.60 and He II $\lambda$ 1640.5 A.

It is interesting to notice that in addition to the main P Cygni
profiles, sometimes narrow features are present, as in the case of CIV
in NGC 6543, where a very outstanding second P Cygni system complete
of absorption and emission parts in both components of the doublet is
observed (cf. Fig. 4 of Perinotto et al., 1989). This system clearly
forms in the nebula, the velocity displacement being consistent with
the expansion velocity of the nebula. The absorption parts of this
narrow system as well as other similar narrow absorption features
occurring in other lines in this star and in BD+30 3639 have been used
to get  information on the chemical abundances of the corresponding
ions in the nebula (Pwa et al., 1984;  Pwa et al., 1986). Other
narrow components have been noticed across the fast wind of A78 by
Kaler et al. (1988). Their interpretation is more involved. In
particular they suggest that one of the narrow absorption feature
corresponds to a density enhancement due to a rebound shock of the
type described by Kahn (1983) and Okorokov et al. (1985).

### 3.1    Statistics of the fast wind

The statistics of the fast wind has been discussed by Perinotto
(1983), by Cerruti-Sola and Perinotto (1985), by Pauldrach et al.
(1988) and by Patriarchi and Perinotto (1989). Cerruti-Sola and
Perinotto from the study of 60 CSPNs observed with IUE at low
resolution, found that in 18 stars the continuum spectrum is not
measurable or is very faint and not clearly of stellar origin. In
these objects there is, therefore, no information on the existence of
P Cygni profiles in the central stars. Out of the remaining stars,
they see in 16 objects no evidence of P Cygni profiles, in 22 evidence
of P Cygni profiles while 4 stars have doubtful P Cygni profiles.
Therefore about half of the observed CSPNs do show a wind, while the
other half do not. In a more recent work, Patriarchi and Perinotto
(1989), studying some 110 CSPNs, reach about the same conclusion on

the frequency of occurence of the fast winds in these stars.

From a study of the occurrence of the phenomenon in the HR diagram, Cerruti-Sola and Perinotto (1985) found that the presence of the wind is strongly influenced by the stellar gravity in the sense that the CSPNs with a gravity smaller than log g ~ 5.2 (cgs units) do practically always show a wind, while in CSPNs with higher gravity the presence of  a wind becomes less and less frequent.

In a subsequent study, Pauldrach et al. (1988) have shown that the presence of the fast wind is not only a matter of stellar gravity, but also of stellar luminosity, an aspect which was not evident in the analysis of Cerruti-Sola and Perinotto.  As far as the star departs from the Eddington luminosity, the wind becomes less frequent (see Fig. 1 of Pauldrach et al., 1988).

These dependences of the fast winds in CSPNs from the stellar gravity and luminosity is in agreement with the expectation from the line-radiation driven theory (see later on).

*3.2    Terminal velocity of the fast wind*

The edge velocities of the P Cygni profiles of the resonance lines of CIV and NV give the most reliable values of the terminal velocities of these winds.  Rather good measurements are now available in about 27 CSPNs, with values of $V_\infty$ ranging from 600 to 3900 km s$^{-1}$ (cf. Patriarchi and Perinotto, 1989).

By plotting $V_\infty$ versus the effective temperature of the central stars,  Heap (1986) has demonstrated the existence of a linear relationship between the two quantities.  This is clearly interpreted in terms of the line radiation driven theory because most of the observed CSPNs can be viewed as stars of a similar mass around 0.6 $M_\odot$ evolving to the left at constant luminosity in the HR diagram (cf. Section 1.3).  This implies a reduction of the stellar radius and therefore an increase of the escape velocity.  Since the theory of line radiation driven winds actually predicts a proportionality between the stellar escape velocity and the terminal velocity of the wind, the previously found linear relationship is in nice agreement with the theory and provides indeed a value for the proportionality coefficient which is close to the one predicted by the theory.

*3.3    Determination of the mass loss rate of the fast wind*

The methods to derive the mass loss rate from the P Cygni profiles can be either "empirical" or "theoretical".  The first require to compare some observables, as the first, plus in case, further moments of the flux distribution or the full shape of P Cygni profile, with the same quantities calculated under some assumptions.

These calculated quantities come from the solution of the transfer equation for the line radiation across the wind, without hypothesis on the mechanism which originates the wind itself.  Among these "empirical" methods, we recall:

1)    The escape probability method developed by Castor (1970) for
      pure   scattering   lines   under   the   well   known   Sobolev

approximation and used to produce the well known Castor and Lamers (1979) atlas of profiles;

2)  the comoving frame method studied by Lucy (1971) under the Sobolev approximation and by Mihalas et al. (1975) free from this approximation. The latter method has been worked out in more detail by Hamann (1980; 1981);

3)  the method of the moments of the flux distribution, developed by Castor, Lutz and Seaton (1981) just for use with the low resolution IUE data. The method has been further studied by Surdej (1982; 1983) and Hutsemékers and Surdej (1987);

4)  the Sei method (Sobolev plus exact integration of the transfer equation), which considers also the effects of collisions in the source function of the line and the presence of some turbolence across the wind. This method has been worked out by Lamers et al. (1987).

The "theoretical" methods also require the comparison of some observables, usually the full P Cygni profile, with the corresponding calculated quantities. In this case, however, the calculated quantities imply a theory for the production of the wind. The theory used so far for CSPNs is the line radiation driven theory first suggested by Lucy and Solomon (1970) and developed by Castor, Abbott and Klein (1975) with various subsequent improvements by Abbott (1982), Abbott and Lucy (1985), Friend and Abbott (1986), Pauldrach et al. (1986), Pauldrach (1987) and Puls (1987).

The first application of these "theoretical" methods to a CSPN was made by Lucy and Perinotto (1987) with a subsequent general application not addressed to specific CSPNs by Pauldrach et al. (1988).

All the present determinations of the mass loss rates of the fast winds in CSPNs have been made with the "empirical" methods, except for the mentioned study of Lucy and Perinotto (1987) of NGC 6543.

Values obtained by different authors in the same objects are quite different to each other. As an example, we report that for NGC 6543 the following mass loss rates have been derived, using practically the same UV lines observed with IUE: $\log \dot{M} = -7.05$ (Castor et al. 1981); $-6.15$ (Heap, 1981); $-6.49$ (Bianchi et al. (1986); $-5.77$ (Hutsemékers and Surdej, 1987); $-7.40$ (Perinotto et al., 1989) (PCL). Also the luminosity adopted by the different authors differs significantly, the dispersion being somewhat more than one order of magnitude.

These determinations are not, however, of similar accuracy, as we will prove. The obtained value of $\dot{M}$ is quite sensitive to the adopted $T_{eff}$ and luminosity of the central star (see e.g. PCL). In the work of PCL, a great effort has been devoted to the best determination of these quantities using a very extended set of selected observations plus non-LTE atmospheric models computed for this purpose (Patriarchi et al. 1989). On the other hand, in the "theoretical" determination of $\dot{M}$ by Lucy and Perinotto (1987), the values of $T_{eff}$ and L of the central star of NGC 6543 follow as best fit parameters (together with $\dot{M}$) to match the observed spectrum. The best values of $T_{eff}$ and L by Lucy and Perinotto are much closer to those of PCL than to those of the other authors. Indeed with the values of L, $T_{eff}$ of the other authors it would have been impossible to reproduce, even approximately, the observed P Cygni line spectrum of NGC 6543. Clearly the two determinations of the fundamental parameters of NGC 6543 by PCL and by Lucy and Perinotto are totally independent.

Since the same procedure used for the determination of $T_{eff}$, L in NGC 6543, has been used also for the stars studied by Cerruti-Sola and Perinotto (1989), we believe that the $\dot{M}$ from this last work and from PCL are the best presently obtained. Anyhow the accuracy remains low, being of about a factor of 3, for the intrinsic difficulty of the analysis. The values of $\dot{M}$ for the seven best studied objects are presented, together with other quantities, in Table 1.

4. *PROPERTIES OF FAST WINDS AND RELATIONSHIP WITH THE NEBULA AND THE CENTRAL STAR*

We summarize in Table 1 some properties of the fast winds as well as of the central stars and of their optical nebulae of the seven best studied objects.

*TABLE 1. Properties of winds in CSPNs*

|  | NGC 1535 | 6210 | 6543 | 6826 | 7009 | IC 418 | IC 4593 |
|---|---|---|---|---|---|---|---|
| Sp | sd O3 | sd O3 | O3f-WR | O4f | cont. | O7 F | O7f |
| $T_{eff}$ | 77 | 90 | 60 | 45 | 88 | 37 | 35 |
| $\dot{M}$ | 1.4-9 | 2.2-9 | 4.0-8 | 6.4-8 | 2.8-9 | 6.3-9 | 4.2-8 |
| $V_\infty$ | 1900 | 2180 | 1900 | 1750 | 2770 | 940 | 1000 |
| $V_N^{exp}$ | 20 | 21 | 20 | 11: | 21 | < 6 | 13: |
| $t_{age}^{exp}$ | 4.7+3 | 3.5+3 | 1.9+3 | 5.4+3 | 6.9+3 | 8.5+3 | 8.0+3 |
| $\dfrac{\dot{M} V_\infty}{(L/c)}$ | 0.01 | 0.05 | 0.6 | 0.7 | 0.03 | 0.02 | 0.2 |

*TABLE 1.  (continued)*

| | NGC 1535 | 6210 | 6543 | 6826 | 7009 | IC 418 | IC 4593 |
|---|---|---|---|---|---|---|---|
| $\dot{M} V_\infty t_{age} =$ | | | | | | | |
| $= V_N^{exp} M_N^{eq}$ | 6.-4 | 8.-4 | 7.-3 | 6.-2 | 3.-3 | 8.-3 | 3.-2 |
| $\dfrac{\dot{M} V_\infty^2 t_{age}}{M_N V_N^2}$ | 0.3 | 0.4 | 3. | 44. | 2. | 7. | 10. |

   The first row gives the optical spectral of the central stars. The subsequent rows give $T_{eff}$ in $10^3$ K, the mass loss rates of the fast winds in $M_\odot$ yr$^{-1}$, their terminal velocity in km s$^{-1}$, the expansion velocity of the optical nebula in km s$^{-1}$, their expansion age in years, the ratio between the flux of the momentum of the fast wind to the same quantity available in the radiation field of the central star (in the single scattering limit) and two more quantities which are now described.

   The first represents the equivalent mass (in solar masses) of the optical nebula, in the case that its momentum would have been provided by and only by the fast wind.  Indeed this equivalent mass is correct as far as the $\dot{M}$, $V_\infty$, $V_N$ have remained constant during the evolution of the object at constant luminousity  and as for as $t_{age}$ represents the true life-time of the fast wind. That $\dot{M}$ remains approximately constant during the evolution of the central star from the onset of the fast wind up to the present epoch, can be considered to be about correct from the predictions of Pauldrach et al.(1988). The terminal velocity, measured at the present epoch, should be instead an upper limit to the one valid for the previous stages because of its inverse proportionality to $T_{eff}$ discussed under Section 3.2.

   On the other hand, the age of the wind, taken equal to the expansion age of the optical nebula, is also an upper limit to the true value, because the onset of the fast wind only occurs (if it is produced by the line radiation mechanism) when the central star becomes bright enough to excite the nebula.  The optical nebula instead expands from quite before, i.e. from the main ejection phase at the end of the AGB evolution.

   A part from possible variations of the expansion velocity of the nebula, we conclude that the equivalent nebular masses quoted in Table 1 are probably upper limits to the true values.  On the other hand, the quoted values are quite smaller than the actual nebular mass of the considered planetary nebulae, which are likely to be close to 0.2 $M_\odot$.  We will come in a minute to the meaning of this result.

   The last row of Table 1 compares the kinetic energy available in the wind during its life-time with the present kinetic energy of the expanding nebula, assumed to have a mass of 0.2 $M_\odot$ in all the considered objects.

## 4.1  *The line-radiation driven wind theory*

The comparison of the momentum of the fast wind with the one available in the radiation field of the central star (see the previous Section) gives a further confirmation that the line radiation driven theory is adequate to explain these winds.

## 4.2  *The multiple-winds interacting theory*

The existence of the optical nebula is viewed as the result of the excitation by the central star of the hydrogen rich stellar envelope ejected mainly during the last phases of the AGB evolution or as the result of the interaction between the previously ejected material and the fast wind, the so called two-wind interacting theory developed by Kwok and collaborators (Kwok et al., 1978;  Volk and Kwok, 1985; Kwok, 1987;  Kahn, 1989).  According to the  latter theory, the optical nebula appears like it appears, because of the compression due to the outer shock resulting from the interaction of the fast wind with the previously ejected slow wind. Instead of two-winds, one can speak of a three-winds (therefore multiple-winds theory) if a distinction is made between the slow "regular" wind ejected during the RG and early-AGB phases and the superwind ejected in the thermally pulsed AGB phase.

The two-winds interacting theory has been criticized by Iben (1984) on account of the fact that the observed momentum in the fast wind seems to be inadequate to account for the momentum of the planetary nebula shell.  This is entirely confirmed by our results in row 8 of Table 1.

On the other hand Kwok (cf. Kwok, 1987) argues that if the outer shock is adiabatic, then the wind and shell momentum are not expected to be equal since the shell is driven by thermal pressure of the shocked region. The relevant quantity in this case would be the kinetic energy.  From the last row of Table 1, the shell and the wind kinetic energies appear comparable, within the obvious uncertainties of the data.

This would confirm the applicability of the two winds interacting theory to the observed planetary nebulae, but only in the adiabatic case.

Where the fast wind plays anyhow a role is into prevent the previously ejected material from moving again inwards due the inner pressure of the nebula shell (e.g. Kahn, 1983).

A further confirmation of the role of the fast wind into causing observable effects in the nebular shell comes from the suggestion of Balik and Prenston (1987) and Icke et al. (1989) that the elliptical and the so called "butterfly" planetary nebulae, including the density condensations or knots that are particularly outstanding at the extremes of the major axis of some elliptical PNe, own their shape and existence (in case of the knots) to non-spherical or bipolar shocks produced by the interaction of a fast spherical wind  with an

300

aspherical or quasitoroidal previously ejected envelope.

The suggested mechanism is quite interesting. One must however explain the asphericity or the equatorial density enhancement of the ejected envelope. This aspect will be discussed later on, in connection with the possible role of the rotation in the CSPN.

### 4.3  *Effects on the evolution of the central star*

The evolution of the central star at constant luminosity proceeds with the consumption of the residual stellar envelope by hydrogen burning in the stellar shell and by mass loss due to the fast wind. The first source is active when the star leaves the AGB while the onset of the fast wind occurs when the central star becomes hot enough to activate a wind according to the line radiation mechanism. Before this fast wind occurs, during the so called proto-planetary nebula phase, again another wind may be present (see Trams et al., 1989). We do not discuss here this phase, because the information is still poor. We refer therefore to the subsequent phase of regular planetary nebula. It can be easily calculated (cf. Iben and Renzini, 1983) that for an hydrogen concentration by mass of $X = 0.75$, the two sources of reduction of the mass of the residual hydrogen stellar envelope are equal for the following $(\log (L/L_\odot);\ \log \dot{M})$: (3.8, -7);  (2.8, -8); (1.8, -9). These values are close to the range of the observed values in Table 1. Therefore the effect of speeding up the evolution due to the presence of the fast wind is likely to be quite significant in many objects. For instance, for a star with $\log (L/L_\odot) = 3.8$ and a residual mass in the stellar envelope of $2 \cdot 10^{-4}\ M_\odot$ (cf. Iben and Renzini, 1983) at the onset of a fast wind with $\log \dot{M} = -7\ M_\odot\ yr^{-1}$, the further evolution at constant luminosity lasts 2500 yr instead than 5000 yr.

### 5.  *EFFECT OF ROTATION IN CSPNs*

We examine now the possible role of the rotation in helping to enhance the mass loss from CSPNs during their life-time.

### 5.1  *Single stars*

Since almost no specific work has been done, to my knowledge, in the field, I feel justified to explore the matter under the very simple minded hypothesis of rigid body rotation. It is known that in the case of the Sun, the solar seismology indicates that the angular velocity in the solar interior is approximately constant except in a narrow zone near to the center and a moderately extended zone in the outer convective zone. In case of the CSPNs, where the size of the outer convective zone is very much larger than that of the core, the rigid body hypothesis may easily break down. Anyhow it may be instructive to look first at this hypothesis.

Considering the critical velocity as the one at which the centrifugal force equates the gravitational force, the ratio between the critical velocity and the actual velocity at the equator is

$$\alpha = \left(\frac{v^{crit}}{v}\right)_{eq} \quad \propto \quad L^{-1} M^{3/2} R^{1/2} (1-\Gamma)^{1/2} \qquad (1)$$

$$\text{where} \quad L = \frac{2}{5} MR^2 \omega \qquad (2)$$

is the angular momentum of an homogeneous sphere and

$$\Gamma = 2.45 \; 10^{-5} \left(\frac{L}{M}\right) \quad \text{(solar units)}$$

is the usual $\Gamma$ factor.

During the evolution from the MS to the white dwarf stage, the stellar mass is practically constant with respect to the variation of the stellar radius. In eq. (2) we may then assume $R^2 \omega \approx$ const, i.e. $vR \approx$ const. Starting from the MS with $\alpha \gg 1$, we see from eq. (1) that the variation of $\alpha$ depends essentially on the variation of R, the factor $(1-\Gamma)^{1/2}$ being of little importance in the relevant cases we are going to consider. These are the progenitor stars with 1.1 $M_\odot$ and 5.0 $M_\odot$ in the MS (see Section 2).

We will consider the theoretical tracks of 1.1 $M_\odot$ by Sweigart and Gross (1978) followed by the track of 0.598 $M_\odot$ by Schönberner (1979; 1981) and the track of 5.0 $M_\odot$ by Becker (1981) followed by the one of 0.8 $M_\odot$ by Paczynski (1971).

In the case of 1.1 $M_\odot$ initial mass on the MS, the corresponding spectral type being around G0, we may assume a rotational velocity at the equator of 5.0 km s$^{-1}$. The initial value of $\alpha$ is then 84. Later on, R increases and so does $\alpha$ up to 1037 at the tip of the RG phase. The radius then falls. Thus at the beginning of the HB phase, $\alpha$ equals 60. Then $\alpha$ increases again up to the tip of the AGB phase. Then it starts to decrease reaching 24 at the maximum $T_{eff}$ of log $T_{eff}$ = 5.18. We mention that a widening of the metallic absorption lines above the instrumental profiles has been observed in various CSPNs by Méndez et al. (1988) with corresponding formally interpreted v sin i ranging between 50 and 100 km s$^{-1}$. Althought the authors "fell more inclined" to interpret this widening as due to turbolence, we underline that such "rotational" velocities would be consistent with the values we obtain for $v_{eq}$ in our schematic analysis of the 1.1 $M_\odot$ initial mass model, when the CSPN is close at his higher temperature phase of its evolution. Then $\alpha$ continues to decrease reaching 1.8 at log $T_{eff}$ = 4.96 and log $L/L_\odot$ = 1.65.

From the above, we deduce that in the rather naive hypothesis of rigid body rotation, in the very late phases of evolution of the considered CSPN and the more later on during the white dwarf phase, the axial rotation would play a role in enhancing the mass loss from the equator.

In the case of the 5.0 $M_\odot$ initial mass, we may start with an initial equatorial velocity of 215 km s$^{-1}$ (Bernacca and Perinotto,

1974). The initial value of $\alpha$ is now 30. The trend is about as before, with values of $\alpha$ quite smaller than for the previous slower rotator, but always greater than 1.0, decreasing from 8 at the maximum $T_{eff}$ of log $T_{eff} = 5.44$ up to 5 in the white dwarf stage. It looks like in this more massive star it would remain difficult for the rotation to become an important source of additional mass loss even in the latest phases of evolution (in the rigid body assumption).

We must also mention that, according to Heap (1983), if the angular momentum is conserved, and the core spins-up, then the outer edge of the core (i.e. the external parts of the CSPN) would rotate at 10 km s$^{-1}$ if the initial mass on the MS is of 1.5 $M_\odot$ and at 800 km s$^{-1}$ if the initial mass is 3.5 $M_\odot$. This would imply that rotationally-enhanced ejection might play a role in CSPNs with massive progenitors. However Heap (1983) reports the above very schematically without giving details of the analysis. It is clearly important to investigate further the matter.

Willson and Bowen (1988) have actually discussed the role of the rotation in red giants and AGB stars under the hypotheses of conservation of angular momentum for unit mass and of rotational decoupling between the degenerate core of small radius and the convective envelope, each of the two assumed to rotate as a rigid body. They suggest that in fact the rotation may play an important role in modulating or even driving the winds from these stars as they make the transition from the AGB to become CSPNs.

We add that a fraction of white dwarfs do have a strong magnetic field (cf. Sion, this Conference). Considering a magnetic field of $10^5$ gauss in a white dwarf of 0.6 $M_\odot$ and scaling with the square of the radius, we infer a magnetic field of 4000 gauss for the corresponding CSPN while at its maximum $T_{eff}$. This magnetic field appears strong enough for the rotating magnetic (plus radiation driven) wind mechanism by Poe et al. (1989) (see also Cassinelli, this Conference and Friend, this Conference) to be active in these stars.

### 5.2  Double sytems

The scenario taking place in case of binaries evolving through the planetary nebula stage is extremely various, depending on the characteristics of the system and the consequent filling up of the Roche lobe from one or both components at certain phases of the evolution of the system. Iben and Tutukov (1989) list 20 different cases.

Very simply, we just mention that when the primary fills its Roche lobe, a common envelope (non conservative case) is formed with an expected density enhancement in the orbital plane (which may or may not coincide with the equatorial plane of the primary). This enhancement is due, we underline, more to the orbital motion than to the axial rotation of the primary. In any case an axial-symmetric density enhancement is expected. Then a spherical fast wind sets on in the primary and may come up with a bipolar morphology, because of its interaction with the previous density enhancement.

This would imply that every bipolar planetary nebula is associated with a binary central star. The large number of known bipolar PNe is indeed consistent with the large number of expected binaries among CSPNs as among any other kind of stars. However the presently detected double stystems in CSPNs are still too few (because of obvious problems of detection) (see Bond, 1989; Méndez, 1989) to confirm the above relationship, althought the small numbers available are in favour of that relationship (Zuckerman and Aller, 1986).

Finally we mention the possibility that some CSPNs behave in a way similar to the RS CVn stars. That means a double star, with a red giant corotating with the orbital motion. The ingredients: rotation, magnetic fields, convective envelope (and therefore turbolence causing waves) would be adequate to have mass ejection with the rotation playing an important role.


*REFERENCES*

Abbott, D.C.: 1982, Ap.J., <u>259</u>, 282.
Abbott, D.C. and Lucy, L.B.: 1985, Ap.J., <u>288</u>, 679.
Balick, B. and Preston, H.L.: 1987, Astron.J., <u>94</u>, 1641.
Bianchi, L., Cerrato, S. and Grewing, M.: 1986, Astron. Astrophys., <u>169</u>, 227.
Becker, S.A.: 1981, Ap.J. Suppl., <u>45</u>, 705.
Bernacca, P.L. and Perinotto, M.: 1974, Astron. Astrophys., <u>33</u>, 443.
Bond, H.E.: 1989, in IAU Symp. n. 131 "Planetary Nebulae", ed. S. Torres-Peimbert, p. 251.
Castor, J.I.: 1970, M.N.R.A.S., <u>149</u>, 111.
Castor, J.I., Abbott, D.C. and Klein, R.I.: 1975, Ap.J., <u>195</u>, 157.
Castor, J.I. and Lamers, H.J.G.L.M.: 1979, Ap.J. Suppl., <u>39</u>, 481.
Castor, J.I., Lutz, J.H. and Seaton, M.J.: 1981, M.N.R.A.S., <u>194</u>, 547.
Cerruti-Sola, M. and Perinotto, M.: 1985, Ap.J., <u>291</u>, 237.
Cerruti-Sola, M. and Perinotto, M.: 1989, Ap.J., <u>345</u>, 339.
Drinkwater, M.J. and Wood, P.R.: 1985, in "Mass loss from Red giants", eds. M. Morris and B. Zuckerman, p. 257.
Fadeyev, Y.A.: 1988, in "Atmospheric Diagnostics of stellar evolution: Chemical Peculiarity, Mass loss and Explosion", Lecture Notes in Physics, ed.K.Nomoto, p.174.
Friend, D.B. and Abbott, D.C.: 1986, Ap.J., <u>311</u>, 701.
Hamann, W.-R.: 1980, Astron. Astrophys., <u>84</u>, 342.
Hamann, W.-R.: 1981, Astron. Astrophys., <u>93</u>, 353.
Heap, S.R.: 1981, in "The Universe at Ultraviolet Wavelengths" (NASA CP-2171), p. 415.
Heap, S.R.: 1983, in IAU Symp. n. 103, "Planetary Nebulae", ed. D.R. Flower, p. 502.
Heap, S.R.: 1986, in "New Insights in Astrophysics"(ESA SP-263,p.291).
Heap, S.R. and Augensen, H.J.: 1987, Ap.J., <u>313</u>, 268.
Holzer, T.E.: 1987, in IAU Symp. n. 122, "Circumstellar Matter", eds. I. Appenzeller and C. Jordan, p. 289.
Hutsemékers, D. and Surdej, J.: 1987, Astron. Astrophys., <u>173</u>, 101.
Iben, I. Jr.: 1984, Ap.J., <u>277</u>, 333.
Iben, I. Jr. and Renzini, A.: 1983, Ann.Rev.Astron.Astrophys., <u>21</u>,271.

Iben, I. Jr. and Tutukov, A.V.:  1989, in IAU Symp. n. 131, "Planetary
     Nebulae", ed. S.Torres-Peimbert, p. 505.
Icke, V., Preston, H.L. and Balick, B.:  1989, Astron.J., 97, 462.
Jacoby, G.H. and Kaler, J.B.:  1989, Ap.J., in press.
Kahn, F.D.:  1983, in IAU Symp. n. 103,  "Planetary Nebulae",  ed.
     D.R. Flower, p. 305.
Kahn, F.D.:  1989, in IAU Symp. n. 131, "Planetary Nebulae",  ed. S.
     Torres-Peimbert, p. 411.
Kaler, J.B., Feibelman, W.A. and Henrichs, H.F.: 1988, Ap.J., 324,
     528.
Kinman, T.D., Feast,M.W. and Lasker, B.M.: 1988, The Astron. J. 95,
     804.
Kwok, S.:  1987, Physics Reports, 156, 112.
Kwok, S., Purton, C.R. and FitzGerald, M.P.: 1978, Ap.J.Lett. 219,
     L125.
Lamers, H.J.G.L.M., Cerruti-Sola, M. and Perinotto, M.:  1987, Ap.J.,
     314, 726.
Lucy, L.B.:  1971, Ap.J., 163, 95.
Lucy, L.B. and Perinotto, M.: 1987, Astron. Astrophys., 188, 125.
Lucy, L.B. and Solomon, P.M.:  1970, Ap.J., 159, 879.
Mallik, D.C.V. and Peimbert, M.: 1988, Rev. Mex. Astron. Astrof., 16,
     111.
Meatheringham, S.J., Dopita, M.A. and Morgan, D.H.: 1988, Ap.J. 329,
     166.
Méndez, R.H.:  1989, in IAU Symp. N. 131, "Planetary Nebulae", ed. S.
     Torres-Peimbert, p. 261.
Méndez, R.H., Kudritzki, R.P., Herrero, A., Husfeld, D. and Groth,
     H.G.:  1988, Astron. Astrophys., 190, 113.
Mihalas, D., Kunasz, P.B. and Hummer, D.G.:  1975, Ap.J., 202, 465.
Okorokov, V.A., Shustov, B.M., Tutukov, A.V. and Yorke, H.W.:  1985,
     Astron. Astrophys., 142, 441.
Paczynski, B.:  1971, Acta Astron., 21, 417.
Patriarchi, P. and Perinotto, M.: 1989, in preparation.
Patriarchi,P., Perinotto, M. and Cerruti-Sola,M.: 1989, Ap.J., 345,
     327.
Pauldrach, A.: 1987, Astron. Astrophys., 183, 295.
Pauldrach, A., Puls, J. and Kudritzki, R.H.:  1986, Astron.
     Astrophys., 164, 86.
Pauldrach, A., Puls, J., Kudritzki, R.H., Méndez, R.H. and Heap, S.R.:
     1988, Astron. Astrophys., 207, 123.
Perinotto, M.:  1983, in IAU Symp. n. 103, "Planetary Nebulae", ed.
     D.R.Flower, p. 323.
Perinotto, M.:  1989, in IAU Symp. n. 131, "Planetary Nebulae", ed.
     S.Torres-Peimbert, p. 293.
Perinotto, M., Cerruti-Sola, M. and Lamers, H.J.G.L.M.:  1989, Ap.J.,
     337, 382 (PCL).
Poe, C.H., Friend, D.B. and Cassinelli, J.P.: 1989, Ap.J., 337, 888.
Pottasch, S.R.:  1984, in "Planetary Nebulae. A Study of Late Stages
     of stellar Evolution", D. Reidel  Publ. Comp.
Puls, J.: 1987, Astron. Astrophys., 184, 227.
Pwa, T.H., Mo, J.E. and Pottasch, S.R.:  1984, Astron. Astrophys.,
     139, L1.

Pwa, T.H., Mo, J.E. and Pottasch, S.R.:   1986, Astron. Astrophys., 164, 184.

Reimers, D., 1975, Mem. Soc. R. Sci. Liège, 6$^e$ Ser. 8, 369.

Renzini, A.: 1981, in "Physical Processes in Red Giants", eds. I. Iben Jr., A. Renzini, p. 431.

Schönberner, D.:   1979, Astron. Astrophys., 79, 108.

Schönberner, D.:   1981, Astron. Astrophys., 103, 119.

Surdej, J.: 1982, Astrophys. Space Sci., 88, 31.

Surdej, J.: 1983, Astron. Astrophys., 127, 304.

Sweigart, A.V. and Gross, P.G.:   1978, Ap.J. Suppl., 36, 405.

Sreenivasan, S.R. and Wilson, W.J.F.:   1985, in "Mass loss from Red giants", eds. M.Morris and B.Zuckerman, p. 261.

Trams, N.R., Waters, L.B.F.M., Waelkens, C., Lamers, H.J.G.L.M. and van der Veen, W.E.C.J.:   1989, Astron. Astrophys., 218, L1.

Volk, K. and Kwok, S.:   1985, Astron. Astrophys., 153, 79.


Willson, L.A. and Bowen, G.H.: 1988, in "Polarized Radiation of Circumstellar Origin", Vatican Observatory Conference, eds. G.W.Coyne et al., p. 485.

Wood, P.R., Meatheringham, S.J., Dopita, M.A. and Morgan, D.H.: 1987, Ap.J., 320, 178.

Zuckerman, B. and Aller, L.H.:   1986, Ap.J., 301, 772.

# WHITE DWARF MASS LOSS, ROTATION, INDIVIDUAL MASSES AND THE IDENTIFICATION OF THE WHITE DWARF REMNANTS OF UPPER MAIN SEQUENCE STARS

EDWARD M. SION
Department of Astronomy and Astrophysics
Villanova University
Villanova, Pennsylvania 19085
U.S.A.

ABSTRACT.    This review focuses upon current observational evidence of white dwarf mass loss, rotation rates and the identification of evolutionary (progenitor) links between white dwarfs and upper main sequence stars. The latter objective rests upon the determination of white dwarf masses through a number of methods, which are discussed. The conclusions are as follows: (1) There is some evidence in the far ultraviolet to suggest that hot DA white dwarfs lose mass via a weak wind. This mass outflow is manifested by the presence of sharp, shortward-shifted absorption lines in the IUE high resolution spectra of a number of hot DA white dwarfs (Bruhweiler and Kondo 1983). There is at present very weak direct evidence of actual mass loss by hot non-DA white dwarfs, except for strong wind outflow from the lower gravity pulsating central star of K1-16. The DO stars, *KPD* 0005+5106 and *PG* 1034+001, may be important exceptions; (2) The general conclusion at present concerning white dwarf rotation rates is that they are slow rotators with established upper limit $v \sin i < 65$ km/s for DA stars and $v \sin i < 135$ km/s for DB stars. Other evidence, presented herein, supports this picture of slow rotation as a general rule for white dwarfs; (3) The identification of quite massive planetary nebula nuclei(PNN) which are the descendants of young massive main sequence progenitors, is demonstrable by comparison with evolutionary tracks on the H-R diagram. These objects are found among the very hot but low luminosity PNN because they evolve very quickly due to their high mass. At this time the hottest known non-DA white dwarfs (the *PG* 1159 stars) appear by this method not to have higher than average white dwarf mass (i.e. $> 0.6$ M$_\odot$) and therefore no link is indicated between these objects and upper main sequence progenitors; (4) At the present time it is not clear whether massive white dwarf progenitors would be expected to leave DA or non-DA remnants. Here, theory and observation provide conflicting indications; (5) The magnetic degenerates tend to have higher than average white dwarf mass and the majority are therefore expected to be the progeny of young, peculiar A and B stars on the upper main sequence. This conclusion is supported by a number of independent lines of evidence; (6) Higher-than-average mass white dwarfs (and therefore the remnants of progenitors more massive and younger than the typical and most populous old disk stars in the solar neighborhood), have been identified through kinematical properties, through



L. A. Willson and R. Stalio (eds.), Angular Momentum and Mass Loss for Hot Stars, 307–329.

gravitational redshift determinations in wide and common proper motion binaries, through kinematical membership in young moving groups, through the gravitational redshifts of the white dwarf members of the Hyades cluster, and through the spectroscopic identification and analysis of white dwarf members of very young open clusters whose turnoff masses to the red giant branch exceed 5 $M_\odot$; A number of evolutionary implications and future prospects are discussed.

## 1. INTRODUCTION

The topic of white dwarfs may seem rather far afield from the general scientific content (and low degeneracy parameter!) of this conference: mass loss and angular momentum loss in hot stars. However I have tried (and it is to be hoped I succeeded) in confining my discussion of the hot stars' degenerate progeny in a way consistent with the overall scientific theme of this NATO Advanced Research Workshop: to summarize our current state of knowledge concerning white dwarf mass loss, rotation and angular momentum loss, and to possibly identify progenitor links with the hot stars on the upper main sequence. While our understanding of these topics is far from being complete or even minimally conclusive, it is nonetheless possible to address these topics in a preliminary fashion and to identify areas of progress and future avenues of research. As a necessary starting point, the spectroscopic nomenclature, surface compositions and effective temperature distribution of types of white dwarfs will be briefly presented (sec. 2.1). The current state of knowledge on white dwarf rotation will be discussed in section 2.2 followed by a review of mass loss evidence in section 2.3. In section 3, the focus is on the identification of massive planetary nebula central stars (hereafter PNN), and white dwarfs with higher-than-average mass (i.e., $> 0.6$ $M_\odot$), both of which are expected to be the progeny of parent hot stars on the upper main sequence with initial masses between 2.5 and 8 $M_\odot$. In sections 4 and 5 it is shown how these objects can be isolated by mass, from their positions relative to evolutionary tracks in the H-R diagram (if distances are reliable), from precise gravitational redshift determinations of white dwarfs in wide and common proper motion binaries or in the Hyades cluster, from membership in moving groups, from kinematical similarity with the space motions of upper main sequence stars, and from white dwarf membership in young open clusters with large turnoff masses to the red giant branch. Conclusions are presented in section 6.

## 2. WHITE DWARF SURFACE COMPOSITIONS MASS LOSS AND ROTATION

### 2.1 White Dwarf Surface Compositions and Temperature Distribution

The white dwarfs, due to their high gravities (gravitational diffusive separation), usually show only the lightest principal atmospheric constituent at their surfaces. They divide into two dominant composition sequences, those with hydrogen-rich atmospheres (denoted DA; see below) and those with helium-rich atmospheres (non-DA). Their spectroscopic properties are determined by the complex interplay of a number of physical processes which control and/or modify the flow of elements and hence surface abundances in high gravity atmospheres: convective dredgeup, mixing and dilution, accretion, gravitational and thermal diffusion, radiation pressure (in the hottest white dwarfs), mass loss, residual nuclear burning, and magnetic fields. Most of these processes remain poorly understood. They manifest themselves in the variety of white dwarf surface compositions exhibited spectroscopically by absorption lines in the far ultraviolet and optical wavelength regions. The spectroscopic classification system described by Sion *et al.* (1983) and summarized in Table 1, provides a physical description of what the spectrum actually contains. The system utilizes a temperature index from 0 to 9 defined by $10 \times \theta_{eff}$ (where $\theta_{eff} = 5040/\mathrm{T}_{eff}$), as well as appropriate symbols for magnetic field/polarization (H, P), variability (V) and spectral peculiarities or unclassifiable spectra (X).

TABLE 1
Spectral Classification and Observed Properties of White Dwarfs
Spectroscopic Types of White Dwarfs

| Spectral Type | Characteristics |
| --- | --- |
| DA | Only Balmer lines; no He I or metals present |
| DB | He I lines; no H or metals present |
| DC | Continuous spectrum, no lines deeper than 5% in any part of the electromagnetic spectrum |
| DO | He II strong: He I or H present. |
| DZ | Metal lines only; no H or He features |
| DQ | Carbon features, either atomic or molecular, in any part of the electromagnetic spectrum. |

A brief summary is presented below, of basic physical properties of the major spectroscopic types and the physical processes presently thought to be responsible for their origin and/or surface chemical abundances.

1. <u>DA Stars:</u> DA degenerates comprise 75%-80% of all white dwarfs hotter than $10^4$K, with the remaining 20%-25% referred to as non-DA. As a result of gravitational diffusive separation, which causes the lightest element to appear at the stellar surface, the optical spectra of DA stars generally show only the Balmer lines of H I. They have essentially pure hydrogen outer layers of uncertain thickness, and occur over a very wide range of effective temperature ($\mathrm{T}_{eff} = 6000$ K to 70,000 K$-$90,000 K). Below 5500 K, the Balmer lines are no longer detectable. Their hydrogen may be either primordial (i.e. remaining from earlier evolutionary stages) or accreted from the interstellar medium (cf. Fontaine and Wesemael 1987; Shipman 1989; Sion 1986). Convective mixing due to penetration of the underlying layer by the hydrogen convective zone may cause DA stars to transform into non-DA objects, but only if the outer hydrogen layer is thinner than $10^{-6}$ M$_\odot$ (D'Antona and Mazzitelli 1979).

2. <u>DO Stars:</u> These carbon and helium-rich objects range from the hottest known degenerate stars (log g $\approx$ 7, $\mathrm{T}_{eff} > 10^5$ K) with carbon, oxygen, and other metals present in their photospheres, down to the coolest known DO stars ($\mathrm{T}_{eff} \approx 45,000$ K$-$55,000 K). In all cases, lines of ionized helium dominate their optical spectra. While their true

hydrogen-to-helium ratios are poorly known due to their high surface temperatures, the coolest DO stars are demonstrably hydrogen-poor and several DO stars with $T_{eff}$ 80,000 K show no evidence of hydrogen based on optical data (cf. Wesemael, Green and Liebert 1985). The origin of DO stars from hydrogen-deficient central stars is discussed in section III-1. The DO stars are thought to be the immediate precursors of the nearly pure helium DB degenerates, although this may not be invariably true. Recent abundance analyses of the hottest (*PG* 1159-035) DO stars (Werner *et al.* 1989b) reveal carbon abundances as high as 55% with helium less than 25%. The determination of these surprising abundances tends to strengthen the confirmation by Sion, Liebert and Starrfield (1985) that the partial ionization of carbon and oxygen near the surface, is the pulsational driving mechanism for these stars. The hottest high gravity (log g = 7) object may be *H* 1504+65, a star in which helium (He II (4686)) is not observed, either because the star is so hot that helium is ionized out or because post-AGB mass loss has stripped the remnant star down to a bare carbon-oxygen core (cf. Nousek *et al.* 1986).

3. <u>DB Stars:</u> The DB degenerates have nearly pure helium atmospheres (neutral helium lines only) with very stringent limits on the amount of hydrogen [N(He)/N(H) $> 10^4$] that can be present without showing Balmer absorption. They lie in the temperature range 12,000 K $< T_{eff} <$ 26,000 K $-$30,000 K (cf. Wegner and Nelan 1986). Below $T_{eff} =$ 12,000 K, HeI lines disappear and DB spectra are no longer recognizable. Approximately 20% of these objects reveal hydrogen spectroscopically and thus are classified DBA (Shipman, Liebert and Green 1987). The most likely explanation for the hydrogen is interstellar accretion. DB stars may be formed as a result of post-AGB late thermal pulses and mass loss (cf. Iben 1984) or from the merger of helium-transfer cataclysmics like AM CVn and G61-29 (cf. Nather, Robinson and Stover 1981).

4. <u>DQ Stars:</u> The DQ (Formerly $C_2$ or $\lambda$ 4670 stars) degenerates generally have helium-dominated atmospheres whose optical spectra show the Swan bands of the $C_2$ molecule but generally no other metals or hydrogen. They lie within the temperature range 6000 K $<$ $T_{eff} <$ 12,000 K and have carbon abundances in the range $10^{-7} <$ C/He $< 10^{-2}$. Their far ultraviolet spectra sometimes reveal atomic carbon (CI) lines. Their photospheric carbon appears to arise from the convective dredgeup of carbon from its equilibrium diffusion tail extending upward from the core ( cf. Koester, Weidemann and Zeidler 1982; Fontaine *et al.* 1984).

5. <u>DZ Stars:</u> The DZ white dwarfs (formerly DF, DG, DK) show metallic absorption features with Ca II H and K absorption dominant, and with carbon noticeably absent or extremely weak in their usually helium-dominated atmospheres. While DZ stars are normally too cool to show absorption lines of the dominant atmosphere constituent, either helium or hydrogen, several helium-dominated DZ stars show trace hydrogen. The hottest examples of this type have $T_{eff} =$ 10,500 K and the coolest examples have $T_{eff} =$ 4500 K. The metals appear to originate from interstellar accretion (cf. Zeidler, Weidemann and Koester 1985; Koester, Wegner and Kilkenny 1990 and references therein).

6. <u>DC Stars:</u> These objects are too cool to show lines of the dominant light element, be it hydrogen or helium. They exhibit featureless spectra to within 5% of the continuum. Most of these objects reveal weak carbon features when examined at high resolution on good signal-to-noise spectra. However, there are apparently still some genuine DC stars in this greatly diminished subgroup.

7. <u>Magnetic White Dwarfs</u>: The Magnetic White Dwarfs (spectroscopically designated DH, DP, DAH, DAP, DXP; see Table 1), of which 27 are known, comprise about 2% of the total white-dwarf sample and are distributed more or less uniformly throughout the magnetic field range $1 < B < 1000$ Megagauss (Schmidt 1987). Although searches for fields weaker than this range are incomplete and detection becomes more difficult in weaker field objects, it is reasonably certain that nearly all white dwarfs do not have detectable magnetic fields (i.e. have fields less than $10^6$ gauss and no more than a few per cent show fields of a million Gauss or higher, cf. Liebert 1988). Like the nonmagnetic degenerates, the magnetic white dwarfs have primarily hydrogen-rich atmospheres based on their analyzable Zeeman patterns. However, the atmospheric compositions of some of the *high field* examples are unknown due to our current lack of knowledge of the Zeeman effect on atoms other than hydrogen.

## 2.2 Mass Loss From White Dwarfs?

### 2.2.1 Mass Loss From DA White Dwarfs?

Among the hydrogen-rich DA white dwarfs, the detection of sharp, shortward-shifted, lowly ionized silicon absorptions in a handful of DA stars below 60,000 K by Bruhweiler and Kondo (1983) remains the only potential indication of wind mass loss but the interpretation of these features remains unsettled. If the interpretation of these shortward-shifted absorption lines is correct, then mass loss rates of less than $10^{-13}$ $M_\odot$/yr may be indicated. Above 60,000 K, the hot DAO1 nucleus of the planetary nebula *EGB* 6 shows strong emission and reveals the presence of a small, high excitation compact nebula close to the star and inside a very large old, faint nebula (Dopita and Liebert 1989 and references therein). However the origin of this compact nebula appears to be most consistent with the disintegration of a gaseous, planet-like body by the radiation field of the white dwarf (Dopita and Liebert 1989). Among the hottest DA stars, $G\,191 - B2B$, $HZ\,43$, and the high gravity DAO central stars of *Abell* 7 and *NGC* 7293, there is no convincing evidence of ongoing stellar wind mass loss.

The best that can be said for DA mass loss at present is that static or slowly expanding haloes have been detected around several hot DA stars which may be manifesting weak mass loss.. The evidence is based upon the detection of sharp, shortward-shifted high ionization absorption features. Their lines typically have velocities between the velocity of the local interstellar medium in the lines of sight to these stars and the Einstein-redshifted photospheric velocity derived from the sharp non-LTE cores of the Balmer lines in these hot DA spectra. (cf. Bruhweiler and Kondo 1983). It is however a distinct possibility that the presence of the high ionization species in the lines of sight to these white dwarfs are manifesting an origin due to a past or ongoing very slow, low velocity mass ejection, perhaps driven by a selective ion wind (cf. Vauclair, Vauclair and Greenstein 1979; Vauclair 1989 and references therein).

### 2.2.2 Mass Loss from DO (Helium-rich) White Dwarfs?

The presence of any stellar wind mass loss from hot helium-rich white dwarfs remains to be established. Among the hot helium-rich DO white dwarfs only the pulsating high gravity DO nucleus of the O VI planetary nebula $K$ 1-16 reveals a fast wind. The other DO

312

degenerates reveal surface metals either as composition relics of prior evolutionary stages (Sion, Liebert and Starrfield 1985) or due to selective radiative acceleration of favored ions (Vauclair, Vauclair and Greenstein 1979; Vauclair 1988). Searches for evidence of mass loss in the far ultraviolet (IUE) range ( cf. Vauclair and Liebert 1986) and for nebulosity or circumstellar shells, in the optical (cf. Kwitter *et al.* 1989) have proven negative.

On the other hand it is not clear that the amount of mass loss present in hot white dwarfs is present at a detectable level. For example Michaud (1987) has pointed out that if wind mass loss rates were higher than a few times $10^{-13}$ $M_\odot$/yr, then abundance anomalies would be manifested at the white dwarf photosphere because such rates would disrupt the equilibrium abundances resulting from radiative levitation in hot white dwarfs. Mass loss rates lower than this value would be extremely difficult to detect in the optical or far ultraviolet wavelength regions because of the exceedingly low densities that would be expected.

This lack of indisputable observational evidence for wind mass loss from hot white dwarfs frustrates the search for ways to produce the very thin (very low mass) hydrogen layers, which appear to be required for most, if not all, white dwarfs in order to be consistent with the observations and theory of white dwarf pulsations (cf. Winget and Fontaine 1982) and with empirical soft X-ray and spectroscopic data (cf. Shipman 1989 for a current review). The residual hydrogen which remains following the planetary nebula ejection phase (cf. Iben and MacDonald 1985 and references therein) is expected to be of order $10^{-4}$ $M_\odot$. It is therefore critical to enlarge and intensify the search for wind mass loss. Despite the frustrating outlook for detecting mass loss, there are some DO degenerates that merit closer scrutiny.

The peculiar DO star, *KPD* 0005+5106 (hereafter *KPD* 0005; cf. Downes *et al.* 1986) stands out as a promising candidate for detecting ongoing white dwarf mass loss, for a number of reasons. Searches for nebular lines and for the red continuum a hypothetical companion star produced negative results and the overall energy distribution was found to be consistent with $T_{eff}$ = 80,000 K. The detection of hydrogen and probable CNO ions in emission makes this DO white dwarf unique among the hottest helium-rich degenerates for two reasons. First the hottest DO objects, the *PG* 1159 stars, and the intermediate temperature DO stars like *PG* 1034+001, reveal no direct spectroscopic evidence for any hydrogen in either absorption or emission, and second, the emission lines of He II and the CNO ions, when present in these stars, is always accompanied by corresponding photospheric absorption lines. (An exception occurs when He II 4686 absorption appears with an emission reversal but He II 1640 absorption lacks an emission counterpart for the same star). It is important to note that the emission reversals reported in the CNO absorption lines of the *PG* 1159 stars have been shown to be due to non-LTE effects (Werner *et al.* 1988) since they appear in non-LTE profiles but are absent in the LTE model atmosphere grids that have been reported. These results would suggest it is unlikely that the emission lines in *KPD* 0005 without accompanying absorption are due to photospheric non-LTE effects though in the absence of optical velocity information one cannot rule out such an origin for the He II 4686 emission reversal.

In *KPD* 0005, the velocity of the lowly ionized interstellar features (e.g. the NI triplet, Si II etc.) is -23 km/s while the far ultraviolet resonance doublets of Si IV, CIV, and NV have an average velocity of -13 km/s. Thus the high ionization ions have a velocity displacement of +10 km/s with respect to the local interstellar medium in the direction of *KPD* 0005. However the widths (FWHM) of the high ionization lines, including the instrumental width of 0.1 Å, are essentially the same as the lowly ionized interstellar lines (0.17-0.19 Å). The absorption line components of the C IV resonance doublet, in the high resolution IUE spectrum of *KPD* 0005, are shown in Figure 1.

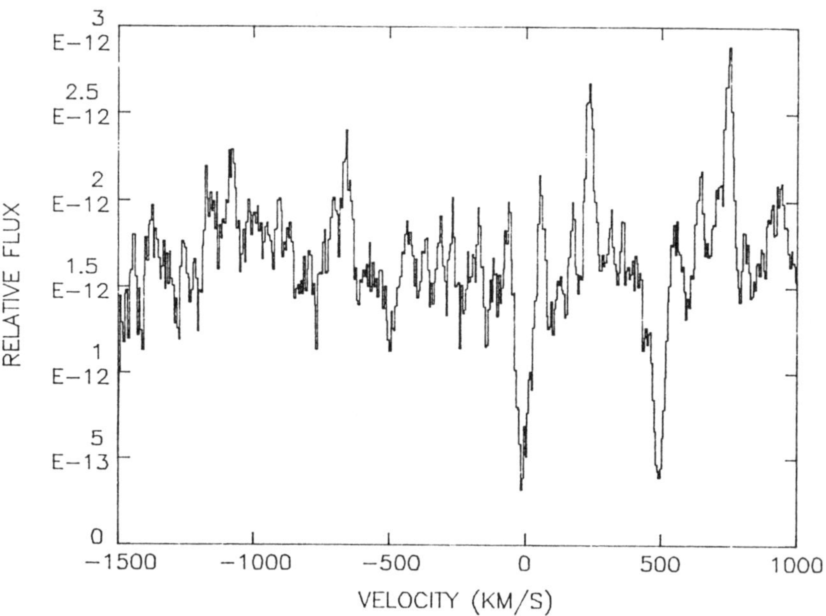


Figure 1. The C IV resonance doublet of *KPD* 0005+5106 plotted in velocity space with zero velocity corresponding to the rest frame of ther C IV 1548.152 as relative flux versus velocity space. From Fritz, Leckenby and Sion (1990).

On the basis of a comparable velocity broadening between the high excitation lines and the low excitation species, it is likely that the high ionization resonance doublets do not originate in the stellar photosphere as suggested in Downes *et al.* (1987) but instead originate in a circumstellar photoionized H II region. The velocity of the high ionization lines in *KPD* 0005 is essentially what one expects for the expansion of an H II region, at the sound velocity in the photoionized gas (Spitzer 1978). A similar interpretation has been advanced for the origin of the high ionization absorption lines in the IUE echelle outburst spectrum of the dwarf nova *SS Cygni* (Mauche *et al.* 1988). Since there is no evidence of ongoing high velocity stellar wind outflow from *KPD* 0005, manifested by *P Cygni* profiles as in some planetary nebula central stars, it is unlikely that the high ionization features arise from a spherical cavity blown into the interstellar medium by a fast wind. *However one cannot rule out that the CNO nuclei in the circumstellar environment of this star were ejected during a prior mass loss stage. Indeed the initial origin of these highly ionized CNO species in the circumstellar environment of this star was very likely due to a prior mass ejection episode.*

The optical emission lines in *KPD* 0005, unlike the emission reversals in the broad photospheric absorption lines of other hot DO stars, may originate above the photosphere and be manifesting ongoing mass loss. Although accurate velocities for these emission features have yet to be measured, it seems apparent from visual inspection of figure 7 in Downes *et al.* (1987) that their wavelength displacements indicate velocities less than 100 km/s. This conclusion is tentative however and attempts to measure accurate velocities of the optical emission lines for comparison with the IUE echelle data, definitely should be carried out.

In the intermediate temperature ($T_{eff}$= 80000 K) DO star, *PG* 1034+001 (Sion *et al.* 1985), the curious appearance of a doubled line structure in the newly detected Si IV line (Fritz *et al.* 1990), is noteworthy. The earlier paper by Sion *et al.* (1985) noted a possible doubled structure in the two components of C IV, which they interpreted tentatively as circumstellar and photospheric splitting. The doubled structure of C IV in the second image of this star (*SWP* 26201) is not as evident but O V (1371) has a pronounced doubled structure with both absorptions equally deep, making it unlikely that the splitting is due to poor signal to noise. It is interesting to consider the possibility, admittedly remote, that these profiles underwent real changes over the time baseline of 2.5 years between the two exposures. This possibility is enhanced by the reported evidence of line profile variations in the shortward-shifted sharp absorption features detected in some hot sdO stars by Bruhweiler and Dean (1984).

The detection of Si IV absorption by Fritz *et al.* (1990), if it arises at the white dwarf surface, appears to be inconsistent with the theoretical radiative levitation calculations of Chayer *et al.* (1989), since they predict that no silicon should be present in the atmospheres of non-DA degenerates with $T_{eff} < 85,000$ K. Fritz *et al.* (1990) suggest that the resolution of this apparent conflict rests in one or more of the following possibilities: (1) the radiative levitation theory is incorrect in its present form; (2) the far ultraviolet resonance doublets do not arise at the stellar photosphere; (3) *PG 1034+001* is hotter than 85,000 K or (4)*some mechanism such as a weak wind (Chayer et al. 1989) is opposing diffusion and radiative levitation in this object. It is possible that the profile changes and line doubling reported above, if real, may be indicating the presence of circumstellar material and therefore some form of weak mass loss.*

## 2.3 Rotation of White Dwarfs

The most general basic conclusion about white dwarf rotation to date is that they are slow rotators. This result is based upon actual determinations or inferences of slow rotation from the effects of rotation on the light curves (frequency spectrum) of the *ZZ Ceti* variables, from variable circular polarization due to changes in the observed magnetic field in magnetic white dwarfs as they rotate and from recent analyses of the very sharp non-LTE Balmer absorption line cores of DA stars. Two examples of the NLTE core of $H_\alpha$ are shown in Figure 2.

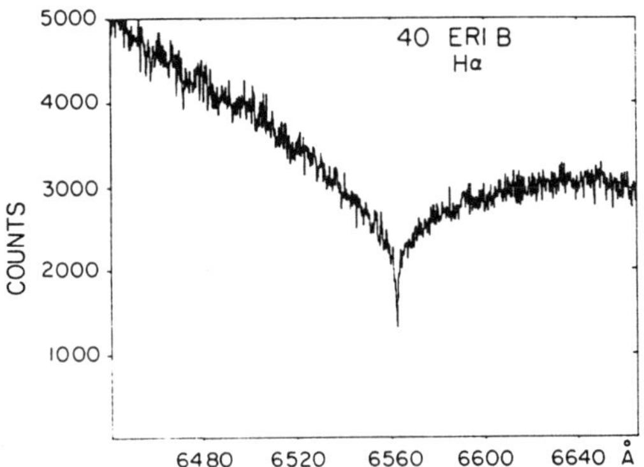


Figure 2. The sharp NLTE absorption core of H $\alpha$ in the spectrum of the DA white dwarf 40 *Eri B*. NLTE model atmosphere fits to such profiles yield reliable upper limit rotational velocities for DA stars. Reprinted from Greenstein *et al.* (1977).

No DA star is known with $v \sin i > 65$ km/sec (Pilachowski and Milkey (1987). For the relatively few DB and DO stars which have been analyzed, the conclusion is the same: slow rotation. Based upon analyses of the sharp neutral helium line cores of several DB stars, all were found to have $v \sin i < 135$ km/sec. The breadth of sharp metallic photospheric absorption lines in the hot DO stars (cf. Sion and Wesemael 1983; Liebert *et al.* 1989) imply a similar conclusion. While the total number of stars which have been analyzed for rotation remains small, there is as yet no example of a rapidly spinning single degenerate. This slow rotation of white dwarfs is unexpected if angular momentum is conserved during evolution from high luminosity, evolved progenitors down to the white dwarf stage. It is possible that either stellar wind braking via exchange of angular momentum between core and envelope during the envelope mass shedding of red giant stages and/or fast post-AGB stellar wind mass loss just prior to the white dwarf stage, may be responsible for the generally slow rotation of white dwarfs.

## 3. IDENTIFYING MASSIVE PLANETARY NEBULA CENTRAL STARS AND THEIR WHITE DWARF PROGENY

### 3.1 The Distribution of PNN and the Hottest White Dwarfs in the Log L - Log T Plane: Comparison with Evolutionary Tracks

If the luminosity L, and effective temperature, $T_{eff}$, are accurately known for a PNN or hot, luminous white dwarf, then their masses could be straightforwardly derived from the core mass-luminosity relation for PNN with active thermonuclear shell sources, of Paczynski (1971):

$$L/L_\odot = 5.9x10^4((M/M_\odot)^{-0.52}).$$ (1)

The evolutionary (nuclear timescale) is also very sensitive to the mass and thus the evolutionary tracks are well separated in the H-R diagram. The maximum excursion to the left (to highest $T_{eff}$) in the H-R diagram is well approximated by the following least squares

316

fit (Tylenda 1989) to the tracks of Paczynski (1971), Schonberner (1981, 1983), Iben (1982, 1984) and Wood and Faulkner (1986):

$$\tau_{nuc} = 1.0 - 4.96(M/M_{\odot} - 1.0) - \log(M/M_{\odot} - 0.52). \qquad (2)$$

This expression provides a good estimate for helium burning PNN as well as the hydrogen burning cases (Tylenda 1989).

The most massive central stars and hence the most likely PNN to have upper main sequence progenitors are expected to be found among the hot but low luminosity PNN because they evolve very quickly while being luminous. Among the leading candidates for being quite massive PNN (and therefore the descendants of originally massive main sequence progenitors), are $NGC\,7027$, $NGC\,2440$, $NGC\,6302$, and $NGC\,2392$, all of which appear to have masses of 0.9 $M_{\odot}$ or greater.

Are there specific spectroscopic types of white dwarfs expected from the evolution of more massive progenitors? Iben (1984) and Renzini (1989) have shown that AGB and post-AGB evolution yield the approximate observed ratio of DA to non-DA stars of 4:1, depending upon whether the central star experiences a helium shell flash during a post-AGB phase or just prior to departing from the AGB for the first time, i.e. when, in the thermal pulse cycle, the AGB star undergoes a helium thermal pulse and whether the PNN remnant suffers a final thermal pulse during its helium shell burning or hydrogen shell burning phase. For the typical white dwarf progenitors of the old disk, Renzini (1989 and references therein) shows that the fraction of the hydrogen envelope mass removed during each post-flash peak is much less than the available hydrogen envelope mass. It is therefore expected that the overwhelming majority of low mass, old disk progenitors (e.g. Mira variables) will yield DA progeny. However for the higher mass progenitors ($3 < M/M_{\odot} < 8$), which are the central topic of this conference, the luminosity of a star at the peak of a thermal pulse should reach the Eddington limit, leading to the hydrodynamic expulsion of the entire hydrogen-rich envelope after a thermal pulse (Wood and Faulkner 1986). If this scenario is correct these objects, according to Renzini (1989 and references therein) should lead to the production of helium-rich planetary nuclei and therefore non-DA progeny. It is thus expected on this basis, that the most massive AGB stars and their upper main sequence progenitors should preferentially lead to the formation of white dwarfs with hydrogen-free surfaces.

Unfortunately the assignment of masses to PNN and the hottest white dwarfs from their placement in the Log L-Log T plane, is seriously hampered by observational uncertainties in determining effective temperatures and in determining distances. The problem is illustrated in Figure 3 where a Log L-Log T diagram from Kaler and Jacoby (1989) is displayed. The filled circles and open circles refer to PNN whose surrounding nebulae have N/O abundance ratios of $> 0.8$ and $< 0.8$, respectively. On the figure are PNN evolutionary tracks from Paczynski (1971) for 0.6, 0.8 and 1.2 $M_{\odot}$ as well as 0.55 $M_{\odot}$ from Schonberner (1981). The 1.4 $M_{\odot}$ track is an extrapolation of these tracks by Shaw (1988). The $T_{eff}$ values are those newly calculated by Kaler and Jacoby (1989) while the upper limit luminosities and their associated uncertainties are given in Table 1 of Kaler and Jacoby (1989).

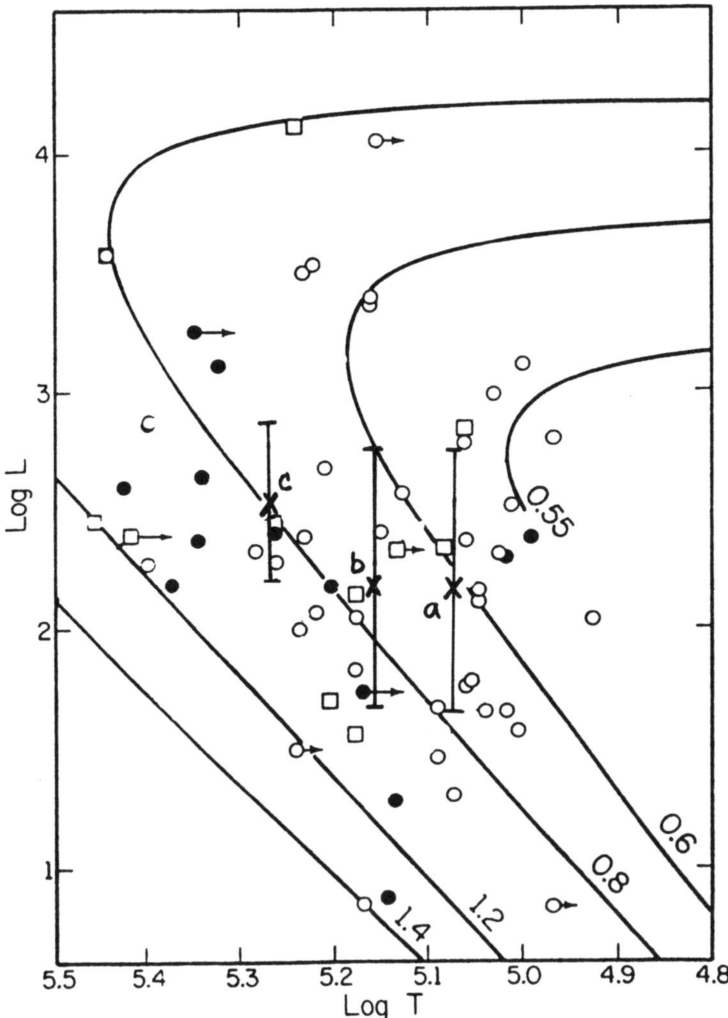


Figure 3. The positions of 62 central stars of optically thick planetary nebula on the Log L - Log T plane with Paczynski-Schonberner evolutionary tracks for various masses in units of $M_\odot$ (from Kaler and Jacoby 1989) . The luminosities are *upper limits only*. The filled circles have enriched surrounding nebulae with N/O > 0.8, open circles N/O < 0.8, and boxes representing nebulae with unknown N/O. The X symbols with vertical error bars in luminosity, represent (a) *PG* 1707+427, *PG* 1424+535 (b) *PG* 1159-035, *PG* 1520+525 and (c) *H*1504+65. The $T_{eff}$ values are those determined recently by Werner, Heber and Hunger (1989) based upon NLTE model fitting while the $T_{eff}$ value for *H* 1504 (180,000 K) is from Barstow and Tweedy (1990), based upon EXOSAT data.

318

On this same diagram, I have plotted the latest and most accurate $T_{eff}$ values for four *PG* 1159 DO degenerate stars, determined recently by Werner, Heber and Hunger (1989a) using their new, non-LTE operator-perturbation model atmosphere code with the inclusion of 169 atomic non-LTE levels. The $T_{eff}$ value for *PG* 1707+427 and *PG* 1424+535 is 110,000 K for both, while for *PG* 1159-035 and *PG* 1520+525, $T_{eff}$ is 140,000 K for both. Also shown is the EXOSAT temperature determination of 180,000 K by Barstow and Tweedy (1990), for *H* 1504+65. Related objects but not shown in Figure 3 are *K* 1-16, which is spectroscopically similar to the *PG* 1159 stars but with a probable somewhat lower gravity, fast wind and surrounding nebular shell, and the nuclei of *Abell* 30 and *Abell* 78 with pure helium inner knots as well as the hot carbon-rich PNN of *NGC* 246. The masses for the four *PG* 1159 stars fall between the tracks for 0.6 and 0.8 $M_\odot$. This result is in approximate agreement with the mass value of *PG* 1159 determined seismologically by Kawaler (see section 3.2.3 below). Note the large uncertainty in the luminosity assigned to these objects largely because of distance uncertainties.

## 3.2 Identification of Massive White Dwarf Remnants from Gravitational Redshifts

### 3.2.1 Methods of Mass Determination for White Dwarfs

Accurate masses of white dwarfs are possible from measurements of the gravitational (Einstein) redshift of their spectral lines. The gravitational redshift of light, due to the gravitationally induced slowdown of clocks in large gravitational fields, can be expressed in terms of the redshift velocity which, for velocities considerably below the speed of light, is given with sufficient accuracy by

$$V_{rs} = GM/cR = 0.635(M/M_\odot)(R/R_\odot)km/sec. \tag{3}$$

If the intrinsic wavelength shift of the spectral lines due to pressure (Stark effect) broadening is properly corrected and if the white dwarf has a known radial velocity (e.g. a distant non-white dwarf wide binary companion of accurately known radial velocity), the gravitational redshift can be extracted and thus the mass/radius of the white dwarf.

The most reliable masses of individual white dwarfs are those obtained from orbit solutions of wide (essentially non-interacting) binaries. These data points are crucial for testing the mass-radius relation for degenerate stars and for testing stellar evolution theory (e.g. the initial (parent) mass- final remnant mass relations). Among the best determinations(see Liebert 1980 for references) are *Sirius B* (1.053 $\pm$ 0.028 $M_\odot$), *Procyon B* (0.63 $M_\odot$), 40 *Eridani B* (0.43 $\pm$ 0.02 $M_\odot$), *Stein* 2051B (0.50 $M_\odot$).

Mass estimates of white dwarfs obtained by model atmosphere analyses of colors (e.g. the gravity-sensitive portion of the Stromgren two-color u-b versus b-y diagram), energy distributions and line profile fits yield mass values of statistical use only. The masses, radii and gravities of large numbers of white dwarfs determined in this way, largely by V. Weidemann and co-workers and by H. Shipman have yielded statistically useful average masses and mass distributions for DA (hydrogen-rich) and non-DA (helium-rich) white dwarfs. The average mass of DA stars is 0.58 $\pm$ 0.05 $M_\odot$ and the average mass of DB stars is 0.55 $\pm$ 0.03 $M_\odot$ (cf. Oke, Weidemann and Koester 1984).

### 3.2.2 Identification of White Dwarfs as Probable Upper Main Sequence Descendants From Einstein Redshift Measurements

The identification of the white dwarf remnants of more massive (upper main sequence) stars is possible by determining gravitational redshifts for wide non-interacting binaries where the radial velocity of a non-degenerate companion is known accurately and the orbital velocity is negligibly small. This technique has been utilized by Greenstein and Trimble (1967), Trimble and Greenstein (1972), Wegner (1981), Koester (1987), Wegner, Reid, and McMahan (1989), and Oswalt (1981) but not yet extensively. These efforts have been confined to relatively small samples of common proper motion systems, having a white dwarf component and even then, of DA spectral type only, because of the less complicated problem posed by the intrinsic pressure shifts of the Stark-broadened Balmer lines. Koester (1987) has presented mass determinations for 10 DA white dwarf members of cpm pairs but no attempted determinations for non-DA stars. Wegner and Reid (1989) have also presented less than 10 determinations, all DA stars, some of which overlap and show good agreement with the redshift mass values determined by Koester. The average mass of Koester's small sample agrees nicely if not unexpectedly with the average value, 0.6 $M_\odot$, of single field degenerates but among his sample are three stars with mass values above 0.7 $M_\odot$. These higher mass objects are almost certainly the descendants of more massive progenitors on the upper main sequence, whose remnant cores had higher masses.

In a related investigation of redshift masses but for single white dwarfs in the Hyades cluster, Wegner (1989) has used the latest convergent point, distance modulus and systemic velocity determination for the cluster, to derive redshift masses for 8 white dwarf cluster members. His results, shown in Figure 4, reveal an average mass of 0.7 $M_\odot$, significantly higher than the 0.6 $M_\odot$ field white dwarf average mass. This result is consistent with the present Hyades white dwarfs being the descendants of main sequence stars whose masses are larger than the turn-off mass of the cluster, which is at least 2.5 $M_\odot$. All but three of the stars in Wegner's sample have mass values between 0.6 and 0.8 $M_\odot$ for a core composition of carbon and oxygen in the Hamada-Salpeter mass radius relation.

### 3.2.3 Pulsations of White Dwarfs and Mass Determination

One of the most remarkable developments in the last fifteen years was the discovery of three distinct classes of pulsating white dwarfs: (1) the DA pulsators (cf. McGraw 1979), whose prototype is *ZZ Ceti*, with effective temperatures near 12,000 K, an instability strip of width, $\Delta T_{eff} = 2000$ K, and multi-periodic luminosity variations in the period range 100- 1200 seconds; (2) the DB (helium-rich) pulsators (cf. Winget 1981), whose prototype is *GD* 358, with effective temperatures near 28,000 K, an instability strip of less certain width but $\Delta T_{eff} < $ a few thousand degrees, and multi-periodic luminosity variations (in 5 of the 6 known pulsators) in the period range 100-1200 seconds; and (3) the DO pulsators (cf. McGraw *et al.* 1979), hottest of the three types of pulsators, whose prototype is *GW Vir* (*PG* 1159-035), with effective temperatures between 110,000 K and 150,000 K,

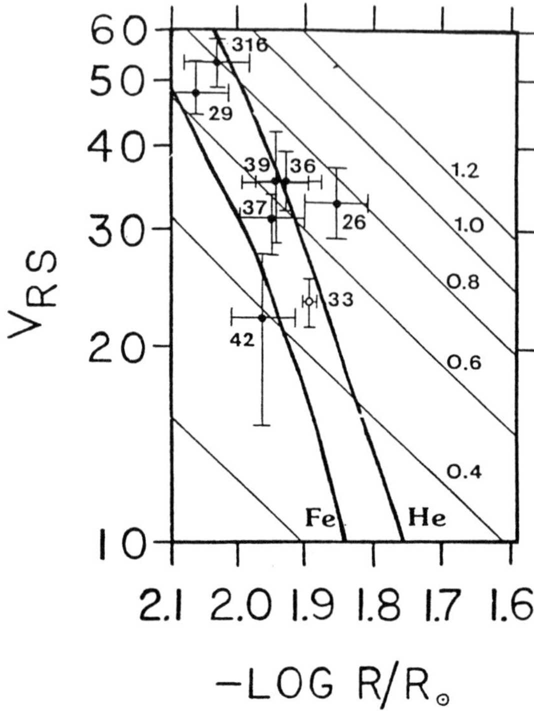



Figure 4. A comparison between the observed gravitational redshift and radii of Hyades white dwarfs and the theoretical predictions of the Hamada-Salpeter (1961) $Fe^{56}$ and $He^4$ mass-radius relations. Courtesy of Wegner (1989).

an instability strip of approximate width, $\Delta\ T_{eff} = 40,000$ K and multi-periodic luminosity variations in the period range 200-2000 seconds. Most of the objects in all three classes have extremely complicated light curves and all white dwarf pulsators appear to be pulsating in non-radial g-modes. The *ZZ Ceti* mechanism appears to require very thin hydrogen layers in order to account for the high temperature boundary (blue edge) and the observed narrow instability strip width. The observed luminosity variations are thought to be due mostly to temperature variations during a global g-mode oscillation. The oscillations are thought to be driven by partial ionization of the most abundant element in the outer layers, hydrogen in the case of the ZZ ceti stars, helium in the case of the DB pulsators and possibly the partial ionization of carbon and oxygen in the *GW Vir* variables although the g-modes may be excited by nuclear shell burning in these hottest pulsating degenerates.

Because the normal modes of g-mode pulsation depend upon the global properties of the star, the periods of the normal modes should change as the star evolves (cools). Since many g-modes are excited in each pulsator and the observed periods are known with extreme accuracy, it is possible to measure the change in these periods in a relatively short time and thus test the cooling theory of hot degenerates (with short cooling timescales) and seismologically probe the interior regions of the white dwarfs.

Of greatest relevance to this workshop are the masses determined from the pulsations of *PG* 1159 stars. Kawaler has found seismologically that the mass of *PG* 1159 is 0.6 $M_\odot$ and one other DO star has been found in this way. Kawaler (1988) has examined the optical power spectrum of the prototypical multi-periodic pulsator, *PG* 1159-035, and found eight statistically significant signals with mean period spacings of either 21.0 seconds or 8.8 seconds. A comparison with evolutionary models reveals that these detected mean periods correspond to a set of l=1 and l=3 modes in a 0.6 $M_\odot$ white dwarf. This determination of mass along with similar determinations for two other *PG* 1159 stars, demonstrates a powerful new means of determining masses of hot pulsating degenerates. The masses of these pulsating white dwarfs, determined thus far, do not indicate a likelihood of their having descended from massive AGB/upper main sequence progenitors.

## 4. SPACE MOTIONS OF WHITE DWARFS

### 4.1. Kinematical Links Between White Dwarfs and Upper Main sequence Progenitors

The most recent and largest compilation of space motions for white dwarfs by Sion *et al.* (1988) confirms the earlier conclusions by Eggen and Greenstein (1967), Sion and Liebert (1977), and Greenstein (1981), that the motions of the local white dwarfs represent an admixture of stellar population subcomponents: the majority of white dwarfs belong to the old disk population subcomponent with typical total space motions of 50-60 km/sec with respect to the sun, while 4-5% per cent have total space motions greater than 150 km/sec, characteristic of the halo and extreme population II subcomponent and several per cent of direct interest to this work, have motions which indicate they belong to the young disk population subcomponent and therefore are associated with young, fairly massive progenitor stars. While the long total stellar ages of white dwarfs and perturbative encounters they suffer during their galactic orbital motions, would tend to smear out kinematical distinctions among the different types of white dwarfs, (e.g. increase their velocity dispersions with age; cf. Wielen 1977), there is evidence that the DQ (carbon-band) degenerates and the magnetic white dwarfs have higher than average and lower than average space motions, respectively (Sion *et al.* 1988).

It is remarkable that the magnetic white dwarfs, with one exception, *G* 195-19, (see section 2.1 above), appear to have low velocity with respect to the sun, a possible indication of youth and evolution from fairly massive progenitors. The kinematical distinction of these objects with respect to the old disk majority of white dwarfs is borne out convincingly by a comparison of their average kinematical properties (see table II (a) in Sion *et al.* 1988). Recent work on *Grw* +70° 7234 by Greenstein (1984) seems to indicate a massive magnetic white dwarf, thus lending additional support to the idea that as a class, the magnetics may have higher than average masses ( $M > 0.6\ M_\odot$). A very recent, even more compelling indication of higher than average mass for the magnetics, comes from an accurate H-R diagram constructed by Liebert (1988), shown in Figure 5, by using well-determined USNO trigonometric parallaxes plus V-I and G-R color indices measured by Greenstein (1984, 1986). The magnetic white dwarfs tend to fall below the parabolic fit to the entire sample of white dwarfs of all spectral class shown in this H-R diagram. The original suggestion that the magnetic white dwarfs are the descendants of upper main sequence stars was advanced by Angel, Borra and Landstreet (1981) who showed that the magnetic white dwarfs were present in approximately the right space densities to be descended from the peculiar A and B stars on the upper main sequence. It is also possible that some of the magnetic white dwarfs had an origin related to pulsars and therefore came from progenitor stars nearly massive enough to have undergone collapse to a neutron star. Some however might even have originated as extinct AM Herculis magnetic cataclysmic variables in which the cool Roche-lobe filling companion has been disrupted or *whittled* down to a very low mass (substellar) degenerate itself, thus terminating magnetically funneled mass accretion and leaving what appears to be a *single* magnetic white dwarf.

In the kinematical study by Sion *et al.* (1988), it was also shown that with the current white dwarf database, and in the absence of radial velocities, there was no convincing evidence of a kinematical distinction between the DA and DB white or that one or the other spectroscopic type appeared to be linked kinematically, with more massive progenitors. This conclusion was contrary to the claim of Guseinov *et al.* (1983), that non-DA stars tended to have higher space motions than DA stars. The removal of color-dependent selection effects and the use of a larger sample with improved photometric and trigonometric parallaxes by Sion *et al.* (1988) appears to account for their differing conclusion. However, it is interesting to explore this question a bit further, by comparing the space motions of DA and DB stars that are thought to have higher than average masses, in the following way. If one adopts, for statistical purposes only, the masses derived by Sion *et al.* (1988) for DA stars, and selects those DA stars having (1) derived masses of 0.7 $M_\odot$ or greater and (2) effective temperatures in the same range (12000 K-30000 K) as the DB stars, then a comparison of their space motions with a similarly selected sample of DB stars can be made. For the DB stars, the best observed, most homogeneous sample is found in Oke, Weidemann and Koester (1984). The results of this comparison revealed what appears to be a possible significant difference between the DA and DB stars so selected, in the sense that the DA stars have a lower average space motion ( $<V> = 43$ km/s) than the DB stars ( $<V> = 56$ km/s) in the same range of $T_{eff}$, thus suggesting that these higher than average mass DA stars may have had more massive progenitors. However this comparison was biased by the fact that the much more numerous DA sample had a high mass tail at $M > 0.7\ M_\odot$, not evident in the smaller DB sample.

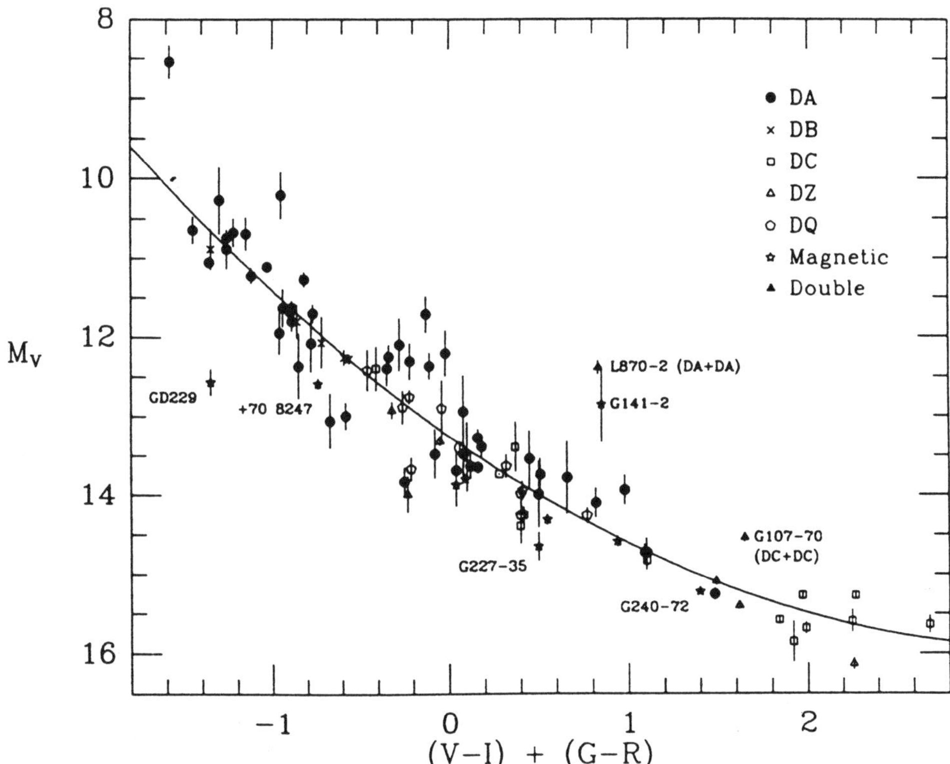


Figure 5. An accurate H-R diagram for white dwarfs with well-determined trigonometric parallaxes and colors measured by Jesse Greenstein (1984, 1986). A parabolic fit to all data is shown. The magnetic degenerates, discovered through either polarimetry or spectroscopy, tend to fall below the fitted curve, implying that they have smaller radii and hence larger mass than the stars on the curve. Reprinted from Liebert (1988).

## 4.2. Kinematical Identification of White Dwarf Progeny of Massive Stars

An important study of the kinematics and ages of wide binaries by Wegner (1981) has also led to the identification of the probable progeny of upper main sequence progenitors. This method (described in section 2.2.2 above) has been used to derive complete space motions for a number of white dwarfs which are members of common proper motion binaries. From the $T_{eff}$ values for each white dwarf component, a cooling age is derived. The distribution

324

of these stars in the UV velocity plane was compared to the positions of the Sirius and Pleiades moving groups and the distribution of the nearby A and B stars. as shown in Figure 6. A number of these objects had to be excluded from membership consideration because their cooling times exceeded the estimated ages of the groups. This resulted in the rejection of all the degenerate stars in Wegner's (1981) sample as members of the very young Pleiades group but several white dwarfs were shown by Wegner (1981) to be probable members of the Sirius group. These white dwarfs, it can therefore be argued, could be the descendants of upper main sequence A and B stars. Additional spectroscopic clues, such as emission lines and peculiar spectra for the non-degenerate companions, were used to strengthen the conclusion that the white dwarf components of some of these low velocity pairs evolved from parent stars that belong to a young stellar population. The increased number of white dwarf companions identified in common proper motion systems, in the years ahead, will allow this method of kinematics and ages of wide binaries, to be very greatly extended.

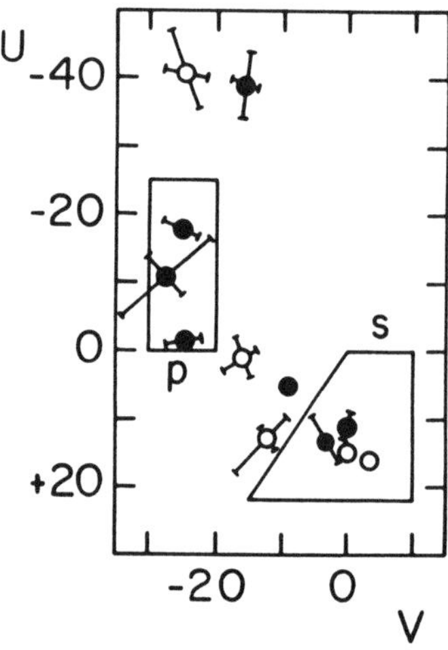


Figure 6. A plot of the U versus V vector components of space motion for white dwarfs with complete space motions from wide binary membership. The boxes in the UV velocity plane labelled *p* and *s* indicate the location of the Pleiades and Sirius group *clumps* in the B- and A- star distributions. Reprinted courtesy of Wegner (1981).

## 5. WHITE DWARF DESCENDANTS OF MASSIVE STARS IN YOUNG OPEN CLUSTERS

White dwarfs have been identified in a number of open star clusters with the largest sample in the Hyades cluster (over a dozen DA stars and one DBA member). The mean mass of the Hyades white dwarfs is larger than the mean mass of field degenerates, thus implying more massive progenitor stars for the cluster white dwarfs. Most remarkably, white dwarfs have been identified in open clusters whose ages are so young that only the massive upper main sequence members of the cluster have had sufficient time to evolve toward the red giant branch (cf. Romanishin and Angel 1980; Koester and Reimers 1985 and references therein). The red giant turnoff mass is 6-8 solar masses in some of these clusters thus implying that the parents of the cluster white dwarfs were at least that massive. The work by Koester and Reimers (1985) utilized a search for faint blue objects on deep UV and red Schmidt plates obtained at ESO/Chile. Subsequent spectroscopic identification of the blue suspects as cluster members, and the determination of cooling times, cluster age and turnoff mass, led to the exciting conclusion that these white dwarfs must ahve evolved from quite massive parent cluster stars. For two of the confirmed DA white dwarfs, *NGC* 2451-1 and *NGC* 2451-5, white dwarf masses of 1.05 $M_\odot$ and 0.78 $M_\odot$ are approximately indicated. The initial main sequence masses of these two white dwarfs must have been $> 8\ M_\odot$ (Koester and Reimers 1985). It can certainly be stated on the basis of these results that larger progenitor masses leave more massive white dwarf remnants. This conclusion is therefore not merely a theoretical expectation any longer. The same argument was advanced earlier in this work to fully account for the observations, physical properties, and origin of the magnetic white dwarfs.

## 6. CONCLUSIONS

This review has focussed upon current observational evidence of white dwarf mass loss, rotation rates and the identification of evolutionary (progenitor) links between white dwarfs and upper main sequence stars. The latter objective rests upon the determination of white dwarf masses through a number of methods. All of these areas are incompletely explored at best and only some very preliminary answers are now available. The main conclusions may be summarized as follows:

(1) There is some evidence in the far ultraviolet (cf. Bruhweiler and Kondo 1983) to suggest that hot DA white dwarfs lose mass via a weak wind. This mass outflow is manifested by the presence of sharp, shortward-shifted absorption lines in the IUE high resolution spectra of a number of hot DA white dwarfs. Mass loss rates are probably less than $10^{-13}\ M_\odot/\mathrm{yr}$. Stellar winds of this magnitude would be extremely difficult to detect and mass loss rates higher than this would produce photospheric abundance anomalies that are not clearly observed. There is at present very weak if any evidence of actual mass loss by hot non-DA white dwarfs. (2) The general conclusion at present concerning white dwarf rotation rates is that they are slow rotators with established upper limit $v \sin i < 65$ km/s for DA stars and $v \sin i < 135$ km/s for DB stars. the presence of sharp absorption lines in the IUE high resolution spectra of hot DO stars constrains how rapidly they may be rotating. Independent evidence from studies of magnetic white dwarfs and pulsating *ZZ Ceti* stars (DA white dwarf variables) supports this picture of slow rotation as a general rule for white dwarfs. (3) The identification of quite massive planetary nebula nuclei which are the descendants of a young massive main sequence progenitors, is demonstrable by comparison with evolutionary tracks on the H-R diagram These objects are found among the very hot but low luminosity PNN because they evolve very quickly due to their high mass. At the present time it appears that the hottest known non-DA white dwarfs (the *PG* 1159 stars) appear by this method not to have higher than average white dwarf mass (i.e. $> 0.6\ M_\odot$) and therefore no link is indicated between these objects and upper main sequence progenitors. (4) At the present time it is not clear whether massive white dwarf progenitors would be expected to leave DA or non-DA remnants. Here, theory and observation provide conflicting indications. Renzini (1989) argues that hydrodynamic ejection of the H-rich envelope during the peak of a thermal pulse in a massive young disk progenitor should leave a white dwarf with a hydrogen-free surface. On the other hand the white dwarfs which have been shown to be the progeny of upper main sequence stars such as the young open cluster degenerates identified by Koester and Reimers, Sirius B itself, are all of spectral class DA. (5) The

magnetic degenerates have been shown to tend toward higher than average white dwarf mass and the majority are therefore expected to be the progeny of young, peculiar A and B stars on the upper main sequence. This conclusion is supported by a number of independent lines of evidence. (6) Higher than average mass white dwarfs (and therefore the remnants of progenitors more massive and younger than the typical and most populous old disk stars in the solar neighborhood), have been identified through kinematical properties, through gravitational redshift determinations in wide and common proper motion binaries where the non-degenerate companion has a known radial velocity, through kinematical membership in young moving groups if the suspected white dwarf member has a cooling age shorter than the age of the group, through the gravitational redshifts of the white dwarf members of the Hyades cluster which has an accurately known distance modulus, convergent point and systemic velocity and through the spectroscopic identification and analysis of white dwarf members of very young open clusters whose turnoff masses to the red giant branch exceed $5 \ M_{\odot}$.

I have tried to demonstrate in this review that it is possible at the present time to isolate a significant number of white dwarfs which are almost certainly the progeny of upper main sequence (e.g. O, B, and A ) stars in the initial mass range $3 < M/M_{\odot} < 8$. These objects have been identified by a diverse array of methods as discussed in this paper. In the years ahead the sample size of these massive white dwarf progeny of the upper main sequence should greatly increase, thus allowing definitive answers to the stellar evolutionary questions addressed in a preliminary fashion in this review.

The research has been supported by NSF grant AST88-02689 to Villanova University. It is a pleasure to acknowledge the kind assistance of Beth Jewell in the preparation of this manuscript. I would also like to thank Drs. Jesse Greenstein, Gary Wegner, James Kaler and James Liebert for their permission to reproduce published figures.

REFERENCES

Angel, R.J.P., Borra, E.F., and Landstreet, J. 1981, Ap.J. Suppl., **45**, 457.

Barstow, M. A., and Tweedy, R. W. 1990, M.N.R.A.S. (in press).

Bruhweiler, F.C., and Dean, C. 1984, Ap.J.(Letters), **274**, L87.

Bruhweiler, F.C., and Kondo, Y. 1983, Ap.J. **269**, 657.

Chayer, P., Fontaine, G., and Wesemael, F. 1989, in White Dwarfs, ed. G. Wegner, (Springer-Verlag: Berlin), p. 253.

D'Antona, F., and Mazzitelli, I. 1979, Astr. Ap. **74** , 161.

Dopita, M., and Liebert, J. 1989, Ap.J., **347**, 910.

Downes, R. A., Sion, E. M., Liebert, J. L., and Holberg, J. B. 1987, Ap.J. **321**, 943.

Eggen, O. J., and Greenstein, J., 1967, Ap.J. **150**, 927.

Fontaine, G., Villenueve, B., Wesemael, F., and Wegner, G. 1984, Ap.J. (Letters), **277**, L51.

Fontaine, G., and Wesemael, F. 1987, in IAU Colloquium **No.** **95**: The Second Conference on Faint Blue Stars, ed. A. G. D. Philip, D. S. Hayes, and J. Liebert (L. Davis Press: Schenectady), p. 319.

Fritz, M., Leckenby, H., and Sion, E. M. 1990, A.J., in press.

Greenstein, J. L. 1976, A.J. **81**, 323.

Greenstein, J. L. 1984, Ap.J. **276**, 602.

Greenstein, J. L. 1986, Ap.J. **304**, 334.

Greenstein, J. L., Boxsenberg, A., Carswell, R., and Shortridge, K. 1977, Ap.J., **212**, 186.

Greenstein, J. L., and Trimble, V. 1967, Ap.J. **149**, 283.

Guseinov, O. H., Novruzova, H. I., and Rustamov, Yu. S. 1983, Ap. Sp. Sci. **97**, 305.

Hamada, T., and Salpeter, E. E. 1961, Ap.J. **134**, 683.

Iben, I. 1982, Ap.J., **259**, 244.

Iben, I. 1984, Ap.J. **277**, 333.

Iben, I., and MacDonald, J. 1985, Ap.J. **296**, 540.

Kaler, J., and Jacoby, G. H. 1989, Ap.J. **345**, 871.

Kawaler, S., and Hansen, C. J. 1989, in White Dwarfs, ed. G. Wegner, (Springer-Verlag: Berlin), p. 97.

Koester, D., Weidemann, V., and Zeidler, K. T., and Sion, E. M. 1982, Astr. Ap. **116**, 147.

Koester, D., and Reimers, D. 1985, Astr. Ap. **153**, 260.

Koester, D., Wegner, G., and Kilkenny, D. 1990, Ap.J., in press.

Kudritzski, R. P., and Mendez, R. H. 1989, in Planetary Nebulae, ed. S. Torres-Peimbert, (Kluwer: Dordrecht), p. 273.

Kwitter, K. B., Massey, P., Congdon, C. W., and Pasachoff, J. M. 1989, A.J. **97**, 1423.

Liebert, J. 1988, PASP, **100**, 1302.

Liebert, J. 1980, Ann. Rev. Astr. Ap., **18**, 363.

Liebert, J., Wesemael, F., Husfeld, D., Wehrse, R., Starrfield, S. G., and Sion, E. M. 1989, A.J., **97**, 1440.

Margon, B., Katz, J. I., and Downes, R. A. 1981, Nature, **293**, 200.

Mauche, C,, Raymond, J., and Cordova, F. 1988, Ap.J., **335**, 829.

McGraw, J. T., Starrfield, S. G., Liebert, J., and Green, R. F. 1979, in IAU Colloquium No. 53,White Dwarfs and Variable Degenerate Stars eds. H. Van Horst and V. Weidemann (Rochester: University of Rochester), p. 377.

Michaud, G. 1987, in IAU Colloquium No. 95: The Second Conference on Faint Blue Stars, ed. A. G. D. Philip, D. S. Hayes, and J. Liebert (L. Davis Press: Schenectady), p. 249.

Mullan, D. J., Sion, E. M., Bruhweiler, F. C., and Carpenter, K. G. 1989, Ap.J.(Letters), 329, L33.

Nather, E., Robinson, E., and Stover, M. 1981, Ap.J., 244, 269.

Nousek, J. A., Shipman, H. L., Holberg, J. B., Liebert, J., Pravdo, S. H., White, N. E., and Giommi, P. 1986, Ap.J. 309, 230.

Oke, J. B., Weidemann, V., and Koester, D. 1984, Ap.J. 281, 276.

Pacynski, B. 1971, Acta Astr., 21, 417.

Pilachowski, C., and Milkey, R. 1987, PASP, 99. 836.

Renzini, A. 1989, in IAU Symposium No. 131: Planetary Nebulae, ed. S. Torres-Peimbert, (Dordrecht: Reidel), p. 391.

Romanishin, W., and Angel, J. R. P. 1980, Ap.J 235, 992.

Schmidt, G. 1987, in IAU Colloquium 95: The Second Conference Faint Blue Stars, ed. A. G. D. Philip, D. S. Hayes, and J. Liebert (Schenectady, NY: L. Davis Press), p.377.

Schonberner, D. 1981, Astr. Ap., 103, 119.

Schonberner, D. 1983, Ap.J., 272, 708.

Shaw, R. A. 1988, in IAU Symposium No. 131: Planetary Nebulae, ed. S. Torres-Peimbert (Dordrecht: Reidel), p.473.

Shipman, H. L., Liebert, J., and Green, R. F. 1987, Ap.J. 315, 239.

Shipman, H. L. 1989, in White Dwarfs, ed. G. Wegner, (Spinger-Verlag: Berlin), p. 220.

Sion, E. M., Fritz, M., McMullin, J., and Lallo, M. 1988, A.J. 96, 251.

Sion, E. M., Bruhweiler, F. C., Mullan, D. J., and Carpenter, K. G. 1989, Ap.J.(Letters), 341, L17.

Sion, E. M., and Liebert, J. 1977, Ap.J. 213, 468.

Sion, E. M. 1986, Pub. Astr. Soc. Pac. 98, 821.

Sion, E. M., Liebert, J., and Starrfield, S. G. 1985, Ap.J. 292, 471.

Sion, E. M., Liebert, J., and Wesemael, F. 1985, Ap.J. 292, 477.

Sion, E. M. and Wesemael, F. 1983, BAAS, 14, 915.

Spitzer, L. 1978, Physical Processes in the Interstellar Medium, (New York: Inter-science), p.55.

Trimble, V. and Greenstein, J. L. 1972, Ap.J., 177, 441.

Tylenda, R. 1989, in IAU Symposium No. 131: Planetary Nebulae ed. by S. Torres-Peimbert (Dordrecht; Reidell), p.391.

Vauclair, G., Vauclair, S., and Greenstein, J. L. 1979, Astr.Ap. 80, 79.

Vauclair, G., and Liebert, J. 1988, in Exploring the Universe with the IUE Satellite, ed. Y. Kondo, (Reidel: Dordrecht), p. 355.

Vauclair, G. 1989, in White Dwarfs, ed. G. Wegner(Springer-Verlag: Berlin), 176.

Wegner, G. 1989, in White Dwarfs, ed. G. Wegner (Springer-Verlag: Berlin), p. 401.

Wegner, G., Reid, N., and McMahan, R. 1989, in White Dwarfs, ed. G. Wegner, (Springer-Verlag: Berlin), p. 378.

Wegner, G. 1981, A.J. 86, 264.

Wegner, G., and Nelan, E. P. 1987, Ap.J. **319**, 916.

Werner, K., Heber, U., and Hunger, K. 1989a, in White Dwarfs, ed. G. Wegner, (Springer-Verlag: Berlin), p. 194.

Werner, K., Heber, U., and Hunger, K. 1989b, preprint (Boulder Workshop on Hot Stars held August, 1989).

Wesemael, F., Green, R. F., and Liebert, J. 1985, Ap.J. Suppl. **58**, 379.

Wielen, R. 1977, Astr.Ap. **60**, 263.

Winget, D., and Fontaine, G. 1982, in Pulsations of Classical and Cataclysmic Variables, ed. J.P. Cox and C.J. Hansen, (Boulder: Univ. Colorado Press), p.46.

Winget, D., Van Horn, H. M., Tassoul, M., Hansen, C. J., Fontaine, G., and Carroll, B. W. 1982a, Ap.J., **252**, L65.

Wood, P. R., and Faulkner, D. J. 1986, Ap.J. **307**. 659.

Zeidler, E-M., Weidemann, V., and Koester, D. 1985, Astr. Ap. **155**, 356.

# EVOLVED STARS AS PROBES OF MAIN SEQUENCE ANGULAR MOMENTUM AND MASS LOSS

M. H. Pinsonneault
Center for Solar and Space Research
Yale University
P. O. Box 6666
New Haven, CT 06511

ABSTRACT.  Models of the spindown of Population I subgiants have been constructed with three different scenarios : rigid rotation, local conservation of angular momentum, and internal angular momentum transport from rotational instabilities; their properties are compared.  The rotation of evolved stars relative to their main sequence progenitors provides evidence for differential rotation with depth.  Different lithium depletion mechanisms result in different total stellar lithium contents even for the same final surface abundance; as a result, lithium in subgiants can be used to distinguish between mechanisms.  No observational evidence for an increase in abundance during the subgiant phase (produced by gravitational settling) or a drastic decrease (required for substantial main sequence mass loss) is seen.

## I.  Introduction.

Evolved stars are powerful probes of angular momentum and mass loss on the main sequence.  In addition, their surface rotation velocities and abundances can be used to directly test post main sequence angular momentum loss and rotationally induced mixing.  During the subgiant phase stellar surface convection zones deepen, and stars that experienced no angular momentum loss on the main sequence will spin down when they develop a deep surface convection zone.  Studying this angular momentum loss, and comparing it with that for lower mass main sequence stars, can provide important clues about the generation of stellar magnetic fields and the nature of stellar winds.  The surface abundances of subgiants also probe the internal main sequence composition profiles of species; on the main sequence itself we can see only surface effects.

More evolved horizontal branch stars are also extremely useful indicators of main sequence properties.  Because they have experienced substantial mass loss on the giant branch, their surface rotation is a sensitive function of the rotation in the deep interior during the main sequence.  In Section II, we discuss the constraints on main sequence properties from the rotation of evolved stars.  In Section III we discuss subgiant lithium abundances as a probe of main sequence lithium depletion mechanisms.



*L. A. Willson and R. Stalio (eds.), Angular Momentum and Mass Loss for Hot Stars, 331–336.*
© 1990 *Kluwer Academic Publishers. Printed in the Netherlands.*

## II.  Rotation of Evolved Stars.

### II.A.  MASSIVE SUBGIANTS (~3.0 $M_\odot$>M>1.5 $M_\odot$)

Stars more massive than about 1.5 $M_\odot$ experience minimal main sequence angular momentum loss. *However, the observed steep drop in the surface rotation velocities of cool subgiants requires angular momentum loss* (Gray 1989; see also Simon and Drake 1989).  We present three cases for the subgiant evolution of a 2 $M_\odot$ star to test for the effects of different degrees of coupling between the surface and interior during this spindown :  1)  Rigid rotation enforced at all times;  2)  Local conservation of angular momentum; and  3)  Case 2, coupled with internal angular momentum redistribution caused by rotational instabilities (Figure 1).  For all sequences rigid rotation was enforced in convection zones and angular momentum loss was applied when the mass of the surface convection zone reached $10^{-4}$ $M_\odot$.  The angular momentum loss model used was initially calibrated on the Sun with the constant in the loss rate multiplied by different factors.  For all three cases, the constant in the angular momentum loss rate must be much larger than the value for the solar calibrated models to reproduce the observations because the slow rotation of the cool subgiants with deep surface convection zones implies a large total angular momentum loss in a short timescale (~ 10 Myr).

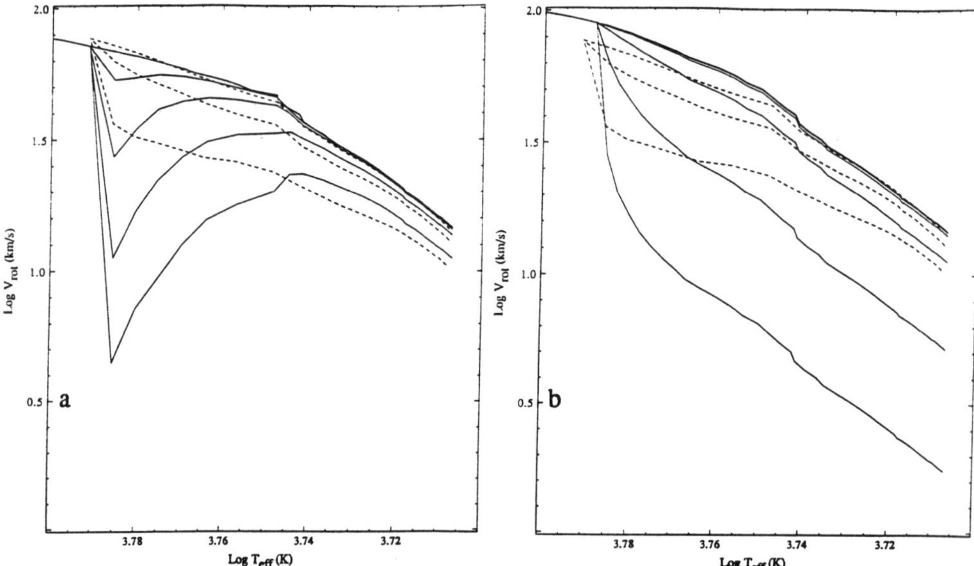



**Figure 1.**  Models of the 3 cases for spindown of a 2 $M_\odot$ subgiant :  Case 1 (solid lines, panel a), Case 2 (solid lines, panel b), and Case 3 (dashed lines, both panels).  The series of lines for each case represent different constants in the angular momentum loss law, with values 0, 10, 100, 1000 and 10000 times the solar calibrated value for Case 1; 0, 1, 10, 100, and 1000 for Case 2; and 10, 100, and 1000 for Case 3.

As we look at progressively higher mass stars they will experience progressively less angular momentum loss because the evolutionary timescale decreases.  Spindown as a function of mass is thus the best constraint on the angular momentum loss rate in massive subgiants.  We can then determine the degree of angular momentum transport from the interior to the surface. Models with local conservation of angular momentum are spun down rapidly when the surface convection zone is shallow. As rapidly rotating material from the inteior is incorporated into the surface convection zone, it spins back up; because of the increase in radius, however, the surface angular velocity is still lower than it was earlier. As a result, such models produce cool subgiants that rotate more rapidly than the other cases. *If the apparent observational lack of rapidly rotating cool subgiants is confirmed, angular momentum must be effectively transported from the interior to the surface.* However, this does <u>not</u> require rigid rotation; rotational instabilites and meridional circulation can efficiently transport angular momentum from the core to the surface even in this rapid phase of evolution.

## II.B. LOW MASS SUBGIANTS (M<1.3 $M_\odot$) AND METAL-POOR HORIZONTAL BRANCH STARS.

Low mass stars experience strong main sequence angular momentum loss. Their subgiant surface rotation velocities therefore depend primarily on the internal main sequence rotation because subgiant loss is minimal. If the entire core rotated rapidly, a drastic drop in rotation period as stars evolved onto the subgiant branch would occur; such a drop is not seen, in agreement with the solar oblateness and estimates of internal solar rotation from inversions of helioseismology data. *However, the rotation periods of subgiants, as inferred from chromospheric activity levels, increase only slowly with decreased $T_{eff}$; this indicates the presence of differential rotation with radius in the interior because rigid rotation would cause a steep increase in period* (Pinsonneault *et al.* 1989, hereafter PKSD).

**Metal-poor horizontal branch stars rotate rapidly, and their main sequence precursors rotate slowly. This combination of observations requires strong differential rotation with depth in main sequence metal-poor stars.** As low mass stars evolve off the main sequence and up the giant branch, they develop deep surface convection zones. As they lose mass on the giant branch, the bulk of the angular momentum (and mass) in the surface convection zone is lost. Therefore, $J_{HB} < J_{GB} < J_{MS}$ ,where $J_{GB}$ is the angular momentum below the maximum penetration of the surface convection zone on the giant branch, $J_{HB}$ is the *total* angular momentum of the horizontal branch star, and $J_{MS}$ is the angular momentum in the core at the turnoff.

If main sequence metal-poor stars rotate rigidly, they do not have enough internal angular momentum to explain the observed horizontal branch rotation velocities. We have constructed models of metal-poor stars which include angular momentum loss and internal redistribution of angular momentum. *These models rotate slowly enough at the surface to satisfy the observed upper limits, and retain enough internal angular momentum to explain the observed horizontal branch rotation velocities.*

## III. Lithium in Subgiants : A Sensitive Probe of Main Sequence Mass Loss and Lithium Depletion.

Lithium survives only in the cool outer layers of stars. The surface lithium abundances of stars decrease as they evolve (Hobbs and Pilachowski 1988, Pinsonneault, Kawaler, and Demarque 1989, hereafter PKD). A number of surface lithium depletion mechanisms have been proposed. Lithium is destroyed in nuclear reactions at the base of the convection zone when it is sufficiently deep. This mechanism must be taken into account in all models, and is important in the pre main sequence and in low mass (M<0.9 $M_\odot$) main sequence stars. Meridional circulation (Charbonneau, Michaud, and Proffitt 1989) and rotational instabilities driven by angular momentum loss (PKSD; PKD) will tend to mix lithium-rich material near the surface down to regions where the lithium is destroyed. Gravitational

334

settling of lithium (Michaud 1986) can also occur on a reasonable timescale for stars with thin surface convection zones. Finally, main sequence mass loss (Hobbs, Iben, and Pilachowski 1989) would progressively expel lithium-rich material until all of the lithium in the star was contained in the surface convection zone. Further mass loss would lead to progressive lithium depletion.

*Because the surface convection zones of subgiants are very deep, their surface lithium abundances allow us to measure the total lithium content of their main sequence precursors. The different mechanisms for surface lithium depletion lead to very different total lithium contents in stars, and can therefore be tested in subgiants.*

Lithium in standard models is nearly a step function, with the mass in the lithium containing region only weakly dependent on the total stellar mass. Models with main sequence mixing will have a lithium abundance that declines nearly linearly with radius across the lithium preserving region. For the same turnoff abundance, models with mixing would predict ~ 2 times lower lithium abundances for cool subgiants than standard models. Much of the lithium lost from the surface by gravitational settling would be stored in the layers below and not destroyed. For the same turnoff abundance, models with settling would have subgiant lithium abundances higher (by up to the apparent main sequence depletion) than standard models. For main sequence mass loss, the total lithium content would be reduced; the limit would be lithium only in the surface convection zone for a total main sequence mass loss in excess of ~ 0.025 $M_\odot$. This would lead to severe subgiant lithium dilution which is a strong function of mass.

We will need a larger sample of observations to test the predictions of each model fully. Based on current observations, however, we can make some preliminary tests. *Main sequence mass loss or mixing will lead to a dispersion in abundance at fixed $T_{eff}$; such a dispersion is seen on the main sequence and in evolved stars (PKD). Observations of evolved metal-poor stars are inconsistent with main sequence mass loss in excess of 0.02 $M_\odot$.* Observations of evolved Population I stars show no evidence for gravitational settling (Balachandran 1989) or lithium retained only in the surface convection zone. If significant lithium settling does occur, the lithium must be transported far enough to be destroyed.

**References.**


Balachandran, S. (1989), 'Lithium in M67 Subgiants', in <u>Proceedings of the 6th Cambridge Workshop on Cool Stars, Stellar Systems, and the Sun</u>, in press.

Charbonneau, P., Michaud, G., and Proffitt, C.R. (1989), 'The Lithium Abundance in Cluster Giants : Constraints on Meridional Circulation Transport on the Main Sequence', submitted Ap. J.

Gray, D.F. (1989), 'The Rotational Break for G Giants', *Ap.J.*, in press.

Hobbs, L.M. and Pilachowski, C. (1988), 'Lithium in Old Open Clusters : NGC 188', *Ap.J.*, **334**, 734-745.

Hobbs, L.M., Iben, I., and Pilachowski, C. (1989),'On Lithium Removal from G Dwarfs', *Ap.J.*, in press.

Michaud, G. (1986),'The Lithium Abundance Gap in the Hyades F Stars : The Signature of Diffusion', *Ap.J.*, **302**, 650-655.

Pinsonneault, M.H., Kawaler, S. and Demarque, P. (1989) (PKSD), 'Rotation of Low Mass Stars : A New Probe of Stellar Evolution', submitted Ap.J.Supp.

Pinsonneault, M.H., Kawaler, S., Sofia, S. and Demarque, P. (1989) (PKD), 'Evolutionary Models of the Rotating Sun', *Ap.J.*, **338**, 424-452.

Simon, T. and Drake, S.A. (1989),'The Evolution of Chromospheric Activity of Cool Giant and Subgiant Stars', *Ap.J.*, **346**, 303-329.

## FIGURE 5

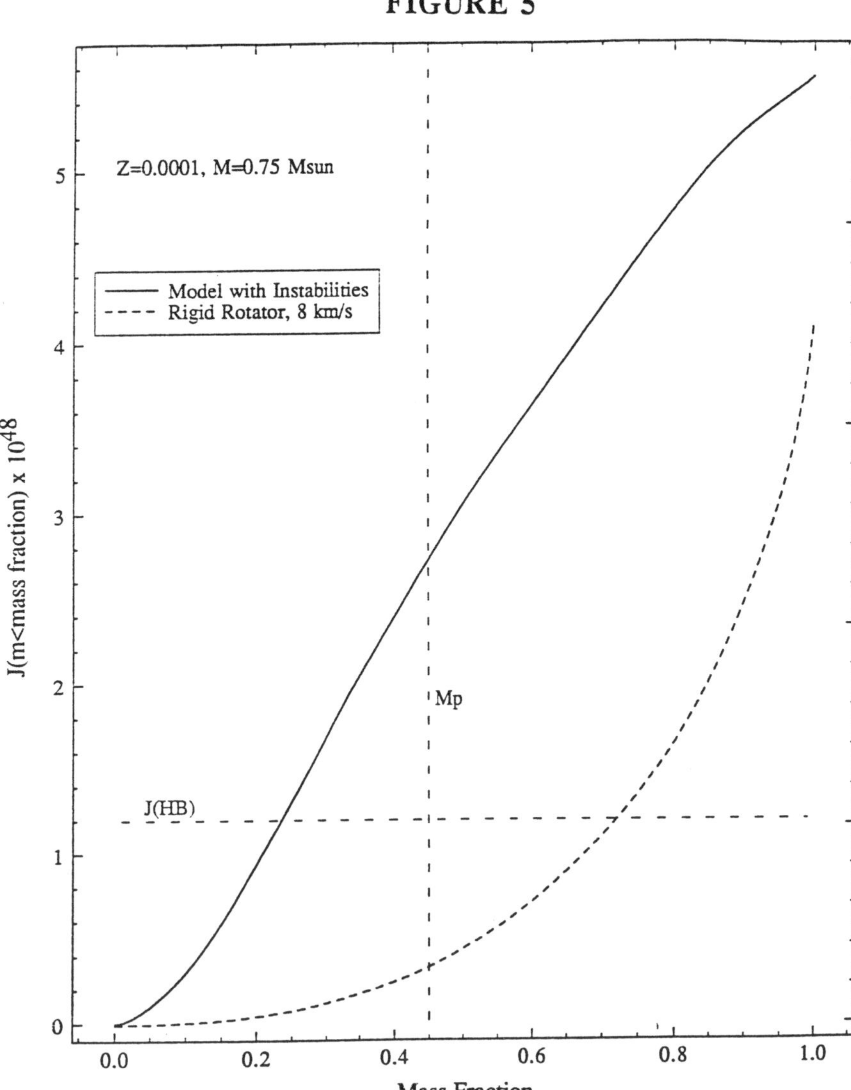



Figure 5 - Angular momentum interior to a given mass as a function of mass fraction for a metal-poor main sequence turnoff star. The solid line is for a model computed with internal angular momentum transport by rotational instabilities. The dashed line is for a rigidly rotating model with $v_{surf} = 8$ km/s, the upper limit from current observations. The degree required for the horizontal branch, and the maximum depth of the surface convection zone during the giant branch, are shown; rigid main sequence rotation is not consistent with the observed rapid horizontal branch rotation even in the unlikely case that the core donates no angular momentum to the envelope on the giant branch.

## FIGURE 6

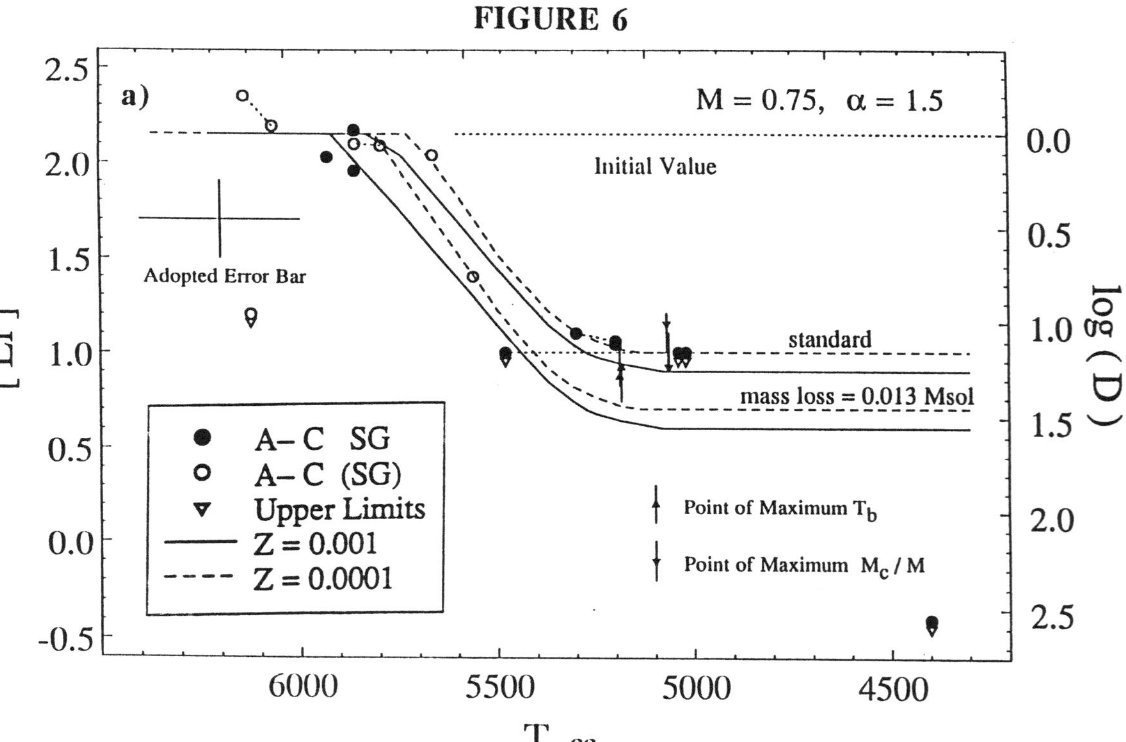


**Figure 6** - Lithium abundances of subgiants compared with theoretical dilution curves for standard models and ones with modest main sequence mass loss (panel a) and for models which have lithium only in the surface convection zone at turnoff (panel b).  Models are computed assuming a 16 Gyr turnoff; modest main sequence mass loss is allowed but higher rates pose serious observational problems.  The open circles are subgiant observations which are moderately metal-poor but more metal-rich than the models.

# THE BIZARRE KINEMATICS OF PLANETARY NEBULA NGC 7009, AND SOME THOUGHTS ON THE TRANSFER OF STELLAR ANGULAR MOMENTUM TO PLANETARY NEBULAE


Heather L. Preston (Astronomy Programs, Computer Sciences Corp.) and
Lee Anne Willson (Astronomy Program, Iowa State University)



ABSTRACT: In an effort to understand the formation of non-spherical morphologies, we are studying the possibilities for efficient transfer of the bulk of a star's angular momentum to its planetary nebula. The kinematics of planetary nebula (PN) NGC 7009 ("the Saturn Nebula") have been mapped using an echelle spectrograph (single-slit mode), with a 3" spatial resolution perpendicular to the slit (1" along-slit). The morphology of this nebula is very unusual, but its kinematics become pathological when an attempt is made to classify it by any of the morphological schemes (e.g., Balick, 1987, refs. therein). There appears to be an inner ring or shell with brightness enhancements aligned with a pair of outer ansae. But on a given "side" (with respect to the nucleus) the inner enhancement and its aligned ansa show velocities in opposing directions, such that if the inner shell were to be modeled as a rotating ring (25 km/sec), the outer ansae would appear to be situated on the edge of another ring, rotating more slowly (10 km/sec) and in the opposite sense. If, instead, the inner shell were modeled as an ellipsoidal wind-blown cavity with a collimation axis nearly in the plane of the sky (as can also be supported by the kinematic data), the counter-sense velocities of the ansae would imply a change in the inclination of the collimation axis of the flow, a phenomenon which is generally only seen in much higher-energy outflows, and which is often associated with precession of the collimation source.


In hydrodynamical models (Kahn and West, 1985; Icke, Preston, and Balick, 1989), a young nebula entering the fast wind phase in the "interacting winds" theory of PN development (Kwok, Purton and Fitzgerald, 1978) may develop a bipolar morphology due to the interaction of the fast wind with a polar-to-equatorial density gradient in the previously lost gas surrounding it, generally expected to be gas deposited by the high mass-loss, low velocity "superwind." In the absence of a match between the percentage of ellipsoidal or bipolar PNe (80%: Balick, 1987) and the percentage of close binary nuclei which might concentrate the ejected RGE mass in the plane (15%: Bond and Livio, 1989), angular momentum transfer from a single-star progenitor has often been invoked in the creation of the desired density gradient. Some mechanisms by which the stellar angular momentum might be efficiently transferred to the nebula are briefly discussed below. Although NGC 7009's central star has not been suggested in the literature as a binary or merged binary, the amount of angular momentum available from a

337


*L. A. Willson and R. Stalio (eds.), Angular Momentum and Mass Loss for Hot Stars, 337–342.*
© 1990 *Kluwer Academic Publishers. Printed in the Netherlands.*

single star is not capable of inducing the kind of rotational velocities which would be required in order that the "counterrotating rings" model fit observations. And the "precessing flow" model, because of its symmetry, seems to demand a binary.

In a previous paper (Balick, Preston, and Icke 1987), the Mach cones of ansae were discussed as a source of kinematic confusion: the expanding ring at each end of an ellipsoidal cavity would be expected to produce an unusual spectral signature as material "piled up" at the intersection of the two contact discontinuities. If such rings were nearly perpendicular to the plane of the sky and if they were quite optically thick, they might produce the observed blue/red pairing of inner shell "cap" and ansa at one end of the major axis, but not the red/blue pairing at the other end.

## 1. Transfer of Angular Momentum

The observations that (1) horizontal branch stars have observable rotation and (2) white dwarfs are slow rotators both support the assumption that transfer of angular momentum from the core to the (to-be-lost) envelope of the star occurs sometime between the red giant or asymptotic giant branch (AGB) and the point at which the star becomes visible again as the central star of a PN. The stars making that transition have deeply convective envelopes, which may overlap regions closer to the core which have been convective (i.e., the first and second "dredge-up" phases of Iben and Renzini (1983); stars that experience a "third dredge-up" have core masses greater than 0.7, so that a third dredge-up may not be counted upon for most PN nuclei). If convective coupling between these two zones in the star occurs more than once, a great deal of angular momentum can be transmitted from the core to the envelope.

Consider the first dredge-up (Iben and Renzini, 1983), which occurs when the base of the convective envelope extends inward to dredge up material that has been involved in hydrogen burning, i.e., material very close to the core. If the core (an n=3 polytrope) is undergoing near solid-body rotation, then the material convecting upwards will have significant amounts of angular momentum and the region of convection will "couple" the core's rotation to the envelope's rotation more strongly than at virtually any time since the star left its (protostellar) Hayashi track.

The next transfer of angular momentum occurs significantly later, after an electron-degenerate core has begun to form (after helium exhaustion). In addition to the normal deepening of the convective envelope, convection begins in the helium-burning shell just outside the core during thermal pulses, due to the combined effects of the sudden increase in luminosity from helium burning (plus release of gravitational potential energy from the contracting core) and the drop in temperature and increase in opacity in the overlying expanding layers. If the helium burning region has become, since contraction began, strongly coupled to the core's rotation, and it later becomes coupled by convective overlap to the envelope, it will act to "spin up" the envelope to whatever extent it can.

The details of the pulsational phase in the late AGB and in the transition to Miras, OH/IR stars, and ultimately PN nuclei are not yet fully understood. However, the combination of convective rotational coupling with heavy mass-loss seems very promising for explaining both points (1) and (2) mentioned above and possibly the relatively clean break that appears to occur in multiple-shell PNe (i.e., major angular momentum transfer episodes from the core to the envelope roughly map onto "shells," with the final, and most efficient, transfer at the end of the high-mass-loss phase).

This convective coupling proposition, and especially the conjecture that the coupling is most effective at the very end of the high mass-loss phase, is supported by existing observations of highly-processed material in the inner portions of PNe, such as the anomalously hydrogen-poor blobs (ansae) found in the central few arcseconds of the otherwise chemically normal nebulae, Abell 30 and Abell 78 (Jacoby and Ford, 1983; also Grauer et al., 1987). Other researchers have proposed that as many as 10% of PNe may be similar to A 30 and A 78 in their chemical development, and may ultimately evolve into R Cor Bor stars (Iben et al., 1983); they explain the depletion of surface H by a combination of deep mixing and a final pulse which turns almost all of the remaining H to He in the envelope. The dynamical mass-loss mechanism, however, appears to change or at least intensify along with the Y, Z enrichment. A 30, for instance, is spherically symmetric in its outer portions, but its H-poor ansae are paired linearly. In this it may be an extreme example of late-transition convective coupling, and research on this mechanism is ongoing in the hope that it might explain the origin of nonspherically symmetric, single-star planetary nebulae without invoking merged binary cores (which almost certainly exist, but which are too uncommon to explain such a large number of nonspherically symmetric nebulae). Note that this picture of the late stages of the high-mass-loss phase does not conflict with the two-winds model of subsequent PN development in any way, rather it is being investigated as a cause of the initial nonspherical symmetry in the density distribution of the low-velocity component of the PN gas.

In the case of NGC 7009, our calculations have shown that a single star could not provide the amount of angular momentum needed to cause either "ring" to rotate at the observed velocity, but the angular momentum present in a merged binary might be sufficient. On the other hand, the apparent counter-rotation of such "rings" argues against their reality. The "precessing central collimation source" model, ugly as it seems, may be the only remaining choice, as it preserves symmetry across the observed object. If the gas around the nebula were the only thing collimating the outflow, then a density perturbation in that gas might change the direction of the flow, but there is no physical motivation for that direction change to be mirrored at the opposite end of the bipolar wind-blown cavity. Unfortunately, there is also no evidence of duplicity in the nucleus of NGC 7009, aside from the circumstantial need for a collimation source which can precess.

## 2.  The Observations

## 2.1.  Kinematics

Kinematic data were obtained in the lines Hα λ6563 Å and [N II] λ6854 Å using the echelle spectrograph and CCD detector on the Mayall 4-m telescope of Kitt Peak National Observatory. The spectrograph was used in single-slit mode, and the particulars of calibration, etc. are presented in Balick and Preston (1987). Velocities discussed in this paper are relative to the systemic velocity of the PN.

The long slit of the echelle spectrograph was placed across the PN image as it appeared on the TV guider at one edge of the nebula, then subsequent exposures were taken by stepping the slit in 3" increments across the image in a direction perpendicular to the length of the slit. Integration times were the same at all offset slit positions. The slit was 1" in width. The seeing was close to 1.5" at all times and the skies were generally photometric.

## 2.2.  Morphology

Images of NGC 7009 (Fig. 1, a) and the rest of the sample of PNe (Balick, 1987) were obtained with the TI 2 direct CCD camera at the f/7.5 focus of the 2.1-m telescope at KPNO in May and November, 1985. A variety of filters was used to give coverage in lines associated with different physical regimes of the nebula. Exposures were attempted in the lines of H$\alpha$, H$\beta$ $\lambda$4861 Å, [O I] $\lambda$ 6300 Å, [N II] $\lambda$6584 Å, [O III] $\lambda$5007 Å, and He II $\lambda$4686 Å. Figure 1(a) presents H$\alpha$ and [N II] images. H$\alpha$ images are contaminated by emission from the [N II] lines at $\lambda$ $\lambda$ 6548 and 6583 Å, and likewise the [N II] image is contaminated by light from H$\alpha$. But the effects are less than 5% and no attempt was made to remove them. Seeing was typically 0.8" to 1.3" and the weather was generally photometric. The integration times for both the kinematic data and the images were adjusted so that the CCD pixels were not saturated by the brightest portion of the nebula. Also the chips used in all but the May '85 observations were preflashed to eliminate low light level charge transfer problems. Bad columns of CCD pixels can be seen in both the images and the spectral data -- the observations were set up to avoid such areas whenever possible.

All images and spectra were corrected for instrumental effects using the standard software available at the telescopes. The corrected data were then processed using AIPS software at the University of Washington's Astronomical Image Processing Facility.

## 2.3.  Data Format

In addition to presenting the kinematic data as a series of images of spectral lines (poster only), we have presented them as a series of "isovelocity images," (Fig. 1, b) similar to the channel maps so familiar to radio astronomers. This is possible because of the way the data are stored. Direct spectra contain information in one spatial  dimension (along the slit: hereafter the $y$ direction) and in one velocity ($v$ ) dimension (radial velocity). By stepping the slit along the surface of the nebula, we obtain information on an additional dimension (the $x$ direction) in the direction of the offset. The resolution in this third dimension is not as good as the resolution along the slit, but the $y$ -dimension data can be smoothed and compressed to an appropriate size. Then complete images of the nebula at velocity intervals corresponding to the $v$ -direction separation between pixels (usually 8.9 km/sec) can be produced. This is done by storing the data as a "cube" -- for each pixel there are three coördinates ($x, y,$ and $v$ ) plus intensity information stored. If the data are displayed as slices along the $v$ axis, they are equivalent to channel maps; if as slices along the $x$ axis, the images are the usual spectral lines, albeit trimmed closely. Isovelocity images are exceedingly useful for attempting to determine the 3D structure of PNe.

## Acknowledgements

The authors acknowledge with pleasure the contributions made to our efforts to understand the evolution of PNs through conversations, collaboration, and/or correspondence with: Bruce Balick, George Bowen, Vincent Icke, Steve Kawaler, Curt Struck-Marcell, and Detlef Schönberner.

Russell Owen and Julie Lutz are warmly thanked for their part in the early stages of the data acquisition, as are the staff members of KPNO and NOAO who helped at every turn and were key to the success of the observational portion of the research, especially Caty Pilachowski.

Financial support (LAW and HP) was received from the NASA grants NAGW1364 and NAG5-707, and from NOAO for HP's transportation and lodgings during observing runs. We are also grateful to NATO and NSF for the support that made HP's attendance at this workshop possible. The Astronomical Image Processing Facility at the University of Washington was made possible by grants from the NSF (AST 83-10552) and the Digital Equipment Corporation.

## Bibliography

Balick, B., 1987 Astron. J. **94**, 671.
Balick, B. and Preston, H. L., 1987 Astron. J. **94**, 958.
Balick, B., Preston, H. L., and Icke, V., 1987 Astron. J. **94**, 1641.
Bond, H., and Livio, M., 1990 (preprint).
Grauer, A. D., Bond, H. E., Liebert, J., Fleming, T. A., and Green, R. F., 1987 Astrophys. J. **323**, 271.
Iben, I., Kaler, J. B., Truran, J. W., and Renzini, A., 1983 Astrophys. J. **264**, 605.
Iben, I., Jr., and Renzini, A., 1983 Ann. Rev. Astron. Astrophys. **21**, 271.
Icke, V., Preston, H. L., and Balick, B., 1989 Astron. J. **97**, 462.
Jacoby, G. H., and Ford, H. C., 1983 Astrophys. J. **266**, 298.
Kahn, F. D., and West, K. A., 1985 Mon. Not. Roy. Astron. Soc. **212**, 837.
Kwok, S., Purton, C. R., and FitzGerald, M. P., 1978 Astrophys. J. Lett. **219**, L125.

Next page: Figure 1. *a.* Direct-CCD images of NGC 7009 (from Balick, 1987). [N II] is on the left, Hα on the right. The long dimension of the frame is 35 arcseconds. The contrast is adjusted logarithmically, in order to show faint features more clearly. East is up, N to the right.

Figure 1. *b.* Isovelocity images of NGC 7009. The set of frames shows the PN as it would appear through a set of very narrowband filters of width and relative offsets of 8.91 km/sec. The final frame in the sequence is a summation of all of the isovelocity images. The middle frame of the individual images is the zero-velocity frame. Blueshifted velocities are to the left. The slit was aligned with the ansae, at PA=77°.

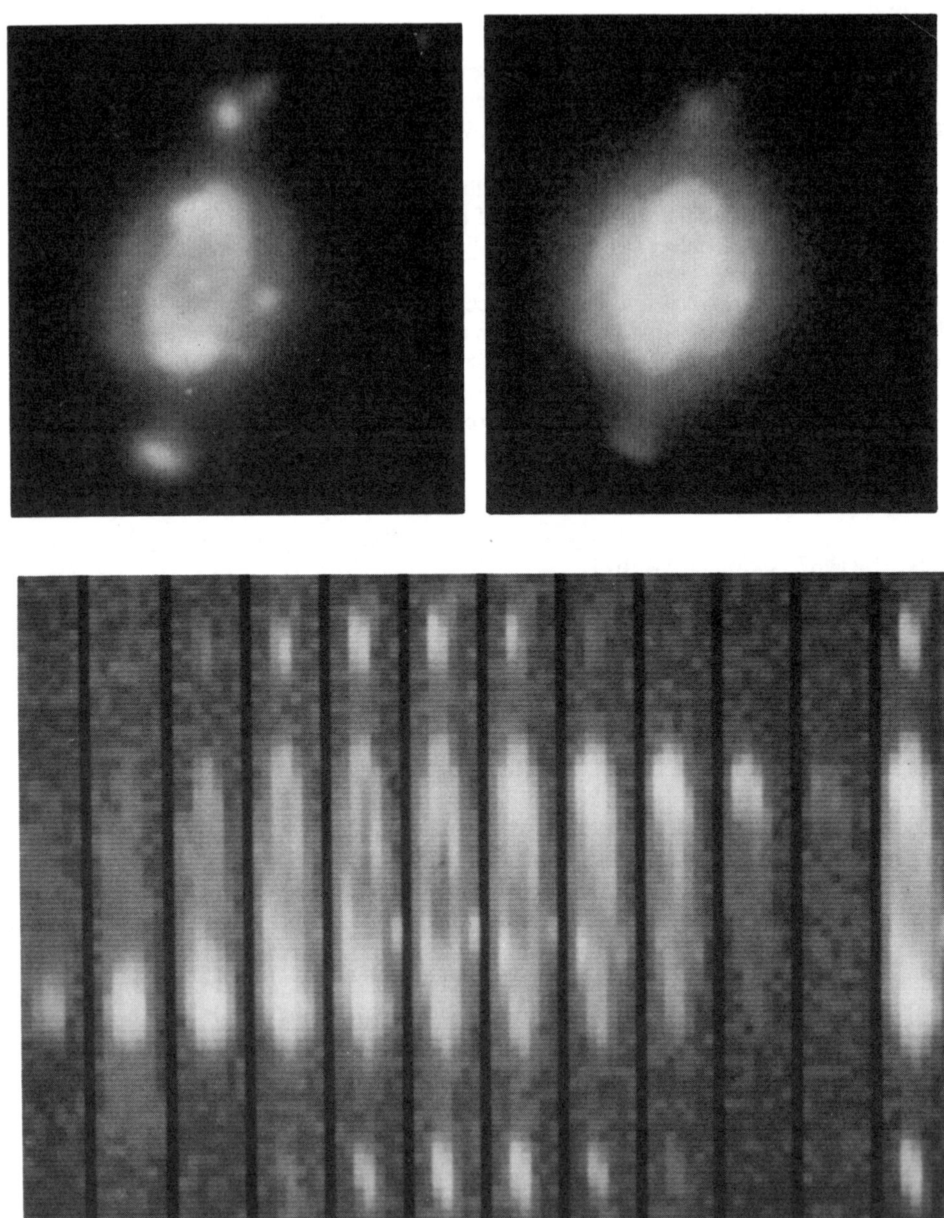

Figure 1.*a*. Direct-CCD images of NGC 7009 in [N II] and Hα. E is up.  *b*. Isovelocity images of NGC 7009.  Successive frames are 8.91 km/sec apart.  Blueshifted to the left.

# CHROMOSPHERIC Hα ACTIVITY IN α ORI

Myron A. Smith and Carol A. Grady
IUE/CSC Observatory
10000-A Aerospace Rd.
Lanham-Seabrook MD 20706

Abstract.    In hot stars the Hα line is formed substantially in the upper photosphere. In cool supergiants it is formed in the chromosphere where it is hot enough to ionize hydrogen. It has often been stated that the Hα line is stationary in α Ori, but evidence is presented herein that it varies on the dynamic timescale for the inner chromosphere. We suggest that this behavior is consistent with magnetic structures that appear and evolve on the surface of the star, and that it provides encouragement for Alfven wave dissipation models for supporting the chromosphere.

## 1. What Supports the Chromosphere?

Chromospheres in α Ori and other M supergiants are known to be distended relative to QS chromospheric models (Hartmann and Avrett 1984; "HA"). The HA model suggests a chromospheric maximum in the region of 1.5R*-4R*, in quantitative agreement with the speckle observations of Hebden et al. (1987). The current consensus is that the most likely way to excite and mechanically support a chromosphere is through the dissipation of either waves or pulsational shocks. HA find that their dynamic wind envelope models, supported by Alfven wave dissipation, predict a rather large terminal velocity. Also, their static models are unstable to perturbations. They suggest that fountains of ejected and returning material can preserve the robustness of dynamic models. Unfortunately this picture, including the assumption that magnetic fields exist at all in α Ori is ad hoc. On the other hand, pulsationally driven models seem to be discouraged because: (a.) the pulsations have a low semiamplitude (1-2 km s$^{-1}$), causing only mild shocks in the chromosphere and (b.) IR spectrophotometry indicates an absence of dust close to the atmosphere, prohibiting a feedback mechanism that permits the dust to amplify the pulsation velocities and hence atmospheric distension, as suggested in Mira stars (Willson 1988).

The question of interest is what drives the chromosphere in M supergiants, magnetic activity or pulsations? We will exhibit evidence that Hα varies on the dynamic timescale for the inner chromosphere. We suggest that this behavior discourages pulsation and favors magnetic activity, and therefore Alfven wave dissipation mechanisms.

## 2. Observations of Hα

As part of the National Solar Observatory's "solar-stellar connection" program, the McMath telescope has been used to conduct Hα observations of α Ori, α Sco, and α Her for five years. From 1984 to 1987 February a Reticon detector was used for the observations; at



*L. A. Willson and R. Stalio (eds.), Angular Momentum and Mass Loss for Hot Stars, 343–346.*
© 1990 *Kluwer Academic Publishers. Printed in the Netherlands.*

344

that time a TI 800×800 CCD detector was substituted. The pertinent spectroscopic quality figures include a S/N of 300-1000 per pixel, a spectral resolution of 40,000, and a dispersion of 81 mA per 15$\mu$m pixel. The observations were conducted at intervals of a few days to a few weeks during the observing season. In each case an observation was accompanied immediately by one of $\alpha$ Tau to provide a wavelength template. As of 1989 April, 93 observations have been made.

## 3. Results.

Smith (<u>et al.</u> 1989) show that the radial velocities of the H$\alpha$ line show almost no correspondence to the RVs of neighboring metallic lines, whether over short or long timescales, over the same period, a result also suggested by less extensive data of Goldberg (1979). In addition, the RVs measured from the inner core of H$\alpha$ correlate with those measured from the lower half of the profile. An examination of the profiles below examines this in detail, and shows that whatever is responsible for H$\alpha$ activity includes a large portion of the profile. Since it also includes regions where the optical depth in the line is low, and the phenomenon extends down close to the photosphere.

In Figure 1 and 2 three H$\alpha$ profiles are overplotted, both for $\alpha$ Ori and $\alpha$ Tau. The small fluctuations in Figure 2 give a realistic estimate of the spectrophotometric errors (including H$_2$O line contamination) for the $\alpha$ Ori profiles since they were observed at virtually the same time and in nearly the same region of the sky.

A complete examination of the individual H$\alpha$ profiles shows two types of variations. The first is a "swaying" of the line relative to its own shoulders and nearby metallic lines. Although these variations generally take a year to develop, on some occasions one can observe systematic 2-3 km s$^{-1}$ shifts in only 6 weeks. Simulations with model profiles show that the fixed position of the "clean" blue wing, while the line core moves, precludes global motions such as from a coherent radial expansion of the chromosphere. Instead they can be more easily modeled by hot, H$\alpha$ emitting centers with an area of perhaps 30-50% of

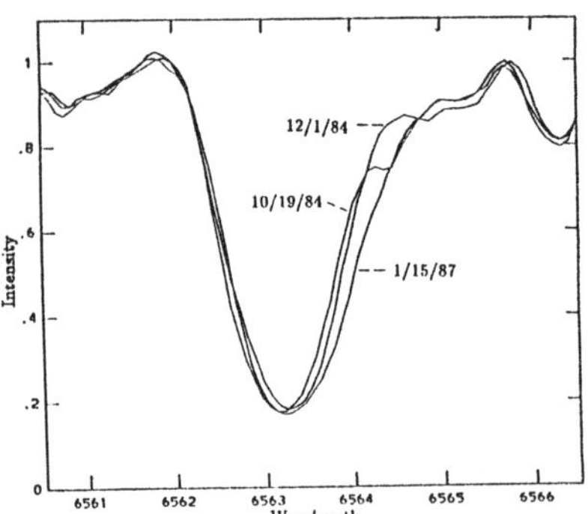

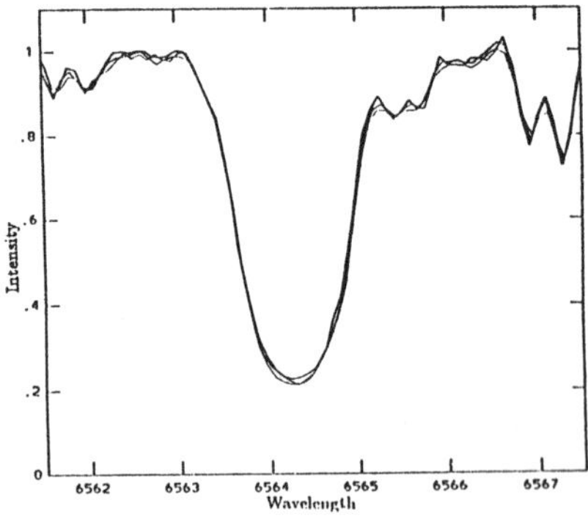


<u>Figures 1 and 2.</u> — Three pairs of observations of the H$\alpha$ profile of $\alpha$ Ori (left) and of $\alpha$ Tau (reference), showing short-term variations along the $\alpha$ Ori profile.

the visible disk, and with an upward velocity slowly varying between 0 and 15 km s$^{-1}$. The origin of this activity, equally apparent in H$\alpha$ observations of $\alpha$ Sco as well, is not understood, but it could be related to reports of large H$\alpha$ structures from speckle observations. It is premature to say "how aperiodic" these variations are, though a single periodicity may be ruled out. Moreover, they do not seem to be responsible for the long-term trend shown in H$\alpha$ RVs. This swaying activity appears to be present nearly all the time. in $\alpha$ Ori.

A second type of activity is an irregular broadening or narrowing of the line profile, generally on the red side, which can extend over an Angstrom. It is generally responsible for the long-term RV pattern shown in Figure 1. When broader, the line generally shows a depressed red shoulder and is also deeper at line center. As remarked by Weymann (1962), it is difficult to decide whether an asymmetry in H$\alpha$ is caused by an excess of emission or absorption at different times. However, because the line is comparatively narrow and symmetric most of the time, we infer that profiles such as those of 1987 January 15 (Fig. 1, see also Fig. 3) result from an anomalous absorption to the red of the line core. On that date the absorption extended out to +1.0-1.5A. On other dates a tapering of the red absorption wing can be followed out to 2 Angstroms.

As demonstrated by the observations of 1984 October 19 and December 1, smaller but credible profile fluctuations can develop and persist close to the line center, even over 6 weeks. In addition, low velocity absorption bumps can form close to the line center over the same timescale, as shown in Figure 3. Clearly, H$\alpha$ shows marked variations up to +50 km s$^{-1}$ over timescales of 0.1 to 1 yr.

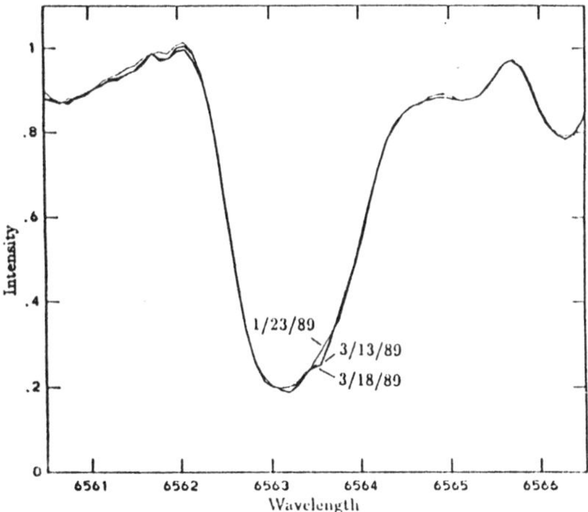


Figure 3. — Three H$\alpha$ observations of the H$\alpha$ profile showing low velocity structure developing over 6 weeks.

4. Interpretation.

Observations of H$\alpha$ alone are not enough to develop a unique picture of the chromosphere. However, our data support the interpretation of Boesgaard (1979) and HA that upwellings of material are injected into and return from the chromosphere. Using a free-fall velocity given by assumed stellar parameters, it would be expected that infall would attain velocities of 50 km s$^{-1}$ if it falls from 2 R$^*$ to the surface. The same parameters correspond to a free-fall time of about one year, consistent with the variability timescale. Our observations are the first direct evidence for aperiodic dynamic motions in the chromosphere. They lend support to Wilson's (1960) initial conjecture that hot eruptions erise "invisibly" and can be observed mainly as cooling infall.

Boesgaard interpreted redshifted Fe II emission as evidence for infall, and attributed it to tossing from convective cells. However, while Smith <u>et al.</u> indeed notice RV metallic line variations on short timescales, they found that these fluctuations did not propagate to the H$\alpha$ line, even to the midpoint of the profile. We argue that the comparatively large velocities required to lift material out to the chromosphere make it difficult to attribute this activity to the mild (subsonic) velocities expected in convective cells. Rather, because of the large optical thickness at these wavelengths, H$\alpha$ and radio continuum variations must denote activity well above the stellar surface. Moreover, because the velocities are consistent with material returning to a point near the surface, we suggest that the chromosphere consists of both high and low density structures. The latter would permit occasional glimpses by an external observer of colliding material near the photosphere/chromosphere interface.

The picture suggested by these temporal variations is probably most consistent with transient magnetic structures that regulate the mean density stratification and the wind flow in the chromosphere. Nearby in the H-R Diagram there is additional support for such activity: density structures that extend to several stellar radii have been observed by the IUE in certain eclipsing late supergiants (e.g., Schroder 1983, Ahmad 1988).

Finally, this evidence seems to strengthen the case for the chromosphere of $\alpha$ Ori being excited and supported by the absorption of magnetic wave energy. The RV variations of metallic lines show that the pulsational amplitudes are rather small. Nonetheless, in advance of HD modeling, it is probably too early to rule out distension of a pulsationally driven "calorisphere" (Wilson 1988). The best evidence for any pulsational coupling to the chromosphere is the brief strengthening of Mg II emission at a phase following the maximum compression (Dupree <u>et al.</u> 1987), suggesting that the inner chromosphere may be affected by pulsational shocks. However, as indicated by the H$\alpha$ variations, it is this very region that is likely to suffer particularly violent penetrations from upwellings. Hence this activity may well dominate pulsations in most of the chromosphere.

5. References.


Ahmad, I. A. 1988 A Decade of UV Astronomy with the IUE Satellite, ed. E. J. Rolfe (ESA SP-281), p. 217.

Boesgaard, A. M., 1979, <u>Ap. J.</u>, 232, 485.

Goldberg, L. 1979, <u>Q. Journal Roy. Astr. Soc.</u>, 20, 361.

Hartmann, L., and Avrett, E. H. 1984, <u>Ap. J.</u>, 284, 238 ("HA").

Schroder, K.-P. 1983, <u>Astron. Astrophys.</u>, 151, L5.

Smith, M. A., Patten, B. M., and Goldberg, L. 1989, <u>A. J.</u>, in press (December).

Willson, L. A. 1988, A Decade of UV Astronomy, <u>op cit.</u>, p. 29.

Wilson, O. C. 1960, <u>Ap. J.</u>, 132, 136.

STELLAR WINDS IN A-TYPE SUPERGIANTS.


A.  Talavera *
ESA IUE Observatory
VILSPA. P.O.Box 54065
28080 Madrid, Spain


ABSTRACT.  A study of the ultraviolet spectrum of the A-type
supergiants based on IUE data allowed us to divide  them  in
two groups:  Group I and Group II.  The Group I contains the
less  luminous  supergiants;   these   stars   show   weak
indications  of mass-loss only in the Mg II resonance lines,
where an absorption  component  of  variable  blueshift  and
intensity  hes  been detected.  On the other hand, the stars
included in the Group II show strong evidence  of  wind  and
mass-loss  in  the  UV resonance lines of Mg II, C II, Si II
and Fe II, and the terminal velocity of the wind from the Mg
II lines is about 250 km/s;  the most luminous A supergiants
belong to this group.

An observational programme to study  the  optical  mass-loss
indicators  in  these  stars has been carried out during the
last year at La Palma and Calar Alto Observatories.  The  Hα
line  has  been observed in a large sample of A-supergiants.
The Hα emission seems to be well correlated with the stellar
luminosity.   The   stars  belonging  to  Group  I  present
symmetric absorption profiles, while the stars  included  in
Group  II  show  asymmetric  profiles,  some of them with an
emission  component  or  P-Cygni  profiles.   We  study  the
relation  between  the  Hα equivalent width and the absolute
magnitude in A-type supergiants.


NOTE.  The work presented in this poster has been  published
in the following articles:

Talavera,  A.   and Gomez de Castro, A.I.  (1987) ' The
    UV  high  resolution spectrum of A-type supergiants',
    Astron.Astrophys., 181, 300-314.

---

* Affiliated with the Astrophysics Division, Space Sciences
Department Estec, Noordwijk



*L. A. Willson and R. Stalio (eds.), Angular Momentum and Mass Loss for Hot Stars, 347–348.*
© 1990 *Kluwer Academic Publishers. Printed in the Netherlands.*

Talavera, A. and Gomez de Castro, A.I. (1987), 'Stellar winds in A-type supergiants', in P. Harmanec (ed.), Pub. Astron. Instit. of the Czechoslovak Academy of Sciences. Pub. No. 70, Prague, pp. 207-210.

Talavera, A. and Gomez de Castro, A.I. (1989), 'Stellar winds in A-type supergiants', in G.Tenorio-Tagle, M.Moles and J.Melnick (eds.), IAU Coll. No.120, Springer-Verlag, in press.

# THE EVIDENCE THAT WOLF-RAYET STARS ARE IN A LATE STAGE OF EVOLUTION


H.J.G.L.M. LAMERS[1], A. MAEDER[2], W. SCHMUTZ[3] and J.P. CASSINELLI[1]
[1]Dept of Astronomy, University of Wisconsin, Madison, USA
[2]Observatoire de Geneve, sauverny,switzerland
[3]J.I.L.A.,University of Colorado, Boulder,USA



ABSTRACT. We present evidence that Population I Wolf-Rayet stars are a late evolution stage of massive stars when most of the hydrogen is lost from the star. In this stage the WR-stars are expected to be Helium stars with possibly a thin H-rich layer in the earliest WR-phase. These arguments are presented here because during various occasions at this meeting it was suggested that Pop I WR-stars are young pre-main sequence objects (cf. A.B. Underhill, this volume).


## 1. The Abundances of Wolf-Rayet Stars

There are two types of WR-stars: the WN stars with strong N lines and weak C and O lines, and the WC stars with strong C and O lines and weak N lines. Both types have strong He and weak or absent H lines. These abundances indicate that WR stars are in a late evolutionary stage.

a) The H/He ratio: The strong He II lines and the very weak or absent H lines clearly show that the H/He abundance of WR-stars is considerably less than the solar value (Conti et al. 1983a, Hamann et al. 1988).

b) The C/N ratio: The most reliable determinations of C/N are those by Hillier (1988; 1989). He finds the following abundance ratios by number: C/N = 0.07 and N/He = 0.004 for the WN-star HD50896 and C/He = 0.5 for the WC-stars HD165763 with an accuracy of about a factor of 3. The non-solar abundances in WN stars are supported by the studies of the ring nebulae around WN stars by Rosa (1989) and Smith (1989).

c) Conclusions about the abundances: The abundances mentioned above are basically in agreement with the predicted abundances of CNO (WN stars) and triple-alpha (WC stars) processed material (see Section 2). Therefore, there is no doubt that we see the products of nuclear burning at the surface of WR stars.

## 2. WR-Stars as Predicted Endpoints of the Evolution of Massive Stars

The comparison of the observed properties of WR stars (cf. Abbott and Conti, 1987) with the results of evolutionary models unambiguously supports the identification of most WR stars as bare cores in post main-sequence evolution. The main arguments are the following




L. A. Willson and R. Stalio (eds.), Angular Momentum and Mass Loss for Hot Stars, 349–352.
© 1990 Kluwer Academic Publishers. Printed in the Netherlands.

ones.

a) The various WR subtypes WNL, WNE (L for late, E for early), WC and WO correspond to different stages of visibility of the nuclear products typical of the CNO-burning and partial He-burning stages. The changes of abundance ratios are quite large: for example, N/C = 1/4 at solar surface, while values of about 50 are predicted for CNO equilibrium and of 0 for He-burning, (cf. Chiosi and Maeder, 1986). The observed and theoretical values agree within the observational uncertainty.

b) From the observed values of the mass loss rates in O-stars and supergiants (cf. average values by de Jager et al. 1988), it is unavoidable that bare cores form at some stage during the He-burning phase of massive stars (initial mass larger than about 40 $M_\odot$). If the WR-stars are not the bare cores, then which stars are?

c) The observed mass loss rates in WN stars are sufficient to lead, at least in a fraction of them, to the visibility of the products of He-burning (WC stars). This provides a natural explanation for the WC-stars.

d) Bare cores with increased mean molecular weight are predicted to be overluminous with respect to their actual masses and to obey a specific mass-luminosity relation (cf. Maeder, 1983). So far there is only one WR-star in a binary system with a reliable luminosity and mass: V444 Cyg with $L=10^{5.1}\ L_\odot$ (Schmutz et al. 1989a) and $M=10\ M_\odot$ (Smith and Maeder, 1989). This agrees with the predicted mass luminosity relation for bare cores by Langer (1989).

## 3. The Number Ratio Between WR and O Stars

a) The typical number ratio of WR/O stars is about 0.1 to 0.2 in the solar neighbourhood (cf. Conti et al. 1983b). It agrees well with predicted values, if the WR stage correspond to a part of the He-burning phase. If WR stars would be in pre-MS evolution, the above ratio should be less than 0.01.

b) The number ratios WR/O and WN/WC change with galactocentric distance and are also different in other galaxies. These number ratios correlate well with metallicity Z (see for example Arnault et al. 1989). Both the observed values and the changes with Z can be well explained by the new evolutionary models, in which the mass loss rates in O and supergiant stars depend on Z as given by the stellar wind models (Kudritzki et al. 1987; Maeder, 1990).

## 4. Transition Objects Between Of and WR Stars

Walborn (1982) and Stahl (1986) discovered stars with characteristics in between those of Of stars and WR stars. These are the Ofpe/WN9 stars. Their optical spectra show emission lines of high and low ionization stages. This indicates that the Ofpe/WN9 stars are in a transition phase between the Of stars and the WN stars. This is supported by the fact that the Ofpe/WN9 stars are enriched in N and He in their atmosphere and in their surrounding nebulae.

The presence of transition objects between the Of and the WN phase and their abundance anomalies (Schmutz et al. 1989b) proves that the Of phase (i.e. evolved O-star phase) and the WN phase follow each other in an evolutionary sequence.

## 5.  A WR-Star in a Binary with a Compact Companion

The WN7 star HD 197406 is in a binary system with a mass function of 0.28 $M_\odot$.
Polarimetric observations indicate a large inclination angle of 67°.  If the mass of the WN7
star is assumed to be 60 $M_\odot$, the mass of the companion is 12.4 $M_\odot$ (Drissen et al.,
1986). The system is at distance of 800 pc from the galactic plane, which indicates that it
is a runaway system.  The large Z-distance suggests that the system had an asymmetric
Super-Nova outburst. If that is the case, the companion must be a compact object, i.e.
black hole.

The existence of a binary sytem consisting of a WR-star and a compact object does not
agree with the idea that the WR-stars are pre-main sequence stars.  On the other hand it is
easily explained if the WR-star is a bare-core, because both the WR-star and the compact
star are late evolution stages of massive stars.

## 6.  Pre-Main Sequence Massive Stars

Wood and Churchwell (1989) discuss the morphologies and physical properties of the
Ultra-compact H II (UC H II) regions that surround young massive stars, and show that
these are massive young stars.  The radiation from the central star is nearly all absorbed by
the circumstellar dust and re-emitted in the infrared near 100 micron.  Wood and
Churchwell present evidence that young massive stars remain in the UC H II regions for
about 15 percent of their main-seqence lifetime.

The WR-stars are not heavily obscured by circumstellar dust.  Their relatively small
near-infrared excess can often be explained as free-free emission arising from the stellar
wind (Hillier et al., 1983).

So young massive pre-main sequence stars have been observed and they are very
different from WR-stars.  This shows that WR-stars are not young main-sequence stars.

## Conclusion

We conclude that there is overwhelming evidence that the Pop I WR stars represent the
late evolution stage of massive stars.

The arguments will be descibed in more detail in a forthcoming paper in Ap. J. by the
same authors.

## References


Abbott, D.C., Conti, P.S. 1987, Ann. Rev. Astron. Astrophys.  25, 113.
Arnault, P., Kunth, D., Schild, H. 1989, Astron. Astrophys. 224, 73.
Chiosi, C., Maeder A. 1986, Ann. Rev. Astron. Astrophys. 24, 329.
Conti, P.S., Leep, E.M., Perry, D.N. 1983a, Astrophys. J. 268, 228.
Conti, P.S., Garmany, C., de Loore, C., Vanbeveren, D. 1983b, Astrophys. J. 274,
302.
de Jager, C., Nieuwenhuijzen, H., van der Hucht, K.A. 1988, Astron. Astrophys.

Suppl. Ser. 72, 259.

Drissen, L., Lamontagne, R., Moffat, A.F.J., Bastien, P., Seguin, M. 1986, Astrophys. J. 304, 188.

Hamann, W.-R., Schmutz, W., and Wessolowski, U. 1988, Astron. Astrophys. 194, 190.

Hillier, D.J. 1988, Astrophys. J., 327, 822.

Hillier, D.J. 1989, Astrophys. J., 347, in press.

Hillier, D.J., Jones, T.J., Hyland A.R. 1983, Astrophys. J. 271, 221.

Kudritzki, R.P., Pauldrach, A., Puls, J. 1987 Astron. Astrophys 173, 293

Langer, N. 1989, Astron. Astrophys. 210, 93.

Maeder, A. 1983, Astron. Astrophys. 120, 113.

Maeder, A. 1990, Astron. Astrophys. Suppl. Ser. in press.

Rosa, M. 1989, in Boulder-Munich workshop: Intrinsic Properties of Hot Luminous Stars, C. Garmany (ed.), ASP Conf. Ser. in press.

Schmutz, W., Hamann, W.R., Wessolowski, K. 1989a, Astron. Astrophys. 210, 236.

Schmutz, W., Leitherer, C., Torres-Dodgen, A. V., Vogel, M., Conti, P. S., Hamann, W.-R., and Wessolowski, U. 1989b, in Physics of Luminous Blue Variables, ed. K. Davidson, H. J. G. L. M. Lamers, A. F. J. Moffat (eds.), IAU Colloq., 113, 289.

Smith, L.J. 1989, in Boulder-Munich workshop: Intrinsic Properties of Hot Luminous Stars, C. Garmany (ed.), ASP Conf. Ser. in press.

Smith, L.F., Maeder, A. 1988, Astron. Astrophys. 211, 71.

Stahl, A.G. 1986, Astron. Astrophys. 164, 321.

Walborn, N.R. 1982, Astrophys. J. Letters 256, 452.

Wood,D.O.S., Churchwell, E. 1989, Astrophys. J. Supp. 69, 831.

# WHY WOLF-RAYET STARS SHOULD NOT BE CONSIDERED TO BE EVOLVED CORES OF MASSIVE STARS


ANNE B. UNDERHILL
Department of Geophysics and Astronomy
University of British Columbia
Vancouver, B.C. V6T 1W5, Canada


Maeder uses three criteria to define the Wolf-Rayet stage of stellar evolution, namely (1) anomalous surface abundances like those in nuclear processed material, (2) an effective temperature higher than 30,000 K, and (3) a dense wind with a rate of mass loss greater than or equal to $10^{-5} M_\odot$ yr$^{-1}$. These are properties which can all be determined from the observations. I think that spectroscopic observations from X rays to radio wavelengths of Wolf-Rayet stars show that Wolf-Rayet stars do not have these properties. Rather the observations show that Wolf-Rayet stars have properties typical of young stellar objects barely arrived on the main sequence and still surrounded by a line-emitting region (LER) which geometrically resembles the LERs of young low-mass stellar objects such as Herbig Ae/Be stars and T Tauri stars.

The evidence that Wolf-Rayet stars are young stellar objects rather than evolved objects is summarized in two papers presented at this conference (Underhill 1990a, 1990b). That the composition of Wolf-Rayet atmospheres is solar is shown by the one-point analyses by Bhatia and Underhill (1986, 1988, 1989). Bhatia and Underhill have used a one-point method of analysis similar to that used by Willis and Wilson (1978) and by Smith and Willis (1982, 1983). The major difference is that Bhatia and Underhill use more realistic model atoms than Willis and his colleagues do and they include in the equations of statistical equilibrium rates for *all* the processes that can occur including photoionization from excited levels of the atoms.

This use of more complete equations of statistical equilibrium and the exploration of a wide range for the parameters are the reasons why Bhatia and Underhill are able to conclude that solar abundances are indicated by the observed spectra of Wolf-Rayet stars. The values of the parameters valid for representing Wolf-Rayet LERs are given in Underhill (1990a). They imply that $T_{eff}$ lies in the range 25,000 − 30,000 K. Such a result is also indicated by the analysis of the integrated fluxes of Wolf-Rayet stars (Underhill 1983b). Other NLTE one-point analyses of Wolf-Rayet spectra (for example Barlow and Hummer 1982; Torres 1987) have used nebular theory which omits photoionization from excited levels. As a consequence they give results like those of Willis and his colleagues who also omitted most photoionization rates.

Large rates of mass loss from Wolf-Rayet stars are usually inferred from the observed radio fluxes measured at 6 cm. These fluxes may arise in three ways: (1) from thermal bremsstrahlung in a spherical wind, (2) from magnetic bremsstrahlung in the parts of the LER threaded by magnetic fields (Underhill 1983a, Underhill 1986, Bhatia and Underhill




*L. A. Willson and R. Stalio (eds.), Angular Momentum and Mass Loss for Hot Stars, 353–355.*
© 1990 *Kluwer Academic Publishers. Printed in the Netherlands.*

1988), and (3) from a possible disk-driven wind (Underhill 1986). There is no need to assume a rate of mass loss greater than about $10^{-6} M_\odot$ yr$^{-1}$ to account for the observed radio fluxes. The fact that the spherical expanding wind models of Hamann, Schmutz and Wessolowski (1988), Schmutz, Hamann, and Wessolowski (1989) and of Hillier (1987, 1989) predict absorption components for He I and He II lines much stronger than what are observed is another reason why it is not desirable to assume that the rate of mass loss from Wolf-Rayet stars is greater than about $10^{-6} M_\odot$ yr$^{-1}$ and that the electron temperatures in the LER are low, less than about 30,000 K.

Bhatia and Underhill (1988) have suggested that the LER is shaped like a disk. Observations showing that a disk with a large hole in the middle and containing a few ever changing, magnetically supported filaments has been presented for HD191765, WN6, by Underhill et al. (1990). The observations of the spectrum of HD50896, WN5, by Ebbets (1979) and as shown in Figure 2 of Hamann, Schmutz, and Wessolowski (1988) indicate that a similar model is valid for the LER of HD50896.

All He I and He II lines in the spectra of Wolf-Rayet stars are longward displaced by 100 – 150 km s$^{-1}$, see, for instance, Underhill, Yang and Hill (1988), Underhill et al. (1990), and Underhill, Gilroy, and Hill (1990). On the other hand, lines from N IV and C IV are shortward displaced by about 60 km s$^{-1}$. Such differing displacements cannot be explained by spherical outflow; they can be accommodated in the model developed by Underhill and her colleagues.

All Wolf-Rayet stars show a significant polarisation, see, for instance, Schulte-Ladbeck and van der Hucht (1989). This indicates that Wolf-Rayet stars are surrounded by a flattened volume of electrons which can be the same as the flattened distribution described as a disk by Bhatia and Underhill (1988), Underhill et al. (1990), and Underhill, Gilroy, and Hill (1990). In the latter paper and in Underhill (1990b) it is noted that the components of the eclipsing binary CQ Cephei do not have the luminosities and effective temperatures of evolved models of Maeder and Meynet (1987, 1988).

On account of all the factors listed above, it is advisable to explore the possibility that Wolf-Rayet stars are massive young stars only now arriving on the main sequence and still buried in the remnant of their natal clouds, rather than to identify them with the evolved core models of Maeder and Meynet (1987, 1988) and of Langer (1989). The statistics of Wolf-Rayet stars relative to other stars of their masses $(10 - 15 \ M_\odot)$ do not vitiate the identification which I propose.

Because, as pointed out in detail by Bhatia and Underhill (1986, 1988, 1989), the observed spectra of Wolf-Rayet stars can be interpreted in a coherent manner using solar abundances, there is no need to assume that Population I Wolf-Rayet stars are evolved objects. The central stars of planetary nebulae which have Wolf-Rayet spectral types are low-mass evolved objects quite different from the Population I Wolf-Rayet stars. It is significant that showing the type of emission-line spectrum called "Wolf-Rayet" occurs both for massive stars and for low-mass evolved objects. This fact indicates that such spectra are generated by processes which act in the LER and are not a unique indicator of the mass and stage of evolution of the star.

# REFERENCES


Barlow, M.J., and Hummer, D.G. 1982, in *Wolf-Rayet Stars: Observations, Physics, Evolution*, ed. C. W. H. de Loore and A. J. Willis, (Dordrecht: Reidel), p. 387.

Bhatia, A. K., and Underhill, A. B. 1986, *Ap. J. Suppl.*, **60**, 323.

Bhatia, A. K., and Underhill, A. B. 1988, *Ap. J. Suppl.*, **67**, 187.

Bhatia, A. K., and Underhill, A. B. 1989, *Ap. J.*, submitted.

Ebbets, D. 1979, *Pub. Astr. Soc. Pac.*, **91**, 804.

Hamann, W.-R., Schmutz, W., and Wessolowski, U. 1988, *Astr. Ap.*, **194**, 190.

Hillier, D. J. 1987, *Ap. J. Suppl.*, **63**, 965.

Hillier, D. J. 1989, *Ap. J.*, **346**, November 1.

Langer, N. 1989, *Astr. Ap.*, **210**, 93.

Maeder, A., and Meynet, G. 1987, *Astr. Ap.*, **182**, 243.

Maeder, A., and Meynet, G. 1988, *Astr. Ap. Suppl.*, **76**, 411.

Schmutz, W., Hamann, W.-R., and Wessolowski, U. 1989, *Astr. Ap.*, **210**, 236.

Schulte-Ladbeck, R. E., and van der Hucht, K. A. 1989, *Ap. J.*, **337**, 872.

Smith, L. J., and Willis, A. J. 1982, *M.N.R.A.S.*, **201**, 451.

Smith, L. J., and Willis, A. J. 1983, *Astr. Ap. Suppl.*, **54**, 229.

Torres, A. V. 1988, *Ap. J.*, **325**, 759.

Underhill, A. B. 1983a, *Ap. J.*, **265**, 933.

Underhill, A. B. 1983b, *Ap. J.*, **266**, 718.

Underhill, A. B. 1986, *Pub. Astr. Soc. Pac*, **98**, 897.

Underhill, A. B. 1990a, in *Angular Momentum and Mass Loss for Hot Stars*, ed. L.A. Willson and R. Stalio, (Dordrecht: Kluwer), Poster Session 1.

Underhill, A. B. 1990b, in *Angular Momentum and Mass Loss for Hot Stars*, ed. L. A. Willson and R. Stalio, (Dordrecht: Kluer), Poster Session 3.

Underhill, A. B., Gilroy, K. K., and Hill, G. M. 1990, *Ap. J.*, **351**, March 10.

Underhill, A. B., Gilroy, K. K., Hill, G. M., and Dinshaw, N. 1990, *Ap. J.*, **351**, March 10.

Underhill, A. B., Yang, S., and Hill, G. M. 1988, *Pub. Astr. Soc. Pac.*, **100**, 741.

Willis, A. J., and Wilson, R. 1978, *M.N.R.A.S.*, **182**, 559.

# KEYWORD INDEX

*The page numbers given are the initial pages of papers for which the authors gave these key words.*